FURTHER PRAISE FOR

WIRED FOR CULTURE

One of the *Guardian*'s Literary Highlights for 2012

Chosen for *Science News*'s "Bookshelf"

"Absorbing." —Olivia Judson, *Wall Street Journal*

"Gorgeously written, elegantly argued, Pagel's *Wired for Culture* demonstrates that genes are only a small part of the human success story; minds and culture are the larger part. A compelling read that allows us to appreciate everything around us with fresh eyes."

—David Eagleman, author of *Incognito: The Secret Lives of the Brain*

"[A] pioneering contribution. . . . Pagel has an interesting take on what is perhaps the deepest controversy among Darwinists: does culture fundamentally change the evolutionary dynamics of our species? . . . This is the best popular science book on culture so far."

—Peter Richerson, *Nature*

"Mark Pagel's *Wired for Culture: Origins of the Human Social Mind* leads readers to a startling view of human evolution. . . . Readers who like being challenged will enjoy those controversial sections and will gladly spend several hours engaging in an intellectual wrestling match with the author." —Fred Bortz, *Dallas News*

"*Wired for Culture: Origins of the Human Social Mind* comes from an evolutionary biologist who takes us back eighty thousand years to a point where humans developed culture. . . . A fascinating exploration."

—*Midwest Book Review*

"Pagel's absorbing account . . . paints a broad picture, impressive for its detail, accuracy and vivacity." —Julian Baggini, *Guardian*

"A thorough, well-researched, and important contribution to understanding the biological and social adaptations in modern humans."
—R. A. Delgado Jr., *Choice*

"[*Wired for Culture* is] a clear and convincing read, and it wouldn't look out of place alongside Pinker and Dawkins."
—Tom Chivers, *Telegraph*

"Pagel's arguments are complex but skillfully assembled, creating a convincing thesis that accounts for the rise of human culture. . . . Crucially, Pagel's arguments steer away from reliance on biological determinism." —Robin McKie, *Observer*

"Readers of diverse perspectives will recognize the timely wisdom in Pagel's concluding reflections on the challenge humans now face in overcoming deeply ingrained ethnic jealousies by developing much more inclusive new conceptions of culture." —Bryce Christensen, *Booklist*, starred review

"An intriguing combination of information on the latest advances in genomics and epigenetics, with an optimistic prediction of a future global society in which inventiveness and cooperation prevail."
—*Kirkus Reviews*

"A far-reaching study of how our species' innate capacity for culture altered the course of our social and revolutionary history."
—*Publishers Weekly*, "Science Picks"

"Pagel examines the evolution of human nature in the tradition of Richard Dawkins's *The Selfish Gene*. . . . Pagel's book is recommended for readers interested in human evolution and human nature."
—Scott Viera, *Library Journal*

WIRED FOR CULTURE

ALSO BY MARK PAGEL

Evolutionary Genomics and Proteomics
with Andrew Pomiankowski (Sinauer, 2008)

The Oxford Encyclopedia of Evolution
(Oxford University Press, 2002)

The Comparative Method in Evolutionary Biology
with Paul H. Harvey (Oxford University Press, 1991)

WIRED FOR CULTURE

Origins of the Human Social Mind

MARK PAGEL

W. W. Norton & Company • NEW YORK LONDON

Printed in the United States of America
First published as a Norton paperback 2013

For information about permission to reproduce selections from this book,
write to Permissions, W. W. Norton & Company, Inc.,
500 Fifth Avenue, New York, NY 10110

For information about special discounts for bulk purchases, please contact
W. W. Norton Special Sales at specialsales@wwnorton.com or 800-233-4830

Manufacturing by RR Donnelley Harrisonburg
Book design by Brooke Koven
Production manager: Julia Druskin

Library of Congress Cataloging-in-Publication Data

Pagel, Mark D.
Wired for culture : origins of the human social mind / Mark Pagel. — 1st ed.
p. cm.
Includes bibliographical references and index.
ISBN 978-0-393-06587-9 (hardcover)
1. Human evolution. 2. Social evolution. 3. Evolution (Biology)
4. Evolutionary genetics. I. Title.
GN281.P32 2012
303.4—dc23

2011044465

ISBN 978-0-393-34420-2 pbk.

W. W. Norton & Company, Inc.
500 Fifth Avenue, New York, N.Y. 10110
www.wwnorton.com

W. W. Norton & Company Ltd.
Castle House, 75/76 Wells Street, London W1T 3QT

1 2 3 4 5 6 7 8 9 0

Contents

PART IV • **The Many and the Few**

Preface

I BEGAN THINKING about the ideas that would lead to this book in the early 1990s while in a remote and barren region of Northern Kenya known as the Chalbi Desert. The Chalbi lies north of the town of Marsabit and to the east of Lake Turkana. Marsabit is about a day's drive north from the town of Isiolo, which sits on the border of a region still called the Northern Frontier District, and that forms roughly half of Kenya. The Chalbi Desert is to some archaeologists the cradle of humanity, the place where the evolving lineage that would eventually lead to modern humans arose. It is a hot and arid region, short of water during the dry season, making it too dry for agriculture, but suited to nomadic pastoralists—people who live by herding animals.

In the Chalbi Desert most pastoralists herd sheep, goats, and camels. A pastoralist's animals are an edible bank account, with varying interest and risk rates. The sheep reproduce quickly but are finicky about what they eat. Goats will eat anything but reproduce less quickly. Camels are the gold bullion of pastoralism: they can survive the harshest conditions but are very slow about making more camels. For the nomadic pastoralists, life is like being an itinerant investment manager. Every day begins with the question of how best to divide one's resources and efforts among their four-legged investment policies: how many sheep should I have, and when I get that number should I trade them for goats and camels?

The Gabbra are a tribe of nomadic pastoralists who live in the Chalbi and descend from Cushitic people who trace their origins to the Horn of Africa. One of these Gabbra, a man called Dido, had seldom been more than about thirty miles from his birthplace, and had spent his life herding his animals and owning what he could fit on the back of a camel when it came time to shift from one area of pasture to another, something these nomads might do several times per year. In addition to his native Gabbra language, Dido could speak English from spending time among missionaries. But Dido also spoke four more languages: Rendille, Samburu, Turkana, and Swahili. I wasn't surprised about Swahili because it is a trade language spoken all over East Africa. But Rendille, Samburu, and Turkana are the languages of other nomadic pastoralists who also live near the Chalbi Desert and who make their living in the same way as the Gabbra. I asked him why he could speak their languages and he replied: "So I can talk to them."

I was asking a different question. What I really wanted to know was why there were four different tribes of nomadic pastoralists all living in the same area and herding the same kinds of animals, whose genetic differences were negligible, and yet who had divided up the land and their lives so exclusively as to speak different languages. Why were there four tribes with different languages, customs, habits, and traditions, and not just one? Why do we humans have a tendency to form into small tribal groups rather than living as one large and homogeneous society? This is a worldwide phenomenon, not confined to this desert region of Northern Kenya. There are currently as many as 7,000 different languages spoken, or 7,000 mutually unintelligible systems of communication in one species, marking out at least 7,000 distinct societies. This is more different systems of communication in a single mammal species—for that is what we are—than there are mammal species. It is 7,000 different ways of saying, "Good morning," or, "Looks like rain today," and means that humans uniquely and strangely among animals often cannot communicate with other members of their own species.

For many anthropologists, culture in its wider manifestations—

including rituals, belief systems, religion, and customs—simply exists and develops its own momentum and directions in separate groups. For them, our cultural diversity arises as a consequence of geography. The wider our species spread after leaving Africa probably sometime between 60,000 to 70,000 years ago, the more different societies and languages would eventually emerge. But this fails to explain a key fact of human societies: that there are many more than we would expect from simple territorial expansion of our species, hinting at more fundamental social and psychological causes. In fact, it is where people are found most closely packed together—the least geographically isolated—that we find the greatest diversity of cultures. There are regions of the northeast corner of coastal Papua New Guinea where a different language is spoken every few miles. I once met a Papuan man from that area and asked him if this could be true. He replied, "Oh no, they are far closer together than that."

This cannot be a simple consequence of geography, and if we are a species with a predilection to form into societies with separate and distinct identities, then this is something that we are going to have to come to grips with in a modern world. It is also something that we will want to understand because it clashes with another feature of our species. The evolutionary biologist Robert Trivers began something of a revolution in our thinking about humans when in 1971 he put forward his idea of "reciprocal altruism." It showed a way that two individuals who were not related to each other could nevertheless benefit from mutual acts of altruism. Here was a way that natural selection could escape Tennyson's "Nature, red in tooth and claw" and promote cooperation among people who in evolutionary terms were, by instinct and temperament, competitors. By the early 1990s evolutionary biologists and anthropologists were coming to appreciate that human beings had taken this even further, having developed the means to behave altruistically toward others and even do so without any expectation of a return from those others. We are capable of great acts of charity, helping others in distress, and of simply being kind, generous, and friendly. No other animal does such things, and so it is a capability we have evolved only relatively recently. But why? So here are two observations that need to be put together—our unmatched ability to get along with each other set against our tendencies to form competing societies often not far from conflict.

Evolutionary biologists are accustomed to recognizing that some kinds of patterns in the distribution of animals and plants and how they live reveal clues about those species' tactics in surviving. It is then a matter of working backward from the patterns to discover what kinds of survival strategies would give rise to them. With humans, we want to ask what is the nature of culture as a survival strategy that it would have this feature of forming us into so many small societies, which seem to act in some respects like an extension of our bodies. We are devoted to them sometimes to the point of self-sacrifice, we cooperate with others inside of them, and we use them to advance our interests. At a psychological level, we display forms of social behavior conducive to living in small groups, such as rewarding cooperation, punishing those who deviate from norms, being wary of outsiders. What kinds of explanation can we give for these features of our lives?

This book is an attempt to answer some of these questions. I am grateful to the following people (in no particular order) who one way or another have given me information, suggestions, and ideas that have been helpful in writing it, some of them so long ago that they won't remember and will be surprised to see their names here: Andrew Meade, Chris Venditti, Andreea Calude, Russell Gray, Rutger Vos, Paul Harvey, Tecumseh Fitch, Irv DeVore, Kevin Laland, Richard Sibly, Ruth Mace, Quentin Atkinson, Jon Wilkins, Mark Beaumont, David Krakauer, Jessica Flack, Matt Ridley, Colin Renfrew, Karl Sigmund, Robert Trivers, and Preethi Chandramohan. I am also grateful to the University of Reading and the Santa Fe Institute for providing me the time and space to think and write. The Leverhulme Trust has supported some of my work on language evolution that I report in a later chapter. My children Thomas and William remind me that life is both simpler and yet more unpredictable than I ever imagined, and therefore worth explaining.

My agents Katinka Matson and John Brockman provided particularly valuable advice and suggestions, and put opportunities my way. Laura Romain and Ann Adelman respectively shepherded and copyedited the manuscript as it moved from typescript to book. And I owe a special degree of thanks to my editor Angela von der Lippe. She carefully read the entire manuscript, making many useful suggestions that improved it greatly.

WIRED FOR CULTURE

INTRODUCTION

The Gamble

That human nature is defined by our response to culture

THE ENGLISH 4TH Baron Raglan, Major FitzRoy Richard Somerset of the Queen's Grenadier Guards, once remarked that "culture is roughly everything we do and monkeys don't." This comment nicely summarizes one of the main messages of this book. Human beings have not always been as we know us now—sentient, big-brained, naked, prolix, artistic, wary, scheming, generous, warlike, forgiving, vengeful, religious, and moralistic. Instead, we were launched as recently as 80,000 years ago when our genes undertook a remarkable gamble. Around that time a species of upright apes, close evolutionary relatives to the chimpanzees, began to perfect a new way of life. Nothing in this species' predecessors would have hinted at what was about to emerge, or that it would have such startling effects. Where previously they had roamed the African savannah for at least a million years, hunting and foraging in small family groups, the new species now came to live in larger tribal societies in which people worked together, customs and systems of beliefs arose, ideas, skills, and technologies were shared, languages evolved, and dance, music, and art appeared. Within a few tens of thousands of years, these tribal groups would spread out to occupy the world as some of them developed the means to live near the sea, others the ability to survive the desert or to inhabit jungles, forests, mountains, or plains. In what was little more than an instant in our long evolutionary history, we had become a single species with a global reach and ways of life as varied as collections of different biological species, and we were soon to become the sole survivor of an evolutionary lineage that had spawned at least six previous human branches.

The world was witnessing the final stages of a shift in the balance of power between our genes and our minds. Human beings had discovered culture. It was not high art and symphonies—those would come—but knowledge, beliefs, and practices acquired from watching, imitating, and learning from others. Today, we take our possession of culture for granted, but it was a development that had to await nearly the entire history of life on Earth. Our world is four and a half thousand million (4.5 billion) years old, and might have been a harsh, rocky place devoid of life for its first 700 million to 1 billion years. Then, from fossil traces buried deep in ancient rocks, we know that life sparked into existence and for the next 3.5 billion years genes ruled, transmitting the instructions that organisms used to survive and reproduce. For most of that time, life consisted of simple one-celled organisms, direct ancestors of today's bacteria; but these gave way around 1 billion years ago to the first multicellular organisms, simple creatures like today's sponges. Five hundred million years after that the first animals with arms and legs would rise up out of the sea and walk on land. These land animals would in turn evolve for yet another 500 million years before the evolutionary lineage that we call the *hominins* came on the scene, a mere 7 million years ago.

Even then, it was only when our species arose within this hominin lineage just 160,000–200,000 years ago that a competitor to the rule of genes finally appeared. Our invention of culture around that time created an entirely new sphere of evolving entities. Humans had acquired the ability to learn from others, and to copy, imitate and improve upon their actions. This meant that elements of culture themselves—ideas, languages, beliefs, songs, art, technologies—could act like genes, capable of being transmitted to others and reproduced. But unlike genes, these elements of culture could jump directly from one mind to another, shortcutting the normal genetic routes of transmission. And so our cultures came to define a second great system of inheritance, able to transmit knowledge down the generations. For humans, then, a shared culture granted its members access to a vast store of information, technologies, wisdom, and good luck. The only other example like this in nature is the lowly bacteria. These simple one-celled organisms cannot exchange ideas, but they have acquired a variety of means for exchanging genes among individuals and even among different species, granting them access to a vast store of genetic technology. And, like us, they have shown great inventiveness and versatility, occupying nearly every environment on Earth.

Our cultural inheritance is something we take for granted today, but its invention forever altered the course of evolution and our world. This is because knowledge could accumulate as good ideas were retained, combined, and improved upon, and others were discarded. And, being able to jump from mind to mind granted the elements of culture a pace of change that stood in relation to our genetical evolution something like an animal's behavior does to the more leisurely movement of a plant. Where you are stuck from birth with a sample of the genes that made your parents, you can sample throughout your life from a sea of evolving ideas. Not surprisingly, then, our cultures quickly came to take over the running of our day-to-day affairs as they outstripped our genes in providing solutions to the problems of our existence. Having culture means we are the only species that acquires the rules of its daily living from the accumulated knowledge of our ancestors rather than from the genes they pass to us. Our cultures and not our genes supply the solutions we use to survive and prosper in the society of our birth; they provide the instructions for what we eat, how we live, the gods we believe in, the tools we make and use, the language we speak, the people we cooperate with and marry, and whom we might fight or even kill in a war.

Most of us assume without reflection that it has always been this way, that human beings have always occupied the world, and that somehow we are the natural and rightful rulers of its domains. But we are new on the scene, and even newer around the world, having only ventured permanently out of Africa probably sometime in the last 60,000 to 70,000 years. Even as recently as 80,000 years ago, our species' continued existence still hung in the balance. An extraordinary degree of similarity in the genes of people from all over the world tells us that we all share a recent common ancestry. In fact, genetic studies now reveal that our ancestors might have dwindled to as few as 10,000 individuals—some say even fewer—making humans as endangered 80,000 years ago as a rhinoceros is today. Then our numbers began to grow and human culture began to flourish, and our species, having come perilously close to extinction, reached a point of no return. Our minds were now firmly in executive control of our fates, and we were showing the adaptability, and producing the artifacts and culture that would propel us out of Africa, and then around the world—specialized stone tools and spear points, carved fishhooks, clothes, shaped blades and instruments, but also sculpted figures, ceremonial burials, musical instruments, and cave art.

The world is now a remarkably different place from what it was throughout the first 99.996 percent of its history. Almost everything around you in your bustling everyday lives is owed to the new evolutionary world in which ideas could accumulate on top of ideas, and most of those ideas were first thought up by someone distant to you in time and space. Having culture is why we watch 3D television and build soaring cathedrals while our close genetic relatives the chimpanzees sit in the forest as they have for millions of years cracking the same old nuts with the same old stones. Even so, having become the first species to throw off the yoke of its genes, our life in the presence of culture would usher in an irony. It is that we have fallen in thrall to the new sets of instructions our cultures provide. This is because to take advantage of culture meant evolving a new kind of mind. It had to be a cultural blank slate or *tabula rasa*, a compliant or docile mind, designed to be programmed by and embrace the culture into which it happened to be born. A wolf brought up by sheep will remain a wolf and soon turn on its benefactors, but a newborn human must be ready to join any cultural group on Earth, and without knowing which. It might find itself living on the Arctic ice, the

Russian steppes, or sailing across Polynesia; it might find itself in the Australian Outback, the deserts of Arabia, on the prairies of North America, or the African savannah, on an island in the Indian Ocean, or fishing along the rich tropical coasts of Papua New Guinea. And so we have had no choice but to evolve to allow our culture to occupy our minds, writing its language and story into our consciousness.

In nearly every other respect for which the great English philosopher John Locke proposed his doctrine of *tabula rasa*, the human brain has been shown to come into the world prepared, and not at all a blank slate. We are primed to learn language, to comprehend shapes and movement, to expect causation, to manipulate numerical quantities, to be afraid of heights, to mimic others, and to favor our relatives. But we are not primed to acquire any particular culture. The one we do inherit is an arbitrary story, an accident of birth, but it is one to which we show a surprising and sometimes alarming devotion. People will risk their health and well-being, their chances to have children, or even their lives for their culture. People will treat others well or badly merely as an accident of their cultural inheritance. If there is a humbling lesson of culture, it is that we do these things even though each of us might have been someone else, with a different internal voice, likes and dislikes, and allegiances. If there is a comparison, it is to ducklings whose parents have been lost and when they hatch from their shells adopt as their parent the first animal that wanders past—even a human. Animal ethologists call this *imprinting*; it is difficult to escape the feeling that we seem to imprint on our cultures, and in a way that is hard to shake off.

Genes are carefully shepherded into our bodies inside small vehicles known as *gametes*—sperm from fathers and eggs from mothers—which are designed to see to it that a body is made that carries a collection of its parents' genes. Part of the imprint of culture is to get us later in life to act as its shepherds. Each of us who has children will have shepherded pieces of our culture into them, some of it from mothers, some of it from fathers, ensuring that they were French, Korean, English, Melanesian, or American, Italian, Russian, or Chinese, and that they were religious or atheist, but also that they spoke a particular language and held certain beliefs about their nation and the rest of the world. We should be aware that it is at least a curious, and surely a compelling, feature of our species that a child born into the world as nothing more than a "blank" human being might be labeled as a Christian or a Hindu, Jewish, Buddhist, Muslim, or Confucian, and that this label—or some other its culture provides—can influence the course of this child's life, as if it were a trait inherited on some gene. There are places all over the world where a child born into one of these religions might peer across a fence at children from another whose parents are sworn enemies of its own, and only then because *their* parents labelled them.

The reason for this shepherding is clear. Human culture has been a development of revolutionary social *and* genetic effect, easily the most potent trait the world has ever known for converting new lands and resources into more humans. Our genes' gamble at handing over control to the new sphere of evolving ideas paid off handsomely. Culture became our species' strategy for survival, a biological strategy, not just some bit of fun and amusement on the side, and it would trump all the wonderful wings and feathers, shells, claws, poisons, acts of camouflage or deception, odors, feats of running speed, long necks or beaks, powerful jaws, and spectacular colors and displays of the rest of the animal kingdom. It didn't have to be this way. Our newly liberated minds might have chosen aesthetic reverie, feckless indolence, jumping off cliffs, debilitating drug use, or mindless warfare. But, for the most part, we didn't. We seem to have followed our ancient genetic instincts for survival, and culture has been remarkably able to oblige.

The question is often asked, What makes us human? Quite apart from its interest to anthropologists and other scholars, it is a question that invades nearly every aspect of our lives, our psychology and behavior. Who are we, and why are we the way we are, so utterly different from other animals? What makes us kind and forgiving, generous and friendly, but also wicked, murderous, and vengeful? Why do we have morality? The usual answers to these questions are that we are made human by virtue of possessing consciousness, or that we have this or that gene, an opposable thumb, or an upright posture and bipedal gait, that we learned to control fire, or that we have empathy, language, or our extraordinary intelligence. For others it is the belief we are made in God's image and in possession of a soul. And it is true, these traits and beliefs set our species apart. But the argument of this book is that these usual answers are the wrong way around. They are the wrong way around because they fail to recognize that it is only because of culture that we have many of these traits. Here is something we will have to get used to: all of us carry around in our minds something akin to a software "operating system" installed without our consent by our parents and others in our societies. It defines who we are and is our internal voice. It frames our social and cultural identities, and fundamentally influences the course of our lives. No other species has such a system. Only when we understand this, and understand how the traits we acquired in response to this new way of life serve our interests, can we begin to grasp what it means to be human.

And here is why. Evolutionary biology teaches us that in a competitive world, if we know something about the environment an animal lives in, we can make some predictions about what it will be like. If an animal lived its life in trees or flying in the air, hunting for insects or swimming in the water, we could expect it to have acquired certain characteristics to promote its survival and well-being—long arms to swing in trees, wings for flight, an acute nose or hearing to detect insects, or a streamlined shape for swimming. Most animals are adapted to a physical environment such as one of these, and are confined to areas of the Earth where that environment is found. But for the last 160,000 to 200,000 years, humans have roamed the Earth conquering its many environments, chauffeured wherever we travelled by the inventive and cooperative tribal societies that are their cultures. And so, we are entitled to expect that, instead of adapting to the demands of any one physical environment, our genes have evolved to use the new social environment of human society to further their survival and reproduction. These are the adaptations that have wired our minds and bodies for culture.

It is a subject that touches the most fundamental aspects of our lives. We will see in the chapters of this book that our responses to culture have produced some of our best and our worst tendencies, creating a species brimming with contradictions. Our possession of culture is responsible for our art, music, and religion, our unmatched acts of charity, empathy, and cooperation, our sense of justice, fairness, altruism, and even self-sacrifice; but also for our undeniable self-interest, our tendency to favor people from our own ethnic or racial groups, wariness of strangers, xenophobia, and predilections to war. But it goes further than this. The nature of our culture will tell us why we alone as a species have language, why it is that we alone can show kindness to strangers, and even to other animals, but also why we can be callous and murderous. It is why we are the only species with morality, but also why we apply it capriciously to suit our needs. Culture equips us with envy, jealousy, and spite, indignation and contempt, but also with friendship, forgiveness and affection, and a conscience. It is why we, and probably we alone, have consciousness, and yet why our conscious mind is often divided between reason and passion, unsure or even in conflict with itself over how to behave. It is why we differ from each other, why we differ so from the other apes despite sharing so many of their genes, why we are shrewd and deceptive, and even why we deceive ourselves. We will see that our cultures can even get us to kill our own children—so-called honor killings—and at the same time can get us to behave so selflessly that we would have to travel all the way to bees in a hive or to the cells in our body to see anything else like it in nature.

True, it would be wrong to suggest we are the only species with culture; it is just that only in humans has the handover been so great and the occupation of our minds so complete. New Zealand's chaffinches, a songbird carried to those islands by homesick Europeans, learn their songs from their parents and thereby produce a surprising range of local dialects. Some chimpanzee troops have cultural traditions in the styles of tools they use to fish insects from the ground, or in the stones they use to crack nuts. Some meerkat colonies living side by side have persistent but arbitrary differences in the times they get up in the morning. There are idiosyncratic hunting styles among some dolphin pods and variety among the songs of some whales. In another dolphin species, females wear decorative sponges on their noses that they have gathered up from the seabed, and some groups of orang-utans make leaf-bundle "dolls." Japanese macaques produce a wonderfully humanlike potato-washing behavior beloved of television documentaries. These cultural achievements are delightful, often entertaining, and sometimes even unexpected. But they bear about as much resemblance to human culture as a gorilla beating its chest or a chimpanzee drumming on a log does to a Bach cantata, scarcely deserving to be compared to the varieties, contrivances, complexities, and intricacies of human science, technologies, language, art, music, and literature.

Still, is the 160,000 to 200,000 years we have been around long enough for traits to have evolved in response to living in the social environment of our cultures? Has there been time enough to become wired for culture? The simple way to answer this question is to look around you. For instance, sometime around 25,000 years ago, people began living above 12,000 feet in the high Tibetan plateau, and they acquired physiological adaptations that allow them to cope with the reduced oxygen at these altitudes. One of these was so advantageous that it might have spread to 90 percent of all Tibetans in just four thousand years. The Dinka tribespeople of Sudan are tall and slim and have unusually dark skin. The Inuit people of northern North America are shorter, of stocky build, and have lighter skin. The Dinkas' spaghetti-like body shape gives them a large surface area for shedding heat, while the Inuits' more spheroid shape reduces their surface area to conserve it. The Dinkas' dark skin protects them from the sun, but the melanin needed to produce it isn't needed in the Arctic, so the Inuit make less of it.

These are all genetic adaptations acquired, in the case of the Tibetans and the Inuits, since our species walked out of Africa: a Dinka raised in the Arctic will not look like an Inuit, and vice versa, and the Tibetan capacity to live with reduced oxygen levels doesn't evolve at low altitudes. If this kind of rewiring of our genes and physiology can take place over such short periods of time, this tells us that other features of our nature, including our psychology and social behaviors, have had plenty of time to evolve since we acquired culture.

Even so, many people hold the view that humans fall outside the grip of Darwinian evolution by natural selection. We are intelligent beyond comparison to other animals, we use language creatively, we have art, music, dance, and religion, and, above all, a free will. But we must be careful. The standard philosophical objection to free will is that we aren't as free to do what we "want" as we would like to think we are, because our current "wants" will always be influenced by our previous wants. And these previous wants form a chain leading all the way back to our birth and early upbringing when we were unable to make free choices. The first of these events over which we had little control might have been the accident of being born into a particular culture.

And to an evolutionist free will isn't even all it's cracked up to be anyway: good judgment should trump free will in most circumstances. Throughout our evolutionary history those of us who behaved in ways that promoted our survival and reproduction, rather than merely doing what we "wanted" to do, will have left the most descendants—descendants who will have inherited these same tendencies. If even just one of your ancestors had decided to give up having children for his or her art, the consequences for you would be no different than had that ancestor been killed—you would not be here today reading this book. Indeed, it is an underappreciated fact of biology that throughout history the overwhelming majority of individuals ever born, hatched, or budded off died long before adulthood. So the world is populated today by a select group of survivors whose ancestors had the dispositions and the wherewithal to survive and reproduce, and this alone tells us there is no particular reason to believe that free will per se has been positively favored throughout our evolution. Survival is a rare thing, far too valuable to be entrusted to what could be a capricious free will.

Many people believe that to allow natural selection a role in defining who we are consigns us to having a selfish agenda, one in which our genes single-mindedly promote their existence. Our genes do that, but it is a misunderstanding of evolution to think that natural selection always favors a nasty and ruthless nature. It is far more creative than that, and nowhere more it seems than in our species. In fact, if the history of biological evolution teaches us anything, it is that natural selection can often achieve the most for its genes by building cooperation among actors or even among genes that avoids debilitating conflict, returning greater gains than could be achieved by competition or a solitary existence. Among the triumphs of modern evolutionary biology is the demonstration that many of the outlines of culture and of our behaviors can be explained as strategies for promoting our survival and reproduction. The influential evolutionary theorist William Hamilton anticipated this some years ago, saying:

> to come to our notice cultures, too, have to survive and will hardly do so when by their nature they undermine the viability of the bearers. Thus we would expect the genetic system to have various inbuilt safeguards and to provide not a blank sheet for individual cultural development but a sheet at least lightly scrawled with certain tentative outlines. . . .

It is those "tentative outlines" we seek to understand.

the rest of this book

THE REST OF this book is about how our cultures came to occupy our minds, what they demanded of us, how those demands have been met, and whether our cultural nature provides useful solutions for living in a modern world. For many people, I think one of the most distinctive and salient features of life in human societies is the sense of belonging to a particular cultural group to which they often feel a surprising attachment and allegiance, one that can even extend in some circumstances to giving up their lives for it. So finely tuned is this tendency in us that even within our societies the cultural subdivisions can acquire a bewildering degree of complexity, as people from different regions detect minute differences in accents, preferences for food, styles of dress, religious beliefs, and manners. To outsiders, these differences may be barely, if at all, detectable. But it is a complexity that seems entirely natural to someone from one of those societies, and the differences that are so small to an outsider can seem large indeed from the inside.

In the Preface I described a tendency throughout our history to form into small *tribal* societies. Some will cavil at this term, thinking it carries bigoted or prejudicial overtones, but I use it far more simply to capture that sense of a group of people, somehow organized around an identity. Even if no one can agree precisely what that identity is at any given moment and who has it or not, most people have a sense of which group they belong to, and just as importantly who doesn't. Our dispositions to form into these groups is a phenomenon that has held throughout our evolutionary history and its effects linger in our behaviors and psychology today. We will see them over and over in this book and so we want to try to understand why we have this particular nature.

I want to call these tribal groups *cultural survival vehicles*. This might seem a cumbersome term, but it is one I have found useful in trying to understand our species. Indeed, it proves so central to understanding what makes us human that it could have been the title of this book. The zoologist Richard Dawkins in *The Extended Phenotype* coined the term *vehicles* to describe structures that carry *replicators*. An example of a vehicle is your body, or the body of a cat or a dog. A replicator, on the other hand, is something that can make copies of itself, such as a gene. Putting these two ideas together, we can see that replicators (think of genes) exert their effects on the external world, and thereby influence the likelihood that they will survive, through the vehicles (think of your body) they build. The distinction between replicators and vehicles is important because it reminds us that an animal's body is merely a temporary structure built by its genes to promote their survival and reproduction. It can be difficult to shake the habit of thinking we are the main players in evolution rather than our genes, but your body is not replicated in your offspring; rather, your genes are, and then again in theirs.

When I use the term *cultural survival vehicle*, it is to capture the idea that our species evolved to build, in the form of their societies, tribes, or cultures, a second body or vehicle to go along with the vehicle that is their physical body. Like our physical bodies, this cultural body wraps us in a protective layer, not of muscles and skin but of knowledge and technologies, and as we will see in the later chapters, it gives us our language, cooperation, and a shared identity. We are the actors that produce this vehicle, behaving almost like individual genes clamoring inside it to exert our effects on the outside world, and influencing *our* likelihood of surviving. Our nature is wrapped up in the strategies we evolved and now deploy to make the cultural survival vehicle work for us. It doesn't matter that it is a shifting and fluid vehicle whose members might come and go, or that we cannot draw clear boundaries around it as a thick layer of hide or skin does. The same could be said of an ant's nest, a lion's pride, a troop of monkeys, or even a herd of wildebeest, and no one doubts that they are also vehicles to promote the survival and reproduction of their inhabitants.

Still, there is a fundamental evolutionary difference between our cultures as survival vehicles and our physical bodies, and it is a difference that will make all the difference. Our genes share a common route into the future, all of them living or dying with their particular body, and this has enforced a degree of agreement among them. So complete is that agreement, most of the time our genes work together seamlessly to build our bodies, not bickering or wishing to go off in different directions. Our societal vehicles are different. Like our physical bodies they have been fundamental to our success, and this is a point that is difficult to overemphasize. Even so, we as inhabitants of these vehicles do not all share a common route into the future. Each of us, unlike our individual genes, is free to reproduce on our own. The essential balancing act of human societies is that they will normally work best when everyone pulls together, but at any given moment what is best for you might differ from that which is best for your group. Our psychology is the outcome of this balancing act. It is the set of temperaments geared toward using our cultural vehicles to promote our individual survival in a world full of others like ourselves.

The topics I shall consider in this book—cooperation, learning, identity, self-sacrifice, language, religion, consciousness, creativity, altruism, deception, greed, and self-interest—have all been studied before and all of us will be familiar with them, having used, experienced, or somehow "participated" in each of them most days of our lives. But if this idea of a cultural survival vehicle can help us to organize our thoughts about humans, it will also help us to see these familiar topics in new ways, and maybe even come up with new explanations for their existence or for why they take the forms they do.

A point that it is difficult to stress too much is that my attempts to understand our traits in Darwinian terms are not meant to promulgate any sort of morality based upon the notion of survival of the fittest. Evolutionists sometimes refer to the failure to distinguish "is" from "ought" as the *naturalistic fallacy*. Just because something has evolved doesn't make it right, even if it has contributed in the past to our survival and reproduction. But if it has evolved for these reasons it is something we might wish to take seriously because it will lurk somewhere in our nature. Equally, to say we have an evolved social nature is not to suggest that our behaviors are determined by genes, but to say we have certain predilections and tendencies. In any given circumstance, the behaviors we produce are some result of genes and the environment of the actor. This is far from determinism, but still revealing of who we are and why we might have the tendencies to behave as we do.

The book is divided into four parts. The chapters of Part I try to answer the question of how our cultures have been able to organize us into the small, closely knit tribal groups that we are calling the cultural survival vehicles. They examine the tricks our cultures have used, what they have extracted from us, and what was in it for us. Part II investigates our cooperative cultural nature, examining the rules we have evolved for making cooperation work among unrelated individuals. This section recognizes that it is not enough to say that our allegiance to our cultures has evolved because they have returned great rewards throughout our history. The cooperative societies from which cultures are constructed are themselves fragile unless tightly controlled by social mechanisms that continue to make cooperation more profitable than unbridled self-interest. Part III examines how life in the presence of culture has sculpted our minds to use our social systems to our advantage. We expect a strong part of our nature to have been influenced by the rewards that come from steering just that little bit more of the cultural wealth our way. Part IV is about the modern dilemma of large nation-states. In countries such as China and India, over 1 billion people fall under the rule of a small elite, and in all of the major countries of the world millions are ruled by a few. That creates a dilemma for the thesis of this book: If humans have evolved a tribal nature that revolves around life in relatively small and exclusive cooperative social groups, how do we explain the enormous social groupings of the modern world—the observation that so many can be willingly led by so few?

PART I

MIND CONTROL, PROTECTION, AND PROSPERITY

Prologue

WHY IS IT THAT we can show such allegiance to the packets of information we call our cultures, and why has this been a rewarding rather than foolhardy and even dangerous thing to do throughout our evolutionary history? For many of us, a slight directed at our culture or even just a piece of cloth that represents it elicits emotions of defensiveness, confrontation, or even aggression. Where do these emotions come from and why do they arise so naturally? The argument of this Prologue—explored in the chapters of Part I—is that culture has worked by coming to exercise a form of mind control over us. We willingly accept and even embrace this mind control, and probably without even knowing it, in return for the protection and prosperity our cultures provide. This is why our cultural identity is so much like a trait that we might have acquired from genes—surprisingly stable and robust to outside influences, likely to be passed on to our children, and to theirs. For instance, asked which team they would support in a match against England, many Scottish soccer fans often reply, "anyone but England," and this despite the fact that Scotland has been a member of the United Kingdom for over three hundred years (although the Scots might say it is *because* of those three hundred years).

There are other cases of animals' minds and behaviors being taken over by an outside force. Evolutionary biologists have a field of study called "parasite manipulation of host behavior." A science fiction writer might more engagingly call it invasion of the body snatchers. When dogs roam in packs they can be menacing and aggressive, but in many parts of East Africa a dog seen roaming on its own excites greater alarm because people know it is likely to carry the rabies virus. The virus manipulates the dog's behavior to roam because a dog that does so is more likely to encounter an uninfected individual to bite. Similarly, there is a well-known fungus called *Cordyceps* that infects a species of carpenter ant. The fungus finds its way to the ant's brain where it controls the ant like a puppeteer, getting it to climb to the top of blades of grass or small plants. Once there, the ant clamps its jaws shut and then dangles like a flag in the wind. Meanwhile, the fungus devours the ant from within and eventually erupts in its brain, flowering out of the top of its head, releasing spores to be carried off to infect some new ant or even some grazing animals. Tiny worms known as brain flukes can do the same.

Successful rabies virus, brain fluke, or fungus genes hijack another animal's body or *phenotype*, including its brains and sensory organs, and use it to walk around, make decisions, and clamp jaws down on things. All of this is in service to the parasite's interests rather than the host's, which is normally killed anyway once it has served its purpose. We instinctively recoil at the specter of one of these parasites infecting us, but we equally instinctively recognize why they can exist. These parasites can live and evolve at our expense because they do not share the same route into the future as the rest of our genes. So long as one of these parasites can get itself transmitted into another body before you die, it will go on to live another day, and this means it can do with you more or less what it likes. The same principle explains why common viruses and diseases can prosper. Thankfully, it also tells us why extremely virulent diseases in general are rare. If you caught a virus that killed you before you could pass it on to someone else, your death would also be its death. This is why, terrifying as it is, the *Ebola* virus—which kills around 90 percent of its victims and frequently within twenty-four hours of symptoms appearing—is quite rare.

With the advent of culture, another sphere of evolving entities arose that does not share the same route into the future as our genes. This new sphere of evolution was the world of ideas. They are cultural replicators that exist by inhabiting our minds, and their "purpose" is to get us to transmit them to other minds. Richard Dawkins coined the term *memes* to describe these units of cultural evolution, which, like the biological brain parasites, will not necessarily evolve to have our interests in mind but *theirs*. Thus, when we recite advertising jingles or tell jokes, or can't get a tune out of our heads, things that are of little or no value to us have somehow commandeered our brains and even acquired mouths and vocal cords to help them invade someone else's mind. This is not to say that all elements of cultural evolution evolve to exploit us, or that all memes are viruses of our minds. Among the most common memes will be those that do us the most good. The success of ideas like how to build a better hand ax or spear, or how to navigate by the stars, skin a newly killed animal, fish, or tie a knot in a rope, comes from these being ideas we *want* to tell others.

Still, the sheer volume of possible memes tells us that competition among them for space in our minds has been intense. This raises the question of whether, just as we cannot defend ourselves against some biological viruses, we might often be at the beck and call of selfish memes that get us not just to sing a tune, but to behave in whatever ways they decree, and then like a biological brain parasite dispense with us. The philosopher Daniel Dennett once quipped that perhaps "a scholar is just a library's way of making another library." Far-fetched? Maybe, but don't forget that some medieval monk scribes devoted their lives to creating libraries full of copies of revered books, even if they didn't go quite so far as replicating entire libraries. And it is difficult to avoid the conclusion that the memes are in control when hearing of yet another religious martyr who has sacrificed his or her life in the name of some religious cause. That act of martyrdom might be very effective at spreading a religious or political idea to others' minds, or at killing minds the meme has not been able to infect, but it surely does nothing for the martyr. Or what of the Christian *Stylites*? These were the religious ascetics who in the early days of Christianity took up residence perched atop tall poles or pillars. What could have possessed them to do this? Some remained on these perches for years: even if this didn't kill them, it is difficult to see how it could have promoted their reproductive success.

Ideas such as these, and others such as celibacy, drug taking, recklessness, birth control, or notions of courage and bravery, can enter into direct conflict with our genes, often damaging our ability to survive and reproduce. Add to these the many silly rituals and customs so common to cultural behavior, and we don't have much encouragement that it is our genes that win against the cultural replicators. In fact, it has become something of a badge of the true believer among those who study memes that there is no reason to expect our genes to win—that there is no reason to expect, as the evolutionary biologist E. O. Wilson maintained, that our "genes hold culture on a leash." It is the theme of films such as the *Matrix* and *Terminator* series, or the computer HAL in *2001: A Space Odyssey*, that the machines we create will ultimately take over and hold us (genes) on a leash. If we scoff at this, we need only think back to the "millennium bug" or the Y2K (year 2000) scare that engulfed the world at the turn of the millennium. Owing to the design of software systems that could not represent a year date beyond 1999, there was a widespread belief (meme!) that computer systems around the world would fail at midnight on the last day of that year. Billions of dollars were spent preparing for this eventuality. If this now seems a long way in the past, it is, but our dependency on machines has only grown since then.

Still, in spite of all this, and in spite of much of the hysteria that can surround the ideas of "exploitative" memes, there is a fundamental reason why we can expect ideas to be rare that directly hurt our chances for survival and reproduction, and for the same reason that brain flukes and rabies are rare. Daniel Dennett has said that "the haven all memes depend on reaching is the human mind, but a human mind is itself an artifact created when memes restructure a human brain in order to make it a better habitat for memes." Dennett is right; memes do depend upon reaching our minds, and they do structure our brains. I might hold the belief that there is only one true and just God. Once that idea has lodged in my mind, it might invite other beliefs such as that people who believe in other gods are a threat to me, and to my way of life. And this might make my mind vulnerable to the further idea or meme that people who stop believing in my god, or people who profess a belief in other gods, should be punished, maybe even killed. These memes are structuring my mind and working together to promote each other.

We know these things happen because we see people being killed for their religious beliefs all over the world. Even so, we shouldn't take this restructuring to mean that our minds are passive in the face of ideas that might act against our good wishes. Our brains are the descendants of the brains of a long line of survivors, and we know this has given them certain predilections, abilities, and biases for dealing with our world. One of these is the ability to make decisions that promote or preserve our well-being. This can give us some hope that rogue and selfish memes that bring us harm can be kept in check. These cultural replicators must contend with the biological ones—our genes—that build the minds these memes need to inhabit. And indeed the memes that promote celibacy, suicidal acts, debilitating drug use, giving your life in battle, mindless sensation seeking, or endlessly playing computer games are not all that successful despite the attention they receive and the number of people they could "infect."

None of this is to say that genes win, and memes lose, but that our genetic selves are not unquestioning havens for memes and their machinations. The invention of agriculture—a set of memes—beginning around 10,000 years ago is a particularly interesting example. Agriculture allows people to produce surplus food and thereby avoid starvation in times of drought or scarcity. Everywhere agriculture has been invented—and it has been invented many times independently around the world—population sizes have increased. But what makes agriculture interesting is that people's health actually declined after it was introduced. Skeletons dug up from some of the first societies to acquire agriculture, for example in parts of Turkey and Egypt, show that people got smaller, their bones were less sturdy, they often had skeletal deformities from the hard labor of grinding corn, and they lived shorter lives. Does this mean that agriculture is a set of exploitative memes? Probably not. On average, people who adopted agriculture left more surviving offspring behind. Agriculture has been good for our genes, and this is almost certainly why it has all but replaced hunting-and-gathering all around the world. Natural selection does not maximize happiness or even well-being, but rather long-term reproductive success.

There is another reason we can expect to have good defenses against exploitative memes or ideas. Richard Dawkins and John Krebs pointed this out some years ago in the context of biological predators and their prey. It goes by the name "the life-dinner principle," a variant of one of Aesop's *Fables*. Aesop knew that when a fox chases a rabbit, the rabbit is running for its life but the fox is only chasing its dinner. This tells us that we can expect natural selection to act more strongly on rabbits to evade foxes than on foxes to catch rabbits. A fox that fails to catch a rabbit can look for another dinner. But there are no more lives for the rabbit that gets caught. An ant that evades a brain fluke saves its life, but the brain fluke can look for another ant to infect. People with psychological defenses against the idea of celibacy preserve their chances of reproducing, but celibacy as an idea can always seek out another mind to infect.

The life-dinner principle gives us a way to understand which of two sets of replicators is likely to "win" when one of them sets out to use the other for its own gain. Winning doesn't necessarily mean driving the loser to extinction, just that on average the winner's adaptations are not bested by the loser's. One rough way to predict the winner in any given arms race is to ask if one side stands to lose more than the other. The life-dinner principle tells us to bet on the one winning which has most to lose if it doesn't. Genes for running fast in rabbits have more to lose if they are not fast enough than genes for running fast do in foxes. A fox gene that is not fast enough to catch *this* particular rabbit can always seek out another. Genes in our bodies might have more to lose if manipulated by a celibacy or suicide meme than the meme has in failing to infect our minds. The memes can always find other brains.

Still, why don't manipulative memes just win outright? They can always evolve new ways to exploit us faster than our genes can evolve psychological defenses to deflect them. But can they? The same could be said of most biological parasites and yet they don't always win. For example, the rapid rate of evolution of the influenza virus is why you are advised to get vaccinated every year, but even if you don't, there is a good chance you won't be infected. In spite of their prodigious abilities to evolve, the viruses don't always win because each of us carries inside our bodies a miniature Darwinian evolutionary system that is evolving in real time. This is our immune system, and it protects us by generating an effectively infinite variety of different immune cells, each one capable of recognizing a different kind of attacker. Natural selection acting inside our bodies favors those immune cells good at defending us, encouraging more copies of them to be made, and allowing the others to fade away. This evolutionary process is going on inside all of us all of the time. It isn't perfect, but it generally serves us well; and those of us with better immune system genes for producing this miniature evolving system will be more likely to survive and pass those genes on to our offspring.

We can imagine that our minds, like our bodies, have a built-in Darwinian "cognitive immune system," its purpose being to protect us from damaging memes that arise spontaneously out of the torrent of idea evolution. Indeed, what could be more important to a species that has handed over much of its moment-to-moment decision making to its conscious mind, instead of relying on instincts or other hard-wired responses to threats from the "outside" world? Our bodily immune systems attack foreign invaders with specialist immune system cells called *T-cells* that attach themselves to the invader and then put up a chemical flag to warn of their presence. Analogously, our minds might have mechanisms for generating a variety of thoughts, hunches, and worries, each used to latch onto an idea (meme) and test it for its usefulness or suitability, attaching warning "thoughts" to some ideas or memes, disarming or rendering inert the really nasty ones. A natural selective process will mean that at any given moment your mind will be seething with a collection of conscious and subconscious thoughts that have proven the most effective in warding off the dangerous or harmful ones—they might even collectively define our "good judgment." The system is not perfect, but just as with our immune system genes, those of us with genes that grant brains capable of forming good judgments will have been more likely to survive and pass them on to our children.

Even so, if we think all of this means we are not in thrall to many of our cultural ideas, or even controlled by them, we would be wrong. This is because the analogy between memes and biological brain parasites breaks down for most aspects of culture. We are not in general locked into a life-dinner "arms race" with culture, continually evolving to avoid its clutches while it evolves to get better at infecting us for its gain at our expense. Where a parasite, suicide meme, or even a silly tune harms its host or at best does it no good, the rules, beliefs, ideas, and customs that make a culture have, on balance, advanced our interests. We have been far better off with than without them. And sure enough, when we look at the products of culture, the list of things that have promoted our survival and well-being is long—tools, spears, arrows, baskets, shelters, methods of hunting and fishing, or even just how to tie a knot in a rope. E. O. Wilson once said that people are "just DNA's way of making more DNA," and what will emerge throughout the rest of this book is that culture—that software collection of ideas, routines, rituals, and behaviors written into our brains—is the most successful way there has ever been of making more people.

But this raises a point of great and fundamental importance to our psychology: we have every reason to suspect that our culture's hold on us is far stronger than that which manipulative parasites or exploitative memes can exact from their hosts. They are things we seek to avoid, and most of us do, even if not always successfully. But we will have been selected to embrace many aspects of our cultures, even allowing them a degree of mind control, because they became the most potent trait we could acquire. We have good reason to expect that human children have been shaped by natural selection to absorb information about their culture from their parents and other teachers, rather than rely on instincts coded by genes. We have to be that way for the simple reason that we rely more than any other species on the accumulated knowledge of our ancestors to survive and prosper. It is true this makes our children vulnerable to people and ideas that wittingly or unwittingly take advantage of them. And yet, we cannot escape the fact that you—for the most part—embrace your culture because it is your ticket into the future, just as your genes—for the most part—embrace your body because it is their ticket into the future.

If we accept that our cultures have promoted our genetic interests throughout our history, then the arbitrariness of our particular cultural affiliation tells us something else: it reminds us that our particular culture is not for us, but for our genes. You could just as easily have received a different set of cultural ideas from the ones you happen to carry around in your mind. This is a shorthand way of saying that our dispositions for culture evolved because they were those that led to the greatest reproductive success. Were our particular culture for us, we might have a choice in the matter of which one we join and which one we might die for. Maybe we have a taste for fish, or are drawn to hot climates. But few of us ever do have a choice because we seem programmed willingly to accept the culture of our birth, and Scotland's soccer fans—not to mention the resilience of culturally defined emotions such as xenophobia and racism to attempts to stamp them out—tell us it is hard to adjust to a new cultural environment once the one we were born into has been installed into our minds.

Still, it is not enough merely to say that culture has been successful. It would have collapsed long ago as an evolutionary gamble if people could have achieved success without succumbing to its control. In return for this allegiance to culture, we are entitled to expect that it will have evolved ruthlessly to promote our existence. Realizing that both of these things can be true is a large part of what this book is about. Chapter 1 is about the prosperity that culture has delivered. Chapter 2 helps us to understand how the mind control that culture exerts on us could have evolved and why it is so important to our societies—an idea we will revisit again in Chapter 4. Chapter 3 is about how culture has farmed or domesticated our many talents.

CHAPTER 1

The Occupation of the World

That humans invented a new kind of phenotype or body, the cultural survival vehicle, *to propel them around the world*

the rise of the genus *homo*

TWENTY-EIGHT THOUSAND YEARS ago, in a large cave at the southern end of Gibraltar, what might have been the last of the Neanderthals died. They hadn't been pushed there by the encroaching Ice Age that would reach its peak in Europe about 8,000–10,000 years later. Instead, this was their last redoubt, having been displaced, outcompeted, or simply killed by modern humans who had relentlessly marched all over Europe following their arrival some 12,000–15,000 years earlier. It must have been a time of terror, confusion, and despair for the Neanderthals, whom we now know were over 99.5 percent identical in the sequences of their DNA to us, and who had been living in Europe for perhaps 300,000 years. The new modern people would have been cleverer and more inventive, more adaptable, more mobile, and certainly more successful. They would have carried a baffling and frightening array of technologies, and they would have been good at using them. Their standard of living would have fallen far short of ours, but to the Neanderthals, life in human society must have seemed luxurious and privileged. And to make matters worse, these modern humans might even have shown up singing and playing instruments, dancing, wearing sewn clothes, producing art and making carved figures. It would have been like a scene from a science fiction story of a people confronted by a superior alien race, except it was really happening.

With the extinction of the Neanderthals, just a single species of human—modern humans or *Homo sapiens*—remained in a lineage that had spawned six or more previous species. That lineage, the genus *Homo* or humans, arose about 2.25 million years ago on the plains of Central–East Africa in the form of *Homo habilis* or "handy man," a small but upright ape whose fossil was discovered by Mary and Louis Leakey in Tanzania in the early 1960s (a recently discovered fossil of a species possibly even older than *H. habilis* has been attributed to the *Homo* lineage and named *Homo gautengenis*). Another lineage of upright apes, the Australopithecines, was alive at the same time. Its most famous member is "Lucy," or *Australopithecus afarensis*, who lived perhaps 3.2 million years ago in and around modern-day Ethiopia. The Australopithecines had smaller brains and were more primitive than *Homo*, and they would later go extinct, possibly the first casualty of the newly evolving human lineage. Both of these lineages of upright apes—*Homo* and the Australopithecines—trace their ancestry back through even earlier species, and eventually back to a chimpanzeelike common ancestor that lived perhaps 6–8 million years ago.

Homo habilis clearly bore the marks of its ancestry to the apes, but not long after its appearance, sometime around 1.8 to 2 million years ago, a species called *Homo erectus* appeared. It stood fully upright and might have reached a height of five and a half to six feet tall. If you were to meet one you would recognize it as different from us, but you would also see in it the first real indicators of what we would become. Gone were the small brains, short legs, and long arms of the Australopithecines and early *Homo* species, and in came the bigger brains, long legs, and shorter torsos of so-called humans. Around 800,000 years ago, possibly even earlier, some *Homo erectus* populations made their way out of Africa and inhabited parts of Eurasia.

Other *Homo erectus*, known by some archaeologists as *Homo ergaster*, remained in Africa. It was these *H. ergaster* populations that by about 800,000 years ago had evolved into a species known as *Homo antecessor*, which in turn gave rise by 500,000 years ago to a species called *Homo heidelbergensis*.

There might have been one or two other premodern or archaic humans around this time with names such as *Homo helmei* and *Homo rhodesiensis*, but there seems to be a growing view that they can all be included within *H. heidelbergensis*. It is difficult to be precise and opinions vary, but *H. heidelbergensis* seems to have spawned two new lineages. One of these lineages would eventually lead to the Neanderthals and a species whose identity was only just confirmed in 2010, and which is known by just a single tooth and a finger bone. Originally designated *X-woman*, this species is now being called *Homo denisovan* after the Denisova Cave of southern Siberia where the tooth and finger bone were found. Dating of the cave suggests these people were still living as recently as 30,000 years ago, meaning they overlapped with both Neanderthals and modern humans. Remarkably, it has been possible to extract ancient DNA from the tooth and finger bone, and this reveals the Denisovans to have been a sister species to the Neanderthals.

The second lineage is the one that would eventually lead through one or more premodern archaic humans to us. Our species, so-called fully modern humans, finally emerged from their archaic *Homo sapiens* ancestors only around 160,000–200,000 years ago. This might have occurred in East Africa, as has long been believed, or possibly in southern Africa, as hinted at by modern genome studies. But wherever we arose, we are still a very young species by comparison to most, like one of those birth-of-a-star events beloved of astronomers. Given our recent appearance, it is not surprising that the first modern humans would have been almost indistinguishable from us today, and one of them brought up in modern society would not be out of place. They had large brains, possibly even somewhat larger than our own, and somewhat bigger bodies; but compared to Neanderthals they were lighter, less robustly built, had high foreheads, and lacked the protruding brow that characterizes that species.

Homo sapiens was distinguished from other *Homo* species, including premodern humans, not just by their appearance but also by showing the first glimmerings of symbolic thinking in the form of art and adornment. It was a revolutionary development in our minds because now one object could stand for another or even for a set of ideas, and a symbol's presence acted to communicate those ideas to other people. Until recently, the earliest evidence for symbolic thinking came from Blombos Cave in Western Cape Province, South Africa where pierced and painted seashells thought perhaps to be early examples of jewelry, and even an engraved stone, have been found and dated to around 75,000 years ago. But excavations not far away at Pinnacle Point in Western Cape Province have now found little beads of red ochre pigment that might be 160,000 years old. Evidence such as this is notoriously difficult to interpret, but these small fragments of ochre are clearly and deliberately scratched as if to make a powder, suggesting that these modern humans were painting something; though whether it was their bodies, caves, or other objects isn't known.

Some modern human groups might have moved north through the great continent of Africa around 120,000 years ago in what was the first "out of Africa" migration. Fossil evidence suggests that some of these people turned eastward and crossed the narrow Bab al-Mandab Strait at the southern end of the Red Sea and into present-day Saudi Arabia. Others continued northward, eventually making their way into the Levant in an area of what is present-day Israel, where archaeological digs show they might have lived alongside Neanderthals. Opinion is divided, but modern humans apparently didn't last in the Levant, or if they did, their numbers were so small that no fossil or archaeological traces remain after around 75,000 years ago. But then back in Africa sometime around 80,000 years ago, maybe somewhat later, the flickerings of symbolic thinking and communication that had characterized our species from its origins gave way to a flowering of culture. Abstract and realistic art appeared, jewelry in the form of threaded shell beads, teeth, ivory and ostrich shells, ochre and tattoos; small stone tools appeared in the form of blades and burins; bone, antler, and ivory artifacts were made; there were tools for grinding and pounding; and improved hunting and trapping technology, including spear throwers, and possibly even bows, and nets.

So imaginative had our species become, archaeologists define the appearance of so-called modern humans—in comparison to the "archaic" modern humans who immediately preceded us—by our artifacts, or the things we made, as much as by any real changes to our appearance. Genetic evidence points to a time around 60,000 years ago, maybe somewhat earlier, maybe somewhat later (the dates cannot be more precise than this), when populations of these modern humans left Africa for a second time. They followed coastal routes across the Arabian Peninsula and into India, thereby acquiring the name "beachcombers." Modern human populations eventually made their way to Indonesia and Papua New Guinea, and then by 45,000 to 50,000 years ago into Australia. The speed with which they got there after having left Africa leads some to believe that these first occupiers might have been descendants of the earlier out of Africa migration into the Middle East, but no one knows for sure. Other modern human migrations up through the Middle East eventually took people into Eurasia, where they replaced the resident *Homo erectus* and X-woman/Denisovan species, and into Europe by around 40,000 to 45,000 years ago, where by 28,000 years ago they had replaced the Neanderthals.

No one can be sure whether modern humans simply outcompeted these other *Homo* species or whether they moved in and killed them off in direct confrontations. As with so many questions like this, the truth is almost certainly some of each. But startling recent evidence reveals that not all of our contacts with these people would have been warlike; some might even have been amorous. A few of the Neanderthal skeletons and the sole Denisovan fragments come from individuals who died recently enough in our past to fall within the range of time that it is possible to extract ancient DNA from their bones. A careful comparison of the complete genomes of these Neanderthals with those of modern humans, and a similar comparison between modern humans and the new Denisovan species, suggests that we might have interbred with both of them. If we did, some might find it reassuring to note that the evidence indicates the contact was limited: modern humans share around 4 percent of their genome with Neanderthals and Denisovans. This is to say that against the background of our genomes being approximately 99 percent or more similar to these species anyway—reflecting our recent divergence from a common ancestor—there are regions whose precise genetic signature gives them away as being even more recently shared.

But there is a further twist. Current evidence indicates that interbreeding with these archaic *Homo* species occurred in two brief episodes. One episode happened possibly in the Near or Middle East shortly after a subset of modern humans left Africa. The second occurred further to the east, probably near to where the Denisovan fossils were found, and only in ancestors of modern-day Melanesian and Australian Aboriginal people. Spectacularly, neither signature of interbreeding is observed in people of African descent. So, we learn that the descendants of modern human populations that migrated out of Africa are mongrels owing to dalliances that occurred while their ancestors were trekking their way around the world. Intriguingly, when the genetic analyses revealed that Neanderthals had red hair, speculation arose as to the origin of red hair in some modern-day Scottish people—it is, after all, an unusual human trait. But the Scots can rest easy knowing that at least this part of their heritage did not come from Neanderthals—the gene causing red hair in modern humans differs from that in Neanderthals.

The era of our occupation of the world might have been a time when sightings of yeti, abominable snowmen, bigfoots, and possibly even hobbits were common, because there would have been at least three and perhaps as many as six distinct human species simultaneously walking the Earth. In addition to ourselves, the Neanderthals, the remnants of *Homo erectus* and Denisovan populations in Asia, and an unnamed archaic *Homo* species in India, there was possibly one other. *Homo floresiensis*, discovered in 2004 and nicknamed "Hobbits" by newspaper editors straining for a headline, stood around three feet tall. These tiny upright apes lived in caves in an isolated pocket deep in the jungles of the Indonesian island of Flores, possibly until 17,000 years ago. Nobody can be sure but they appear to be a dwarfed and small-brained descendant species of the *H. erectus* populations that had left Africa. Features of their anatomy, at least, indicate they are a separate branch of *Homo* evolution to our own. They almost certainly survived longer than the other competitors to modern humans by staying out of sight—there is no evidence that modern human populations ever made contact with them.

After around 28,000 years ago the only places left on Earth for modern humans to occupy had never before seen human species of any kind. By around 18,000 years ago, groups of what we would now think of as Siberian people moved north and east into a large landmass known as Beringia that had been exposed when sea levels dropped during the last Ice Age. Beringia connected what is now present-day Russia and Alaska, allowing these Siberian people to walk into the Americas. They quickly colonized the northern and southern landmasses of this large continent that spanned nearly the entire north-south axis of the world. People might have reached Cape Horn at the southern tip of South America—nearly 9,000 miles in a straight line from where they first entered North America—by as early as 15,000 years ago, although the evidence is controversial. To get there would have meant moving out of cold polar regions through the temperate climes of North America, across the dry deserts of Mexico, through Central America, and then down the rugged coasts of Patagonia. Eventually people would move all the way to the harsh, cold, wet, and windy environment of the islands of Tierra del Fuego that lie just off the southern tip of South America.

The last unoccupied region of the world was now the Pacific Ocean. Even though it lapped up against the shores of South America, that continent's new occupants evidently were not sailors. Instead, the occupation of the Pacific would come out of the geographical Far East (or from the vantage point of the people in South America, the West). But it would have to wait until around 6,000 years ago when other descendants of Siberian populations moved out from present-day China and Taiwan. They migrated southward through the Philippines and then sailed eastward out into the vast Pacific, carried by newly developed and increasingly sophisticated boat technologies. One by one, thousands of far-flung islands would be occupied as these Lapita people, so named by archaeologists for their distinctive pottery designs, learned to navigate by the stars using "star maps" made from knotted string. They would do this thousands of years before the Vikings made their tentative journey a few hundred miles across the North Atlantic to what is modern-day Britain.

The settlement of the Pacific was a breathtaking and daring accomplishment that can really only be appreciated by going to one of these remote Pacific islands to get a sense of the scale and danger of that ocean. Most of the islands would have been small and well out of sight distance from one another, and no one had the slightest idea what if anything was out there. Still, something pushed these Austronesian people on, and their boats eventually reached the far-flung shores of Easter Island, and then somewhat later they finally made contact with South America, probably somewhere in present-day Chile. There they might have had a surprise because unlike the thousands of islands they had occupied throughout Oceania, this piece of land had people on it. They might also have felt a strange sense of déjà vu because these were people from whom they had separated maybe 20,000 or more years earlier, perhaps somewhere in the Altai Mountains of present-day southwestern Mongolia, and whose broad facial features they might have shared. Polynesian people went on to discover New Zealand in the south and Hawaii in the north only around 1,000 years ago. Reaching those islands, given their remoteness and the technology that was available, must surely rank alongside going to the Moon, and even then we knew the Moon was there before we set off.

The modern-human occupation of the world was now complete, all within a few tens of thousands of years after leaving Africa; and most of this within the first 20,000. It was an occupation that had begun back in Africa when as few as several hundred to several thousand people left that ancestral continent, so that today, remarkably, and in such a short period of time, all of us on Earth trace our ancestry back to this small and intrepid band. By the end of the Polynesian expansion, humans now inhabited deserts, savannah, prairies, marshes, rain forests, and ice. They lived on tops of hills, in valleys, islands, and most places in between. They spoke thousands of distinct and mutually incomprehensible languages. They had evolved a variety of different mating practices such that sometimes men had more than one wife, other times women had more than one husband, and in other societies people practiced monogamy. In some societies men would have to pay fathers bride prices for their daughters in marriage, in other societies fathers had to pay a dowry to get their daughter married. Some societies transmitted their wealth—in the form of belongings and territories—down paternal lines, while others transmitted it down the maternal side of the family. They constructed all manner of different kinds of shelter, ate everything from seeds to whales, and engaged in bizarre belief systems and behavioral practices. A single species had acquired a global reach, and specializations, lifestyles, and beliefs as varied as collections of different species.

Archaeological and fossil remains show that as humans occupied the final uninhabited regions of the world, they came with the same voracious momentum that had enabled them to displace, outcompete, or simply kill off premodern human populations. In North America large mammals such as the saber-toothed tiger disappeared soon after our arrival, and in New Zealand and Polynesia the huge flightless Moa birds and giant turtles survived only a few centuries in our presence. Having never before encountered our kind, they were not prepared for the highly efficient and hungry conquering machines that they met. It is a pattern that continues today. Wherever humans are found in large numbers the ecological world around them has been stripped of its diversity, with most of the animals and many plants having been driven out or made extinct. Vast swathes of Northern Europe are devoid of nearly all the mammal species that would have been present when modern humans first appeared. It is a trend we have continued right up into the present day. The title of Rachel Carson's famous paean to ecology, *Silent Spring*, is a lament to the near eradication of songbirds in America.

imitation, social learning, and culture

THE RISE of the genus *Homo* is principally the story of a single species—our own, *Homo sapiens sapiens*, as it is sometimes called to distinguish it from the Neanderthals, or *Homo sapiens neanderthalensis.* Where all the other *Homo* species in our history, including the Neanderthals, were confined to the environments their genes adapted them to, modern humans were able to spread around the world by producing technologies that allowed them to transform the world to suit their needs. The puzzle and irony of this difference is in the names. *Homo sapiens* means "wise man," but as we have seen, the genetic differences between ourselves and the Neanderthals are slight, amounting to less than one half of 1 percent in the sequences of our genes, and yet only one of these wise men occupied the world.

What made it possible for our species, the "wise wise man," to survive in and exploit nearly all of the world's habitats, where so many others, including the other wise man, had failed or in most cases never even ventured? The flowering of culture, communication, and symbolic thinking that distinguishes our species from all others can probably trace its origins to a development that was almost certainly minuscule on a genetic scale, but vast in its potential. Modern humans seem, uniquely among animals, capable of something that psychologists and anthropologists call *cultural* or *social learning.* Social learning can be fiendishly difficult to define but has—when applied to humans—two important features that together make the entirety of our cultural achievements possible and effectively open up an unbridgeable gulf between ourselves and all other animals. One is that we are capable of sophisticated copying and imitation of *new* or novel behaviors merely by watching or observing others, and without the need for specific training or rewards. We can then transmit these new behaviors faithfully to others. The second feature is that humans act as if they know what they are copying and why, and so they can choose to copy the best from among a number of alternatives, and even attempt to improve on it.

Of course, all animals can learn, and there are numerous examples of animals behaving in ways that give the appearance of social learning, at least for some for highly specific tasks. For instance, in the early 1920s, residents of the village of Swaythling in the south of England began to notice holes poked into the foil tops of the milk bottles that had been left on their doorsteps by the milkman. They soon came to realize that a local bird, the blue tit, closely related to the American chickadee, had learned to peck at the bottles to obtain cream off the top. These birds even recognized that the different colored foil tops on the milk bottles identified some as having more cream than others. This behavior spread throughout southern England for the next twenty years, and is often quoted as an example of one bird learning from another by social learning. Other examples are the washing of food at the seaside by Japanese macaques, and the well known "fishing" by chimpanzees in which they use slender sticks to draw ants or termites out of their underground nests.

These behaviors are compelling to watch, but on close inspection it is apparent that the animals are not so much copying each other in any exact way, as having their attention drawn in some very general way to something they can do. Equally, none of these behaviors seems to improve over time—the birds don't get better at obtaining cream from milk, and the macaques don't get better at washing food. By comparison, if you were a hunter-gatherer living in our distant past and you witnessed for the first time two people making hand axes by chipping or "flaking" them out of larger pieces of stone, you would probably understand or be able to work out what they were for, and you might even recognize differences between them that meant one of these hand axes was going to work better than the other. Later, you might try to imitate the chipping actions of the person who made the better one, and you might wonder how to improve on it. Maybe you want to make it sharper or perhaps less prone to breakage. Even if initially the two axes seemed the same to you, you might later notice that one of them seemed to be sharper or easier to hold, or you might notice that one of the two people was able to skin that large animal they had just killed better than the other. In either case, it would be obvious which one to copy.

Obvious to you at least because, incredibly, we seem to be the only species that can do this. We can learn new behaviors according to the old saying, "Monkey see, monkey do," but the surprise is that monkeys, and other animals, for the most part, cannot, or when they can, it is often limited to highly specific behaviors that already exist in their repertoire of behaviors anyway. Thus, if you were like other animals watching the two people make a hand ax, you might not pay any attention at all, and if you did, you might make nothing of it. Instead, their actions might do little more than draw your attention to stones, and you might even pick two of them up and bang them together, because this is something you do anyway. One of the stones might even crack and chip off a piece that you might then pick up and later discover, again by chance, that it was useful for chopping. Biologists and anthropologists who study animal learning call this *stimulus enhancement*, and it is distinct from true social learning. The distinction is subtle but important because, unlike true social learning, learning from stimulus enhancement doesn't seem to translate into new or purposeful behaviors that faithfully copy or pick up where others have left off. Rather, it just makes use of old or even hard-wired behaviors already in an animal's repertoire, even if sometimes in a slightly new context. Pecking is something that birds do anyway—it is for many birds the principal way they obtain food. Pecking at the tops of milk bottles, then, might be less a case of precise social learning than of stimulus enhancement.

Battles rage among scientists who study animal behavior over whether some specific action is a case of social learning or stimulus enhancement. But if we apply a simple test most examples provide a clear answer: does a species show any evidence of the behavior becoming more sophisticated and refined over long periods of time? This test is crucial. If an animal lacks true social learning, each new generation will have to rely on trial and error, catalyzed only by a little push of stimulus enhancement, to discover for themselves how to perform some action or use some tool; they do not seem to learn directly from others. For example, chimpanzees are often lauded as the champions of culture in the rest of the animal world. Anthropologists have documented that chimpanzee groups living in different parts of the forest have about thirty different cultural traditions in such things as how they fish for ants and termites, or how they use stones to crack open nuts. But these differences among chimpanzee societies almost certainly owe more to the vagaries of local circumstance than to any real design on the part of the animals, because there is no evidence that the chimpanzees, or for that matter any other animal, get better at using or producing these tools—they don't build better nut-cracking devices or invent better ways to fish for termites. Instead, being surrounded by other chimpanzees that use the sticks this way seems to make it more likely that a naive one will pick up a stick and poke and prod things with it—things chimpanzees do anyway. Then, just by chance this might lead to acquiring a few ants or termites to eat and this reward seems sufficient to keep the behavior going.

What this means for most species is that any new innovations or improvements seem limited almost entirely to what an individual can produce on its own, because, they don't seem to recognize and then acquire them from others the way we do: they don't seem to be aware of the innovations, much less whether they are useful or not. Even if there were a chimpanzee-Einstein, its ideas would almost certainly die with it, because others would be no more likely to copy it than a chimpanzee-dunce. And this means that, lacking social learning, there is no real cultural ratchet that leads to improvement over time, no shared reservoir of accumulated ideas, skills, and technologies. Instead, each individual chimpanzee is left to come up with its own rules and own particular styles of nut cracking, or termite fishing. Indeed, were we to go away for a million years, upon our return the chimpanzees would probably still be using the same sorts of tools to fish termites from the ground. The same is true of the birds pecking at milk bottles or the macaques washing their food. This simple difference creates a vast difference between the other animals and us. Just imagine if each generation we had to learn for ourselves how to make fire, flake hand axes, make bows and arrows, sew clothes, navigate by the stars, or build shelters or hunt game, not to mention how to build printing presses, computers, and spacecraft.

The reason animals don't seem to move beyond stimulus enhancement or having their attention drawn to things is that they don't seem to put themselves inside the minds of others, or, as Michael Tomasello argues in his *Cultural Origins of Human Cognition*, they seem to lack a "theory of mind," or the ability to adopt another's point of view. They don't seem to assume as we do that someone is doing something *for some reason or purpose*. Lacking a theory of mind is why a chimpanzee can be taught to paint, but the animal is not really "painting," just spreading paint onto a surface, and chimpanzees will go on doing this, aimlessly painting over what they have done until they grow tired of the activity or until someone takes the brush away. They don't seem to "get" the bigger idea of what they are doing or why. Indeed, a chimpanzee can be taught to use a saw to cut wood, push a broom to sweep a floor, drink tea from teacups, exchange plastic tokens for food, and even wash dishes, given a sufficient reward. But before you think of employing one around your house, be aware that it will as happily wash a clean dish as a dirty one. This is because it is not washing dishes to clean them, as we do, but to get a banana.

Even in humans, a theory of mind emerges only sometime around a child's third or fourth year. A procedure known as the Sally-Anne test shows a child two dolls: Sally and Anne. Sally has a marble that she (with the help of a human experimenter) puts in her basket. Then Sally leaves the scene. While Sally is away, Anne takes the marble from Sally's basket and puts it in her box, which differs in appearance from Sally's basket. Sally then returns and the child is asked where Sally will look for the marble. Children younger than three to four say she will look in Anne's box (intriguingly, many people who suffer from autism also respond this way). But older children realize that Sally can have beliefs that differ from theirs, and they correctly say she will look in her basket.

There are many varieties of these "false-belief" tests, and they can be difficult to interpret, but they all point to three to four years of age as being a critical period during which children's awareness of others' minds develops. To be fair to the other animals it must be allowed that some of them, especially some birds, behave as if they have an awareness of what others are thinking. For instance, if the small bird known as the nutcracker (and some other jay and crow species) sees another bird watching it while it hides its food, it will return alone later to hide the food in a new spot. What matters for this discussion, though, is that no other animal ever seems to get as far in their understanding of others' minds as a four- to five-year-old human. Even among the Great Apes, any theory of mind seems to be no more advanced than that of a human two-year-old.

It is staggering and baffling to us that other animals could be so dim-witted. It is not that they are stupid: a chimpanzee is better at being a chimpanzee than you are. It is just that they lack social learning, and this small difference has made all the difference. But what of our more recent ancestors, such as the many now-extinct species in the *Homo* lineage? The African Rift Valley is an angry tear in the Earth's crust that stretches for thousands of miles. It was in a part of the Rift Valley in Tanzania called the Olduvai Gorge that the pioneering archaeologist Louis Leakey discovered objects over 1 million years old that appeared to have been deliberately and intentionally shaped by hands. Later work established that these were *Homo erectus* hand axes, and they were often found near to bones with cut marks on them made by the same hand axes. For many archaeologists this time in our history, perhaps 2 million years ago, is one of those defining moments, a time we can look back on romantically as being the moment that creative thinking arose. These early humans, it seems, had acquired not only a compulsion to make things but also the insight that they could alter and improve them. It is that "light bulb" moment in Stanley Kubrick's *2001: A Space Odyssey* when the ape throws a bone up into the air. We watch it tumbling end over end, and as it falls back to the ground the bone transforms into a spaceship on an interstellar mission.

But the archaeological record tells a more prosaic story about our ancestors' creativity. Remarkably, from careful sifting through layers of strata, archaeologists have been able to determine that our *H. erectus* ancestors living on the African savannah stubbornly chipped the same hand axes out of larger stones for nearly all of their 1.8 million-year history, without making any serious alterations to its form or function. For tens of thousands of generations of parents and their offspring, and the individuals watching them, this species produced the same basic tool. Their culture—their toolkit—wasn't evolving, and this is not what we expect of an animal with social learning. It is not even clear that the Neanderthals possessed the capabilities for social learning, or if they did, they were not nearly as sophisticated as ours. The Neanderthals would have been recognizably similar to us, yet hauntingly different, being stocky and muscular, with large bulbous noses. But despite having brains at least as big as ours, the Neanderthals lacked most of the outward signs of sophisticated culture so common to modern human archaeological sites of the same period. The Neanderthals did not produce any art, they didn't have musical instruments, and there is no evidence that they carved figures.

Chauvinism? Some will say yes, and the Neanderthals have their apologists. But at a time when modern humans were overflowing with sophisticated artifacts, there is no evidence that Neanderthals could engrave or shape bones, they had no sewing, no weaving, no bows and arrows, and no spear throwers, even though they would have been able to observe all of these things among the talented newcomers who had moved in right alongside them. Some Neanderthal archaeological sites yield shaped pieces of shells that might have been used as jewelry, and other sites suggest that they added symbolic objects such as flowers to graves. (Burial itself should not be taken as any sort of religious, spiritual, or symbolic act. Dead bodies decay rapidly and attract flies and predators, so burial is simply a prudent thing to do.) But even these practices seem only to appear when Neanderthals had prolonged exposure to modern humans, and very recent evidence suggesting that Neanderthals actually produced and wore pieces of jewelry is now being reinterpreted as the work of *Homo sapiens*. The paleontologist Chris Stringer has even speculated on a BBC radio program that if there were slight changes in some Neanderthals' capabilities late in their history, this might reflect brain genes they had acquired from their interbreeding with modern humans.

This description of the Neanderthals is not what we expect of a species with true social learning. No one can be sure, but we can only guess that the mental life of the Neanderthals was, and still is for all other animals, a plodding, inflexible, literal, and unimaginative existence, at least compared to ours. While we were spreading around the world, the Neanderthals' limited technologies meant they were confined almost exclusively to the environments of Western Europe, parts of the Middle East, and southern Siberia. While we were using sophisticated spears and arrows to hunt large mammals, the Neanderthals were close-range hunters with short spears for jabbing, or who relied on clubbing or stoning their large prey—and each other—to death. The Neanderthals' famously robust and muscular physique probably speaks volumes about their lack of cultural complexity, while our gracile and refined appearance trumpets our virtuosity at substituting tools and clever thinking for brute physical force.

The Neanderthals' stocky build made them well adapted to the cold climates of much of Europe and Eurasia, but the irony is that our species—whose tall and slender bodies were certainly not cold-adapted—replaced the Neanderthals during the Ice Age that engulfed these lands. It seems the Neanderthals simply could not adapt their lifestyle of hunting for large game rapidly enough to the declining populations of large animals that the encroaching ice would cause. But we could, and the difference is probably down to social learning. Or think of it this way. Twenty-eight thousand years ago, the Neanderthals sat in Gibraltar going extinct while gazing across the straits to the warmer climes of Africa clearly visible only eight to ten miles away, but they were unable to make boats to carry them there.

cumulative adaptation and cultural survival vehicles

THE NEANDERTHALS' plight reminds us that each of the many biological species on Earth exploits its particular environment, but for the most part it is *only* that environment that it can occupy. This is because biological species are vehicles built by sets of genes that have evolved together over millions of years to be good at solving the problems posed by a particular environment. For instance, woolly musk oxen are the product of a coalition of genes that natural selection has roped together to produce a vehicle suited to surviving the cold temperatures of Siberia. A different coalition of genes gives rise to camels, a vehicle good at surviving even the scorching deserts of the Sahara; monkeys are vehicles adapted to climbing trees; and the coalitions of genes we call penguins produce a fishlike bird vehicle that can survive the Southern oceans. A camel would make a poor musk ox and a penguin a poor monkey. A cross between a musk ox and a camel—were one possible—probably wouldn't be much good at being either a camel or a musk ox.

The lesson we learn from this is that there are no real shape-shifters in nature, nor anything like children's Transformer toys that can change what they are. Being limited to what their coalitions of genes evolved to do, no one species can do everything. That was, of course, until humans came along and rewrote all the rules that had held for billions of years of biological evolution. Here was a single biological species using just a single coalition of genes that was nevertheless able to adopt different guises and forms in different places. In one place we could be like a heron able to pull fish from the sea, in another like a lion able to bring down large prey, in another like a camel able to survive in the desert, and in yet another we could float on the water like a duck or a seagull. Our cultural survival vehicles were built not from coalitions of genes but from coalitions of ideas roped together by cultural evolution. This meant that for the first time a single species was able to spread out and occupy every corner of the world. Where all those species that had gone before us were confined to the particular genetic corner their genes adapted them to, humans had acquired the ability to transform the environment to suit them, by making shelters, or clothing, and working out how to exploit its resources.

It was social learning that made our shape-shifting possible because social learning is to ideas what natural selection is to genes. Both are ways of picking out good solutions from a sea of variety. Natural selection builds complex adaptations like eyes and brains from the successive accumulation over millions of years of many small genetic changes, each one of which improves on its previous form. Equally, social learning builds complex societies by a process of cumulative cultural adaptation as people select the best from among a range of options, improve on them, or blend them with others—what Matt Ridley in his book *The Rational Optimist* calls "ideas having sex." And so our knowledge, ideas, technologies, and skills accumulate and build increasingly complex objects. When someone noticed that a club could be combined with a hand ax the first hafted ax was born. When someone tied a vine to the ends of a bent stick, the first bow was born and you can be sure the first arrow soon followed.

The analogy with genetical evolution is deeper than mere words: just as genetical evolution brings together the sets of genes that produce a successful biological species or vehicle for a particular environment, cultural evolution brings together the sets of ideas, technologies, dispositions, beliefs, and skills that over the millennia have produced successful societies, good at competing with others like them, and well adapted culturally to their particular locale. These are our cultural survival vehicles, and it is important to see them as not different in principle from biological vehicles, it is just that the information on which they are based takes a different form: it resides in our minds rather than in our genes. Thus, when people walked into the Arctic and survived, it was because they had acquired the knowledge and technology to make clothes suitable to that harsh environment, to build shelters out of ice, and to fish in the cold Arctic waters. At a later time and different place, when Polynesian people invaded the Pacific, it was because they had acquired the technology to produce seagoing boats, and the knowledge of how to navigate by the stars.

Indeed, we can think of our differing cultural survival vehicles as playing the same ecological role as different biological species. Just as a camel would make a poor musk ox, a Polynesian would not be well equipped to survive the Arctic. But of course our cultures can adapt on the fly and without having to wait for genetic changes to come along, and so the rapid spread of our various cultural species around the world after we left Africa is like a tape of biological evolution speeded up a millionfold or more. Almost everything around us today in our modern world can be attributed to social learning and the cumulative cultural adaptation it propels.

This is not to say our genes played no role in our occupation of the world, just that it was our cultures that took us to its various environments to begin with. When people walked into the Arctic, they began to evolve genetically to have a stocky build that made them better at retaining heat, but it was their culture and not their genes that took them to the Arctic to begin with. Similarly, the Polynesians would also adapt genetically to their hot and sunny environment by becoming leaner and darker-skinned, but again it was their culture that got them there.

Modern genomic studies of large numbers of people are discovering many small genetic differences among human groups that confer some sort of advantage in their environments. For example, in some European and African societies with a long history of dairying, adults have acquired the ability to digest milk. We have seen how some Tibetan people have acquired an extraordinary capacity to extract oxygen from the air at high altitudes, and how some Han Chinese have an unusual ability to metabolize alcohol. Hunter-gatherer groups exposed to more starch in their diets produce more salivary amylase—an enzyme that begins the process of digestion while food is still being chewed—than those whose diets contain less starch. Differences in facial appearance around the world might be related to arbitrary preferences in the choice of mates.

These are just some of the many small genetic differences among human groups that have arisen as a result of being thrust into environments that our cultures opened up to us. And it is remarkable how quickly we have adapted. The 60,000 to 70,000 years since modern humans spread out of Africa is little more than the blink of an eye when stacked against the 6 to 7 million years that separate us from our Great Ape ancestors. The presence of these genetic differences, however small, tell us that we have had a habit of keeping to ourselves as we spread out around the world, because had we not, our genetic differences would have become blurred. This is not to deny that human groups have always traded with each other, intermarried, fought wars, and traipsed across each other's territories. But it is only by having a tendency to maintain our identities in separate cultures or tribal groups that natural selection could have sculpted our many differences, and have done so in just the few tens of thousands of years since we walked out of Africa. Then again, we might have guessed this was the case: how else but through a tendency to keep to ourselves in our cultural survival vehicles can we explain a single species that speaks at least 7,000 mutually unintelligible languages?

But why do we behave this way? Could it be that our cultural survival vehicles have evolved tendencies to protect the knowledge and wisdom to which they owe their success?

cultures carve up the landscape—linguistically

A WALK along the northeast coast of Papua New Guinea will bring you into contact every five to ten miles with a tribe speaking a different language: in that part of New Guinea you could encounter Korak speakers, quickly followed along the coast by Brem speakers, who in turn are followed by Wanambre speakers, and none of these more than ten miles apart. Each of these tribes is a distinct group of people, making their living alongside each other in the dense forests of that island, and speaking mutually unintelligible languages. If we were to encounter this diversity of languages inside an area of a typical medium-sized town, we might expect to find not just three or more different languages spoken, but three or more distinct groups of people, brought up speaking a different language and living separate lives, each having carved out a portion of the town to live in!

The density of languages in Papua New Guinea strains credulity, but recall how the Papuan man asked if it could be true that the societies which spoke these different languages were this tightly packed together replied, "Oh no, they are far closer together than that." And it is true, an astonishing figure of over 800 different languages, or about 15 percent of all languages found on Earth, are spoken in the mere 312,000 square miles of the island of New Guinea—with many having only a few thousand speakers. This is an area slightly bigger than the state of Texas. Languages are even more tightly packed in the tiny Polynesian island archipelago of Vanuatu, northeast of Australia. Vanuatu's islands cover just 4,100 square miles, and yet over 100 distinct languages are spoken on them, each one by an average of just 2,000 speakers. Even this gives a more sedate picture than becomes apparent when one is on the ground in these regions. For instance, the Vanuatu island of Gaua covers 132 square miles, and like so many of the islands in this region, it is the roughly circular remnant plug of an ancient volcano. Gaua is just twelve to thirteen miles in diameter, but this speck of an island supports five languages—Lakon or Vuré, Olrat, Koro, Dorig, and Nume. This is a density of languages about tenfold higher than that of Papua New Guinea.

Language is one of our defining traits as a species, but we are probably the only animal in which two of its individuals plucked from different places—even right next door—might not be able to communicate with one another, almost as if they were two different biological species. Sometimes, even speakers of the same language can confuse one another: a young English boy I know, travelling in America, was told by someone who overheard him speaking, "I can tell from your accent that you're from somewhere in Europe." By comparison to our linguistic isolation, you could take a gorilla from its troop and put it in any other troop anywhere gorillas are found, and it would know what to do. There would probably be some fighting over territory, and attempts at establishing who is dominant over whom, but for the most part life would be routine. The new gorilla would communicate as all gorillas communicate, fight as gorillas fight, make the same kinds of nest, and eat the same kinds of food. There is nothing special about gorillas. This experiment could be repeated with donkeys, or ducks, or goldfish, or frogs, and get much the same outcome.

So, why is it that groups of people in New Guinea, or more generally just about anywhere in the tropics, all more or less living the same lifestyle, divide up their territories so exclusively as to evolve different languages, and sometimes every few miles? What makes this division even more peculiar is that, where different biological species specialize at exploiting different features of the environment—what biologists describe as a species' *niche*—in any given area the humans are all occupying more or less the same niche, save for one: human societies seem to have a disposition to acquire their own linguistic niche and then maintain it. The anthropologist Don Kulick describes how

> New Guinean communities have purposely fostered linguistic diversity because they have seen language as a highly salient marker of group identity . . . [they] have cultivated linguistic

> differences as a way of "exaggerating" themselves in relation to their neighbors. . . . One community [of Buian language speakers], for instance, switched all its masculine and feminine gender agreements, so that its language's gender markings were the exact opposite of those of the dialects of the same language spoken in neighboring villages; other communities replaced old words with new ones in order to "be different" from their neighbors' dialects.

Kulick also relates an account from another linguist of a New Guinean village of Selepet speakers. One day, the community met and collectively decided to change their word for "no" from *bia* to *bune.* The reason they gave was that they wanted to be distinct from other Selepet speakers in a neighboring village, and with immediate effect. They have spoken differently ever since. We can only sympathize with the confusion someone would have felt who had gone away hunting for a few days.

There is speculation that humans might be innately programmed to recognize and prefer people who share our language, or that if not innate, the preferences arise very early in life, even before we can speak. By five to six months, infants prefer to look at people whom they have heard speaking their native language. Katherine Kinzler and her colleagues note that

> Older infants preferentially accept toys from native-language speakers, and preschool children preferentially select native-language speakers as friends. Variations in accent are sufficient to evoke these social preferences, which are observed in infants before they produce or comprehend speech and are exhibited by children even when they comprehend the foreign-accented speech. Early-developing preferences for native-language speakers may serve as a foundation for later-developing preferences and conflicts among social groups.

Neighboring communities also of course distinguish themselves in customs, beliefs, art, dance, weaponry, costumes, singing, music, and architecture. For instance, among the nomadic pastoralists of Northern Kenya, the Gabbra people dress simply in muted colors, while their next-door neighbors the Samburu wear vivid red robes, and the nearby Turkana favor dark colors and, among the women, copious amounts of metal jewelry and neck rings that can give them a daunting appearance.

It is a pattern seen all over the world. In the first years of the nineteenth century, the explorers Meriwether Lewis and William Clark made their long trek from the Missouri River all the way to the west coast of America and then back. The lands they walked through were uncharted, and their diaries show they were struck by the sheer number and variety of the Native American tribes they encountered. Dayton Duncan and Ken Burns, in their account *Lewis and Clark: The Journey of the Corps of Discovery*, write that

> the dizzying diversity of Native American life is one of the clearest (though unspoken) images to emerge from [Lewis and Clark's] journals. The West through which the Corps of Discovery traveled was neither an "uninhabited wilderness" nor a single "Indian world." Indians thought of themselves as many different people, not as one monolithic group—and understandably so. They were as varied as the western landscape itself. . . . [Some] roamed the Plains following the buffalo herds, living in tepees that could be moved at a moment's notice; people who were farmers . . . lived in permanent villages of rounded earth lodges; people who lived in stick wickiups and dug for roots; people who fished for their food and dwelled in large houses made of wooden planks. Some measured their wealth in horses; others had no horses at all. Some were predominantly tall, or wore forelocks of their hair pushed up as a sign of distinction. Others were shorter, stouter and saw beauty in a forehead flattened by boards.

Lewis and Clark were encountering just a small number of the many different cultural survival vehicles that had evolved in the interior of that vast continent. Prior to the arrival of Europeans, there were over five hundred distinct Native American languages spoken in what is now the territory of Canada and the United States, and these are just the ones that linguists and archaeologists have been able to document.

The human tendency to separate into distinct societies has given human language a geographical mosaic on which to play out its evolution, and we expect new cultures and languages to form naturally as people spread out. The puzzle is that human groups appear to do just the opposite: it is where people are most closely packed together, as in Papua New Guinea or Vanuatu, that the greatest number of different societies is found. A colored-in linguistic map of the tropics looks as though a patchwork quilt has been laid over the landscape. By comparison in the northern regions of North America, there are only a handful of societies, each one occupying a huge area, in the entire west to east expanse of that vast continent. People need to move over large areas in this sparse landscape just to eke out a living, and this tends to blend culture and language as people continually trade, marry, and talk to each other. Remarkably, it is a pattern human societies share with biological species. Where dozens of species of bats, similar numbers of small mice and rat species, and hundreds or thousands of species of insect live in the tropics, the frozen wastelands of the polar regions of North America can only support a handful of different mammal species, such as the caribou, wolf, and polar bear.

Human cultural groups have historically partitioned the landscape among themselves almost as if they were separate biological species. But why speak a different language every few miles? Why not in regions such as the tropics form one giant cooperative society? We seem confronted with the idea that human groups have had an innate tendency throughout history to divide and form into new groups, distinct from others, and just as soon as the environment will allow it. Is it to establish an identity and to protect our knowledge and wisdom from those who might eavesdrop or, worse, subvert our group? If so, it is a risky thing to do because smaller groups are more easily overrun by others, and more vulnerable to bad luck, the loss of key people, and knowledge. But if a group can split off and survive, its members will be able to compete with and maybe even displace other groups looking to use the same lands. The advantage of forming a new group is that now *your* offspring rather than someone else's come to inherit the lands around you, and they, in turn, can use them to have even more children. Our tendency to form into distinct societies might have its origins in our most basic instincts to promote copies of our genes.

cultures restrict the flow of genes

EVEN IN our modern developed world with roads and other links, people can often differ culturally and linguistically over a few tens of miles, as any trip around the shires of England, the French *Départements* or the Swiss Cantons will reveal. These differences tell us that we are a species with a long-term history of staying put, or at least of not moving very far. It is easy to dismiss this as a simple consequence of a lack of mobility; but why does that lack of mobility exist? We must remember that we are the species that occupied the world, and we managed to do so before trains, motorbikes, cars, wagons, roads, or even footpaths were invented. So, our apparent lack of mobility really tells us that, historically anyway, once we get to a place we have tended to stay there, and maybe even slow the pace of others seeking to move in.

There is an ancient moor about five miles north of the city of Oxford in England, called Otmoor. The narrow road that winds around the roughly circular moor is about fourteen miles long and runs through seven villages, all of which date back more than 1,000 years, into Anglo-Saxon times: Oddington, Charlton-on-Otmoor, Fencott, Murcott, Horton-cum-Studley, Beckley, and Noke. In the 1960s, the biological anthropologist Geoffrey Harrison, working at Oxford University, became aware that the seven villages had kept detailed parish records of births, deaths, and marriages dating back at least four hundred years. Harrison and his colleagues realized that they could use these records to track the movement of people among the villages. They quickly established that there was very little mobility and that marriage was one of the few ways by which people moved to a different place on the moor. But the bigger surprise was that prior to the eighteenth century, people often did not venture any further than the neighboring village in their search for someone to marry. In fact, because there was no route directly across the swampy moor, people had been moving from village to village around it in a circular pattern for centuries.

This surprising lack of mobility is not a phenomenon confined to small villages in a rural part of England. Walking through New York City's Little Italy and Chinatown in Manhattan, it is easy to stand in the middle of the street that divides these two communities and hear Italian spoken by third- or fourth-generation descendants of Italian immigrants on one side and Chinese spoken by third- or fourth-generation descendants of Chinese immigrants on the other. If cultures throughout our history have tended to keep to themselves, avoid each other, or even erect barriers, such as the geographical patterns might suggest, then this should be seen in our genes. Anyone can spot genetic differences between the Chinese and Italian groups, but there are more subtle differences even among people who otherwise "look" the same.

Some years ago the statistician Robert Sokal measured a large number of background genetic markers in samples of people from all over Europe. These markers do not influence how we look or feel or what we are like. They are called *neutral markers*, and they merely identify people who have been separated for some time. Sokal applied a method called *wombling* (named after the statistician W. H. Womble) to measure the rate of change in genetic markers between these different locations. If people gradually diffuse over an area such as Europe, then no boundaries of abrupt genetic difference are expected. But Sokal discovered thirty-three boundaries in Europe that separated areas of especially sharp differences in the genetic markers. Not surprisingly, most corresponded to physical barriers such as the Alps or the English Channel. But for eleven of them, there was no physical or political barrier to the movement of people. Instead, in nine of these eleven places it was language differences that kept people from mixing. It seems humans prefer to have sex with people they can talk to!

We think of Europe as one of the most crisscrossed, settled, and resettled areas of the planet, and so genes should be thoroughly mixed. But even in Europe knowing a person's genetic profile can often place them within a hundred or so miles of their birthplace, or if not them, their parents' birthplace. In fact, plotting measures of genetic similarity among people of European ancestry produces a pattern that resembles the shape of Europe. Clusters on the lower left of the grid are from Spain and Portugal, the United Kingdom samples fall in the upper left, with French samples in between, those from Scandinavia and Germany fill the upper right, and samples from Romania, Bulgaria, and the nations of the former Yugoslavia fill in the lower right.

If we measure large numbers of neutral genetic markers from populations around the entire world and then use them to form clusters, we get back groupings that bear a striking resemblance to what have conventionally been recognized as the major divisions of people on the planet: Europeans and Western Asians, Africans, people from the Americas, Eastern Asians, and Australasians. But this is merely a statistical statement and should not be used to say that there are "races" of people with abrupt or clear genetic boundaries between them—there are not. All of humanity shares the same genes and we can all happily and successfully interbreed. And contrary to the pronouncements of some well-known public figures right up until recently, there is not a shred of evidence that human groups differ in the genetic factors that cause intelligence or even in general cognitive abilities. But the existence of distinct clusters of genetic similarity and dissimilarity seen all over the world tell us that we are a species that has made a habit of maintaining our differences.

cultures slow the flow of information from "outside"

THE ENGLISH famously lost an important battle to William, Duke of Normandy, at Hastings in the year 1066. William the Conqueror's reign and the Norman conquest of England that he brought with him carried with it many elements of the French language. French was even the official language at the English court for around three hundred years, and the effects can still be heard in the words of government like "parliament," "legislature," "executive," "judicial," "bureaucracy," and of course "government." But the English language, like its people, stubbornly refused to be overrun. It is true that English today boasts a large vocabulary because about half of its words derive from the Romance or Latinate languages, of which French is one. But the other half proudly shows off its Germanic ancestry, and its core vocabulary words are more likely to be of Germanic than French origin. For example, English speakers say "good" (from German *gut*) not *bon*, "mother" and "father" (from German *mutter* and *vater*) not *mère* and *père*, and "milk" (from the German *milch*), not *lait*. On the other hand, this large mixed vocabulary gives English a freedom and variety of expression other languages will not enjoy. English speakers sitting down to dinner (from the French *dîner*) can refer to "beef" (from the French *boeuf*) or "cow" from the German *kuh*. And there is pork and swine, mutton and sheep.

It can be revealing of our sentiments to see how cultures treat instances of traits coming in from "outside." When cultural traits are transmitted vertically from parents to offspring or teachers to the young, no one takes much notice. But traits acquired from other cultures are far more likely to be regarded with suspicion or even indignation, the home population being seen as an unwilling and impotent victim. The French are convinced their language is now suffering the reverse of what happened to English after William the Conqueror. France, in a spasm of linguistic isolationism, now devotes a government ministry to slowing or banning what is portrayed as an overwhelming march of English and American words, customs, and phrases into the French language and culture. Phrases like *le weekend* or "fast food" just won't do for this ministry. The result is that while the other nations of the world work away on their computers, the French resolutely sit at their *ordinateur* (and the Swedes at their *dator*). The British are alert and often piqued at what they perceive to be Americanisms invading their language. But one review of the English language by the *Oxford English Dictionary* researchers revealed that English had admitted at least 90,000 new "meanings" (being what a word refers to) over the past century, but only 5 percent were acquired from outside British culture. If change is an enemy, it resides within, but it would appear that cultures like to shoot messengers.

So sensitive are cultures to their use of language that George Bernard Shaw once purportedly remarked that Britain and America are "two nations divided by a common language." Some attribute the quote to Oscar Wilde or even Winston Churchill, but regardless of who said it, it is not hard to find examples. Older British hotel staff might still ask visiting American guests what time they would like to be "knocked up" in the morning. An American family hosting a British visitor might be disappointed to hear their guest describing their house as "homely," which to Americans means unprepossessing. American visitors can elicit similar reactions in their British hosts when they comment on a person's trousers by saying "nice pants" (suggesting to the uneasy Britons that the Americans have somehow gained knowledge of their underwear). Or as so often happens in a pub, an American guest might request two drinks at the bar by holding up two fingers to the bartender—a dangerous gesture in the UK at best, but plainly rude when it is the back of your hand that faces someone.

These anecdotal accounts can be placed upon a firmer footing. If cultures routinely swap traits with other cultures, they should tend to share the most traits with those of their nearest geographical neighbors. If, on the other hand, cultures are more closed than this and tend to acquire their traits from previous generations, then they should tend to have the traits of the cultures from which they descend; that is, of their ancestral culture. Geography and ancestry are normally correlated, but measures that can separate the two show that traits related to hunting and fishing practices, family structure and kinship, are often handed down over generations rather than acquired from neighbors. Techniques and patterns of Iranian rugs and even Native American longhouse designs tend to be handed down from generation to generation rather than acquired from neighbors.

Cultures can and do acquire ideas and skills from those nearby, but neighboring societies often differ far more at a cultural level than might be expected because they resist such influences—just think of Québec and the rest of Canada, or the outpouring of cultural diversity that occurred when the Soviet Union collapsed: in Central Asia alone, Turkmenistan, Uzbekistan, Kazakhstan, Chechnya, Tajikistan, Moldova, Kyrgyzstan, and Dagestan reappeared, all differentiated by culture, ethnicity, and language. Even under the influence of close geographical neighbors, or oppressive regimes, cultures can remain stable and coherent units. Cultural evolution is not the free fair exchange of ideas it could be.

culture and the missing years

IF THE genetic changes that defined our species were in place by 160,000–200,000 years ago, and social learning has been the force we think, why then did it take until 60,000–70,000 years ago before we were able to leave Africa and successfully colonize the rest of the globe? These are sometimes called the missing years for modern humans, and so staggering is the space of time for a species not to make use of its full capabilities that many authors suggest there was a very late genetic change, perhaps even as late as 40,000 years ago, that finally made us "fully" human. Forty thousand years ago is identified as the likely candidate time period because this was the time of a great flowering of human cultural innovations, including ornamentation and painting, but also tools and other aspects of what archaeologists call "material culture," or the things we made that leave a trace in the archaeological record—for example, fine hand axes, small and complicated blades, art and decoration and arrowheads.

But such a late genetic change is unlikely since it implies that the change occurred after modern humans had left Africa. Then to explain how all modern humans came to have the same capabilities requires awkward scenarios in which, for example, some genetically superior group arises perhaps 40,000 years ago in Europe, or Asia Minor, or Australia, or New Guinea, and then resettles the entire world; or worse, that they don't, implying some groups are inferior to others. Another suggestion is that the same mysterious late genetic change occurred repeatedly in separate groups, at all the different places in the world we were to be found at that time. Neither of these scenarios is correct. The genetic evidence shows that we all trace our ancestry back to African populations that lived probably 60,000 to 80,000 years ago, not to some more recently inhabited part of the world where a supposedly superior group arose.

Instead, our missing years might be the product of a phenomenon known as *random drift* that can cause small populations to lose information, and thereby slowed the pace of cultural evolution early in our history. In small populations, chance or random events from one generation to the next can strongly influence whether something or someone survives. This can slow the rate at which they can adapt if the effects of drift oppose useful changes. Its consequences can be worked out precisely using mathematical arguments, but the idea readily surrenders to a verbal one. Imagine you live in a small island tribe of around fifty people and there are five among you who carry a gene that improves celestial navigational skills. I do not suggest that there are genes for celestial navigation, but if you have ever tried to use the nighttime stars to navigate, as the Polynesians did to dramatic effect in occupying the Pacific, you will recognize that any genetic differences among people that made some better at this than others would have been strongly favored. Suppose now that all five of these people set out in a large boat on a seagoing journey, and during that journey a terrible storm blows up and they all drown at sea. In one stroke of bad luck this valuable gene has been driven to extinction, at least in this society.

But now, instead of fifty people in the tribe, let's consider there are five hundred, and that as before 10 percent of the people or fifty carry this special gene. It would take a particularly extreme spell of bad luck—or perhaps a very big boat—for all fifty of these people to find themselves in the same boat on the same stormy night. There is no reason to restrict the argument to genes. In a small group, ideas, technologies, and skills can easily be lost owing to the effects of random drift. The reasons could be bad luck, or perhaps there are not enough models for others to imitate, key people might die, there are fewer people to come up with new ideas, and fewer people to correct others' mistakes. We must also bear in mind that learning by watching and imitation is difficult, and so knowledge and skills are prone to drifting away from their starting points. Most of the time when you try to imitate someone doing something complicated, you get the task or the behavior wrong the first time around: just watch someone fly-fishing or hitting a golf ball and then try to repeat the action. Now imagine someone is trying to learn that same skill by watching *you*.

In his book *Guns, Germs, and Steel*, Jared Diamond describes how isolation and small population sizes can even cause societies to lose traits directly related to survival and prosperity. Diamond tells us how when Europeans first landed on Tasmania in the eighteenth century, they sent back descriptions of a culture that had lost many of the technologies originally taken there from the Australian mainland. People had reached Tasmania by 34,000 years ago but became cut off from mainland Australia when sea levels began to rise around 24,000 years ago. By the time the European explorers arrived tens of thousands of years later, the Tasmanians had no bone tools, no fishhooks, no hafted tools or spears, and no spear throwers. They had even lost their appetite for fish! Having lost bone tools they could not sew clothes. They had resorted—as did the Tierra del Fuegians of South America—to smearing their bodies with seal or other marine mammal body fat to preserve their own body heat.

If you think these examples say more about Fuegians or Tasmanians than they do about you, ask yourself if you could make fire without matches, or tell the difference between the edible and inedible plants in your local wood or forest. Dependence of knowledge and learning on the size of the group is surprising to us because we enjoy the benefits of societies that write things down, or draw pictures, or take photographs. Even so, we still see the value of groups today as when we get together to play card or other games. Often no one individual knows all the rules, but by pooling everyone's slightly different portions of the knowledge the game can usually be reconstructed.

demography and the "rule of two"

IT IS not enough that our species could use social learning to acquire the skills to move into most of the environments on the planet. And neither is it enough to say that we acquired the psychological dispositions to protect and keep intact our cultural survival vehicles. For our species to occupy the entire globe, our populations would have had to be expanding. Had they not been, we would either have stayed put, or when we moved to new lands we would have vacated the territory we left behind. But this is not what happened: the combination of human culture and social learning has meant we have repeatedly produced excess numbers of people, enough in fact to occupy the entire world. Indeed, human beings are distinguished in the biological world as having broken a hallowed rule of demography that we can call the "rule of two," and to have done so over long periods of time.

The significance of this achievement is appreciated when we recognize that any proper history of life on Earth is a history of death, and the reason is the rule of two. The rule gets its name from the fact that throughout history, females—of plant and animal species—have left, on average, just two offspring that will survive long enough to do the same. Some have left more, others fewer, but the average is roughly two. It is a surprising statistic because, for example, a female rat has a prodigious ability to make more rats: she reaches maturity at about thirty-five days old, she can produce a litter of up to twelve pups, wean them in a month, and then start all over again, breeding year round until she dies, typically at two to three years. This high output is true of most small animals. Indeed, the rabbits' impressive reproductive potential is immortalized in the phrase "breed like rabbits." But even then, a typical female rabbit leaves just two surviving offspring.

A larger animal like a female elephant takes longer to reach maturity—around ten years—and when she does reproduce it is one at a time, and it takes her far longer to rear her offspring before she can reproduce again. But female elephants reproduce into their sixties, and so they also tend to leave about two surviving offspring. The same is true even of those wildly fecund organisms the trees. An oak or chestnut tree that lives for centuries and rains acorns and chestnuts down in our forests and on our lawns and streets could produce millions of offspring in its lifetime, but oaks and chestnuts on average leave just two surviving offspring trees. Go outside and stare at the vast trunk of one of these trees—some weighing hundreds of tons—and all its many branches. It is a sobering thought that all of that effort in making the wood, and in producing all of the tree's bark, branches, and leaves over so many years, comes to so little.

This rule of two turns out to have a simple explanation that tells us it could not be any other way. For every offspring of a female, a male has been involved (the rule of two becomes the "rule of one" if we have in mind asexual species that reproduce on their own). If a male and female produce two surviving offspring before they die, those two replace this male and female. If this male and female were to leave behind even just three offspring, and each of these three in turn produced three that survived to reproduce, and so on, the numbers of this species would increase without end. Consider just a population of fifty males and fifty females in which each of these females produced three surviving offspring. In the first generation, the 50 females would produce 150 surviving offspring (three each), increasing the population size by 50 once both the parents had died. These 150 would in turn become 225 when each of the 75 females in this generation left behind three offspring. It is easy to see that the world would quickly become covered in layers of rats, or rabbits, and even only slightly less quickly in a layer of elephants if they could break the rule of two. If oaks and chestnuts could leave more than two, our world could become forests of these great trees. The capacity of common bacteria to reproduce is so great that if their growth went unchecked we would in a matter of days (or less time) all be standing up to our waists in a mat of bacteria that carpeted the entire world.

We learn three lessons from this. One is that for most species the difference between the numbers of offspring they produce and the numbers that survive is so large that it is not much of an exaggeration to say that all offspring ever born die before they get a chance to reproduce. Such startlingly high levels of mortality are the same as saying that competition for survival is fierce. It is this competition that lies behind the nineteenth-century philosopher Herbert Spencer's summary of Darwin's evolution by natural selection as "survival of the fittest." Those few of us who have survived are well adapted: we are the rare descendants of a long line of other rare survivors who were our ancestors. The genes in our bodies and those of every other organism are those that have survived for millions or even in some cases billions of years because they were good at producing successful *vehicles*, while uncountably greater numbers have died trying. This means we can expect the genes we see today to be very good at promoting their interests, and they will do so by means of the ways they vary the bodies they produce. But even with all this fine-tuning, the average female still produces just two surviving offspring.

The second thing we learn is that different organisms have adopted different tactics for trying to break through the two barrier. Some—like oak trees and rabbits—go all out. Others, like elephants and whales, show more restraint, but put more effort into each offspring.

The third thing we learn is that all those different ways of producing offspring, some as rabbits, others as trees, are just different but approximately equally good ways of making vehicles for transporting genes into the next generation. All that time, bulk, energy, and trillions of individual cells required to make an adult elephant yield the same number of surviving offspring averaged over long periods of time as a rabbit, or even a single-celled yeast (of which there are, technically, not males and females, but two mating types called α and *a*). Nearly every cell that resides inside a complex organism like a tree or ourselves never sees the light of day, laboring away instead to propel a small number of others into the future. It is even starker than this. The egg of a female and the sperm of a male are single cells. We could say that the trillions of cells that make up our bodies spend a lifetime devoted to seeing just two of their kind escape into the next generation.

It is easy to read this as demonstrating that animals act for "the good of the species," holding back so as not to overpopulate. But the truth is nothing of the kind. Occasionally, a species will break the rule of two for short periods of time. If a more fecund female came along who could on average leave three surviving offspring, or four for that matter, natural selection would favor her: her greater number of surviving offspring would gradually come to dominate the population in which she lived, and eventually all females would be of her kind, able to trace their ancestry ultimately back to her. But if this happened, this species' overall numbers would rapidly increase and two things would follow. One is that at some point the species would reach what is called its "carrying capacity," a number that attempts to describe how many individuals of a species the environment can support. If the population expands above the carrying capacity, some of the excess individuals will die of starvation. The other is that this species' increased numbers would mean that its predators would come to enjoy a bounty of prey and their numbers would thereby increase. The combination of running out of food and the extra predators would reduce the average number of surviving offspring from the superfemales back to two.

Some species can break the two barrier for short periods of time when they have just evolved or when they are introduced to a new area. A newly evolved species that consisted of just a single male and female would have to break the barrier ever to increase in numbers. So, the surviving species we see around us have broken the barrier at some point in their history. But these species will now be at their carrying capacity and leaving on average just two surviving offspring. When rabbits were introduced to Australia, they bred like rabbits. The Australian environment had not had rabbits before, and it is likely that the diseases that kill other small Australian animals did not affect them. But the growth of rabbits was soon contained by introducing a virus that controlled their numbers by killing some of them and making others weak or ill.

Now another newly introduced creature—the cane toad—is eating its way across Australia. It seems unstoppable because its poisonous skin either kills or repels the native Australian predators. These cane toads will eventually reach their carrying capacity, and other animals are already discovering how to avoid their poisons. Some field biologists report that the kookaburra has learned how to flip the cane toad over onto its back before eating it, to avoid the toxic skin. If this strategy succeeds, kookaburras will also probably leave more than two surviving offspring, at least for a while, and Australia will ring to the sound of kookaburras even more than it currently does.

Nature is never quite as tidy and predictable as these examples suggest, but the rule of two is what we often call the balance of nature, and it is how things have worked for billions of years. That is, until a species came along that discovered how to break this rule and do so over long periods of time. Once again, that species is human beings, and for at least the last 80,000 years or so we have carpeted the planet with our excess offspring, and continue to do so. Our discovery for breaking the rule of two was to build cultural survival vehicles. The Earth had not seen the likes of this before or since, and this is the sense in which we saw in the Introduction that culture became our species' biological strategy. Here was a force that could not only deploy technologies such as fire, clothes, and shelter to adapt different environments to it but has been able throughout its history repeatedly to produce innovations that reset the world's carrying capacity to hold more people in a given area. Plagues, wars, and droughts, and the occasional collapse of civilizations, have at times slowed our march but as yet not stopped it.

We didn't break the rule of two only by altering the carrying capacity: modern human women achieve a higher birth rate than other large Great Apes. As a rough estimate a wild chimpanzee female might give birth once every four to six years. The comparable figures for gorillas and orang-utans are once every four to five and once every six to nine years, respectively. By comparison, human women living in hunter-gatherer groups might have had a baby about every three to four years, and a two-year gap is common in modern societies. Human women also maintain a longer *reproductive lifespan* than these Great Apes, reproducing for around thirty years of their lives. This is nearly double that of a gorilla, ten years more than a typical chimpanzee, and perhaps five years more than an orang-utan.

Our rapid rate of reproduction might owe something to a peculiar feature of our species. Human women have a long period late in their lives when they don't reproduce, called the *menopause.* Many people simply assume that the menopause is a consequence of our living longer, and that in our "state of nature" women would not have lived long enough for it to occur. But this idea has in more recent years given way to an intriguing suggestion. It is that the menopause might have evolved as an act of nepotism or help directed at relatives. Natural selection might have favored women who ceased their own reproduction late in life to help their daughters or their daughters-in-law to reproduce, rather than compete with them. Those who advance this idea, known as "the grandmother hypothesis," suggest that having an extra pair of hands around would have meant that the daughters or daughters-in-law could reproduce more quickly. Natural selection would have favored this period of menopause if by helping her daughter or daughter-in-law this grandmother eventually gained more grandchildren than she would have had she chosen to continue to reproduce herself.

Another possibility is that human women have been able to maintain a higher reproductive rate than other Great Apes and still provide for their young simply because modern human societies from early in our history have been more efficient at providing food, shelter, protection, and other resources for people. Whatever the reasons for our higher growth rate, reproduction is to the growth of populations what interest is to money. Populations grow by compounding themselves as the babies born now grow up themselves to have children. So, this higher total reproductive output of our species along with our ability to control our environments has meant that wherever human groups ventured, they would likely have filled up their space and found themselves in constant and intense competition with other human groups doing the same. This tells us that competition among cultural survival vehicles throughout our history has been intense, and just as is true of our genes, those that have survived will have acquired traits that make them good at promoting their inhabitants' survival. For nearly all of our history up to sometime around 80,000 years ago, we were like other animals and we barely increased in numbers. The species that immediately preceded us, *Homo erectus*, the Neanderthals, and even so-called premodern or archaic *Homo sapiens*, struggled to replace themselves and then went or were driven extinct. As recently as 20,000 years ago, our numbers may have amounted to just a few millions, maybe fewer, in the world. But our growth has been rapid, and especially so in the last 10,000 years, with over 6 billion of us today. It is all down to social learning and our distinct cultural survival vehicles.

CHAPTER 2

Ultra-sociality and the Cultural Survival Vehicle

That even a disposition to die for our cultures can be adaptive, just as it can be to fight to the death for your own body

visual theft

WE HAVE SEEN that by sometime around 160,000–200,000 years ago our species might have acquired the capability to learn new behaviors from watching and imitating others. This put us on a trajectory of cumulative cultural evolution as ideas successively built and improved on others. It is something no other species has achieved, and it continues today at ever-increasing rates because the sheer volume of cultural knowledge acts as a vast crucible for innovation. We need look no further than the chairs we sit in, the televisions we watch, the books we read, the cars we drive, the computers we work on, the spaceships and high-energy physics laboratories we produce, or even the food we eat, to see its effects. And so, to most commentators social learning is "job done," "end of story"—our species could make things, so we have prospered in a way that other animals didn't. But in fact our acquisition of social learning was just the beginning of our story as a species because it would create a social and evolutionary crisis, the resolution of which would lay the foundations of our psychology and social behaviors and determine the future course of the world.

Here is why. Social learning is visual theft. If I can learn from watching you I can steal your best ideas and without having to invest the time and energy that you did into developing them. If I watch which lure you are using to catch fish, or how you flake your hand ax to give it a sharp edge, or secretly follow you to your hidden mushroom patch, I am benefitting from your knowledge and ingenuity, and at your expense because now I might even catch the fish before you do. Social learning really is visual theft, and in a species that has it, it would become positively advantageous for you to hide your best ideas from others, lest they steal them. This not only would bring cumulative cultural adaptation to a halt, but our societies might have collapsed as we strained under the weight of suspicion and rancor.

So, beginning about 200,000 years ago, our fledgling species newly equipped with the capacity for social learning had to confront two options for managing the conflicts of interest social learning would bring. One is that these new human societies could have fragmented into small family groups so that the benefits of any knowledge would flow only to one's relatives. Had we adopted this solution we might still be living like the Neanderthals, and the world might not be so different from the way it was 40,000 years ago when our species first entered Europe. This is because these smaller family groups would have produced fewer new ideas to copy and they would have been more vulnerable to chance and bad luck.

The other option was for our species to acquire systems of cooperation that could make our knowledge available to other members of our tribe or society even though they might be people we were not closely related to—in short, to work out the rules that made it possible for us to share goods and ideas cooperatively. Taking this option would mean that a vastly greater fund of accumulated wisdom and talent would become available than any one individual or even family could ever hope to produce. That is the option we followed, and our cultural survival vehicles that we travelled around the world in were the result.

We take this cooperation for granted in our modern world, but it rests on a psychology and social behaviors new to evolution, and unique to our species because no other animal has confronted the crisis of visual theft. Consider that even the simplest acts of exchange among unrelated people wobble on an unstable tightrope, because now my instincts to take advantage of you will not be held back by the usual bonds of family ties. This is because when I help a relative I help a little genetic bit of myself, and so natural selection favors my nepotism so long as the help I provide to them is not too costly. It also means I have less incentive to cheat that relative, or for them to cheat me. When we watch a streaming mass of ants mount a suicidal charge out of their nest to take on some foe, we admire their unflinching courage, but we recognize these ants are dying to save their brothers and sisters, and especially their queen. She is the source of additional copies of their shared genes, produced in the form of more brothers and sisters. The same logic of what evolutionary biologists call *kin selection* tells us why your skin cells are happy to die to protect you from the penetrating and deadly rays of the sun. These skin cells are all genetic clones of each other, only too happy to die if this makes it more likely the body they inhabit will reproduce one day.

But the hallmark of modern human tribal societies was that they were not limited to relatives, and this meant that evolution had to confront the problem of having a potentially unruly mob of individuals on its hands, each looking out for their own well-being. The difference is everything. Now anything I do for you might benefit you and your genes, but at my expense. In fact, once people started living together in groups, natural selection would have favored a raft of *selfish* psychological ploys for taking advantage of others' good nature, because there would have been a continual conflict between what was best for you and what was best for your group. I might take more than my share from the dwindling grain store when you are not looking, or attempt to convince you that I am hungrier than you when food is scarce. I may "forget" in future to return a favor, I may try to escape your attentions, I might return less of a favor, or plead poverty. I might lag behind out of harm's way in battle, or I might get angry if you try the same, and spread rumors that you are not to be trusted.

Left unchecked, these ploys would have caused our societies to collapse before they got off the ground. To make our societies work, then, we had to acquire the social and psychological systems that could somehow overcome and tame selfish instincts born of millions of years of evolution by natural selection to cheat, exploit, dupe, and even murder one's rivals. The solution was simple in principle but profound in its effects: natural selection found ways that made it possible for individuals to align their interests with those of their group. If the benefits of the cooperation that might flow from this alignment could exceed the returns from acting on pure self-interest, cooperation, even with non-relatives, begins to make sense. Maybe you show great courage in battle and this makes it far more likely your group triumphs over an aggressive foe. If as a consequence you are also more likely to survive, this apparent altruism is a good strategy for you. Or maybe by virtue of having some skill that you share with other members of your group, like being good at making spears or at navigating on the open seas, you acquire a value to the group or a reputation that makes people treat you more charitably. If the collective action you inspire, or the benefits you can bring to the group, can return more to you than behaving selfishly, then your apparent altruism is really a case of enlightened self-interest, and the usual conflict of interest between what is best for you and what is best for the group can vanish.

In fact, the monumental and even sometimes terrifying achievement of human culture has been to discover how to get groups to act together in a coordinated way. It is monumental because by unlocking the psychological means to pool our efforts and skills, it granted our societies a formidable degree of shared purpose that could be put to use in solving the problems of survival. At the same time, aligning individuals' interests with those of their groups could be terrifying in making our cultural survival vehicles formidable competitors against other groups that might be competing for the same territories. Now, someone could forage while another hunted, someone could mend sails while another plotted a course, two could stand guard with sharpened spears while a third steals a competing tribe's animals, and my army could cross the valley and attack yours, or repulse it when you attack me.

As I mentioned in the Introduction, the distinctive and salient feature of much of our social existence is the sense of belonging to a cultural group toward which we feel an allegiance that we often do not easily extend to others outside of that group. That sense is the emotion natural selection has kindled in us to get us to behave as a group with a shared purpose. The unusual psychology it brings us can even extend to getting us to engage in costly altruistic acts that rival those of the social insects—the ants, bees, wasps, and termites. Who, for example, can forget images of Japan's fabled World War II Kamikaze pilots, or the warriors in World War I streaming out of the trenches "over the top" to die in battle? You will take this cooperative psychology entirely for granted because it has been wired deep into your DNA, but no other animal does anything like this. You will never see a group of horses or a group of chimpanzees streaming out "over the top" to die for each other. No lion or zebra holds doors for one another, no ape ever politely stood in line; they don't look after the elderly or help those in distress. It is true, elephants are sometimes described as "grieving" for a dead member of their group, but this behavior is normally directed at relatives.

If the social insects are sometimes described as *eusocial*, or truly social, humans have uniquely among the animals achieved a *hyper-* or *ultra-sociality*. This label acknowledges that our altruism has broken free of acts aimed merely at helping relatives. In this chapter we will even see how our evolved cooperative psychology can admit a disposition toward suicidal self-sacrifice. Surprisingly, confusingly, and seemingly paradoxically, it will be shown to be in an individual's self-interest to have this disposition. At the same time, a troubling feature of the way this disposition evolves is that it can cause us to treat people from other societies, and sometimes from even our own, crudely and violently. That is the fragile nature of our sociality and psychology, and it arises because our cultures are cooperative vehicles for the survival of unrelated people, and their genes.

vehicles and the discovery of cooperation

THE ORIGINS of human cooperation can be traced to developments in the earliest replicators that populated the Earth beginning perhaps 3.8

billion years ago. The earliest replicators were probably *RNA* molecules or *ribose nucleic acids*, a simpler form of the DNA molecule or *deoxyribose nucleic acids*, whose twin strands elegantly intertwine in a twisting helical shape. Nearly 4 billion years ago, the biotic world of the very young Earth may have comprised little more than naked replicating segments of RNA floating in a warm primordial soup. Molecular biologists call this the *RNA-world*, and RNA may be the ultimate or *ur*-ancestors (meaning the original or earliest form) of all life on Earth.

One of the more remarkable discoveries during the early years of molecular biology in the 1960s was that strands of RNA all on their own can have distinctive shapes or what biologists call *phenotypes*. Whereas all DNA molecules have more or less the same shape, strands of RNA that differ in their chemical makeup of nucleic acids fold and twist into different forms. This discovery about RNA gave molecular geneticists a mechanism for the early evolution of life on Earth. It turns out that the different shapes can influence the survival of one strand in competition with others. Some shapes are, for example, more resistant to being pulled apart by water. An RNA strand with the right chemical makeup to adopt one of these shapes will live longer and therefore tend to accumulate in competition with RNA strands more easily pulled apart. The RNA strand itself becomes a kind of phenotype, and its own survival vehicle.

The early biotic world was probably one of competing strands of these simple replicators of RNA, and whichever strand was lucky enough to escape the forces of nature and find enough chemicals to replicate itself would have dominated. But there would have been only so many different shapes. At some point, two strands of RNA—competitors for the same chemical resources—might have discovered they could physically combine to make a new kind of shape. This transition would have produced the first *vehicle* comprised of more than one replicator, and it would have increased the complexity of life. But the question is why would this new cooperative venture work? On their own, each of these strands could replicate whenever it wished, but together they would have to give up some of this freedom. Why, for example, should I join forces with you to pick apples when I can pick them on my own, or better yet steal from you?

One of the great insights of evolutionary biology has been to understand how entities that would otherwise compete can be tamed or domesticated to form alliances that serve both. John Maynard Smith and Eörs Szathmáry have called these the "major transitions" in our evolution, and two RNA strands joining together might have been the first of these major transitions. Perhaps the two strands of RNA could help each other to duplicate or copy themselves, or perhaps they could better avoid being pulled apart by water. Living longer would have given this new joint vehicle more chances to replicate itself. We can see from this a reason to give up some of your freedom and to cooperate with a former competitor: your joint enterprise can work if the payoffs more than offset the loss of the freedom to act alone. I should join forces with you in picking apples if we get more than twice as many together as we do separately. Even so, why shouldn't I wait until we have accumulated a large number of apples and then run off with them?

The first of the evolutionary transitions unfurled the first partnership, and the world of evolution never looked back. Once two or more replicators can combine to produce a vehicle that gives them better returns than they would get from competing, their fates become linked, and it is this linkage that can tame their instincts to compete with or exploit each other. Now the answer to why I should not run off is clear: to give up on the partnership is to give up some of its riches from future returns. Once fates are linked, replicators acquire a new incentive: to become better and better at what they do because now they have less reason to fear betrayal. They can specialize in ways that promote the partnership even more. Maybe if we join forces to pick apples and you are stronger than me, I should specialize in standing on your shoulders and you should specialize in supporting me.

Later on in the course of evolution, partnerships of genes moved beyond shapes to devise even better ways to influence their survival. Collections of genes joined forces in cells that housed the genes and protected them from hot or cold or acids or salt, or from other predatory bits of RNA that might pull them apart chemically. Eventually, collections of individual cells came together in the big multicellular bodies such as our own that are very good at surviving, often living for many years. These large bodies were partnerships of billions, maybe trillions of cells, all clones of each other and specialized into different roles as hearts or muscles or kidneys, livers or brains. These large cooperative vehicles were a success because, on their own, nearly all of the individual cells would have died. The incentive for these cells to join forces is clear: all they had to achieve out of their partnership in forming a body was to improve on their stark individual fates.

A new kind of evolutionary transition occurred when the genes residing in separate individuals learned to contribute to a shared vehicle that acts something like a large body itself. The Australian compass termites are famous for building tall, monolithic, skyscraper-like structures that can be taller than an adult human. Areas in which they are prevalent resemble a sort of haphazardly laid out graveyard with a collection of unusual tombstones. But far from marking graves, these mounds provide a safe and warm environment for thousands, maybe millions, of individual termites—brothers and sisters who cooperate to construct and maintain the mounds and who rarely if ever reproduce themselves. Instead, their collective actions are really not so different from the collective actions of the cells in your body; they are simply more loosely organized. Like the cells in your body, the vehicles these termites produce serve their reproductive interests by promoting the queen's—their mother's—reproduction. The same is true of the other social insects—the ants, bees, and wasps. In fact, in each of these societies the queen plays a role not different from the special cells in our bodies we call our *germ line*—the source of our eggs and sperm. Her immense reproductive output more than pays for her offsprings' sacrifices for her. On their own, their chances of survival are nearly zero.

slime molds, suicide, and tribal minds

THE LAST of the great evolutionary transitions was the transition to human societies. Now groups of human individuals acquired the abilities—some learned, some no doubt cast in our genes—to construct a shared cooperative vehicle. But unlike beehives, ants' nests, and termite mounds, human societies are constructed around unrelated people, all of whom are seeking to further their own reproductive interests, not those of a single mother or queen they all have in common. The old rules of nepotism that natural selection had so skillfully exploited to make the shared vehicles of the social insects would have to be thrown out the window. Now any help you might provide to your cooperative group could benefit someone else. How, then, did we manage to construct the systems of cooperation that would allow us to form these shared survival vehicles we call our cultures or societies?

There is an organism that can provide some clues, and its solution to this same problem has profound implications for understanding its nature and ours. The slime molds or social amoebae are a species of single-celled creatures that live on the forest floor. Most of the time they lead a solitary existence resembling tiny drops of jelly. But when they suffer from starvation, something wondrous occurs. One after another sends out a chemical alarm summoning them to unite. From all around, the amoebae converge, eventually forming into a streaming multicellular saffron-colored carpet. This society of strangers then oozes across the ground. At a suitably sunny point the carpet stops and then the amoebae cooperate to build a physical tower or stalk. It is composed of their individual bodies and it rises from the forest floor as they climb up over each other, in effect standing on each other's shoulders. Some climb to the very top of this tower, where a fortunate group makes its way into a bulbous cluster. This cluster acts as a launching pad for spores that will be carried on the wind or the backs of passing animals to better lands, where they will become the progenitors of the next generation of amoebae. The rest will die, having quietly given their lives for other amoebae they did not know and had probably never seen.

The social amoebae have long been a puzzle. Why do so many give their lives for so few in a supreme act of altruistic self-sacrifice? Were the amoebae relatives, as is true of the ants, wasps, termites, or skin cells in our bodies, we would have a ready answer—that they were dying to promote copies of their genes residing in those relatives. But the amoebae are not all relatives, and so the puzzle of their sociality is how natural selection could ever favor the cooperative individuals over selfish ones who never give their lives, given that an altruist runs the risk of helping others who have no intention to repay the kindness. We can begin to see a solution to this puzzle in two choices that confront an amoeba. One is to join in the building of the tower and risk helping others; the other is to remain solitary. Joining the tower gives an amoeba at least some chance to reproduce because it might just find itself getting into the launch pad cluster at the top. Remaining solitary on the forest floor takes away even this small chance. This means that the cost to the amoeba of joining the tower is negligible or even zero, because there is nothing else it could use its time and energy to do that would improve its chances of reproducing—remember, it is starving.

Natural selection has favored the altruistic disposition to build a tower, then, despite the risks of helping others, because over long periods of evolutionary time amoebae with this disposition will, on average, have left more offspring than those who acted alone. It might seem odd, but this statement is true even though the most likely outcome to any one amoeba that helps to build the tower is to die, and to die "childless." The strategy works because building the tower at least holds out the chance, even if a small one, of entering the sought-after cluster of cells at the top, and sending your spores wafting off into the breeze. When that happens, some other amoeba will have given *its* life for *you*.

For the amoebae, the choice to behave altruistically is a stark and lonely one because the rewards are so rare. But the amoebae's actions and the societies they produce, even if temporary, illustrate the fundamental ingredient needed to get altruism to evolve. It is that altruism can thrive if altruists can surround themselves with other altruists. This ensures that selfish cheats are excluded from enjoying the benefits of altruistic acts, and means that any given altruist is just as likely as any other to be helped. It also means that in the long run altruists receive more benefits than they would by acting alone. A collection of "like-minded" individuals can even produce more benefits than simply adding up everyone's individual help. In the case of building a tower, more individuals acting together means a taller tower, and taller towers are better at dispersing spores.

Without perhaps realizing it, we have discovered something fundamental about the amoeba's disposition that will prove relevant to trying to understand humans. It is that the amoeba's altruism is not one of "expecting" a return from some other particular amoeba it has helped. Rather, it is a disposition merely to grant assistance, to club together to form a "mutual aid society." An important and surprising aspect of this kind of altruism is that individuals can acquire tendencies to behave in ways that are costly or even deadly (remember that most amoebae die), and yet those tendencies can, paradoxically, evolve. The paradox is resolved when we realize that natural selection promotes replicators, not the temporary vehicles such as you and me or an amoeba that merely carry them. It is something that you might find peculiar, especially in this context, but read on.

Vehicles—the individual amoeba in the case we have been discussing—are just the ways that genes act on the outside world to get themselves transmitted. The tower-building amoebae each carry a copy of a gene we can call the altruism gene. That gene, by causing a disposition to build a tower, is more likely to get itself replicated than a gene for remaining solitary, because the solitary amoeba will find itself alone on the forest floor. The advantage to the tower-building gene is slight, but it is better than the solitary choice. This is true even if some or even most of the individual amoebae in which that altruism gene resides die in spite of building the tower. The key point is this: the death of these amoebae promotes copies of this same gene that reside in the other amoebae in the stalk, even though in every other respect those others are not relatives, and could even be "strangers."

Once we understand how the amoebae's altruism works, the problem of getting altruism to evolve seems easy to solve. But the real challenge of making altruism work lies in coming up with some way that altruists can identify and then associate with other altruists, and thereby exclude those who lack the altruistic gene. How can I possibly know whether, if I help you, you won't take my help and run? The social amoebae solve this problem by building the shared multicellular body or vehicle, the carpet that eventually forms the tower. The mere presence of another amoeba in the carpet automatically identifies it as one that carries the altruistic disposition to build a tower. The tower is itself, then, the collection of "like-minded individuals."

Geneticists would describe the tower-building social amoebae as related to each other at a single *locus*. This is just a technical term for saying that at a single place in their genomes these amoebae share identical versions of the altruism or tower-building gene (as a comparison, parents and their offspring are normally related at 50 percent of their many thousands of genetic loci). It is extraordinary that a single gene can produce such profound effects against the background of all the amoeba's other genes, but it can do so because building the tower benefits not just the gene for that disposition but all of these other genes. Those genes have no reason to oppose the building of the tower, even though for most of them the fate is death, because they can do no better than try their luck in the tower. The towers align individual with group interest—the amoebae have a shared fate—and so what might seem like foolish and gullible behavior is really their best bet for getting transmitted into a future generation of amoebae.

We set out in this chapter to understand a puzzling side of our nature—our tendency to help others, even when that help might not be directly reciprocated. If this altruism extended no further than to holding doors for people or giving up seats on trains, we might be tempted to dismiss it as just a charming side of our nature, although even doing that would simply raise the question of why we have evolved minds that find helping others an attractive thing to do—don't expect this of an orang-utan or a gorilla. But we've seen that our behaviors go well beyond this, including risking our health and well-being or even sometimes giving our lives in war. That surely cannot just be a charming side of our nature; or if it is, we need to know how, in a Darwinian world in which the fittest survive, people willing to give their survival away can have prospered.

But if our journey into the details of amoebae sociality has done its work, you will be able to see the answer in the reflection of our cultural survival vehicles in their towers. For at least the last 160,000 to 200,000 years humans have resided in small, close-knit cultural societies that develop strong identities, often around their common language, and restrict the flow of people, ideas, and technology. The amoebae show us that these are just the behaviors that can favor the evolution and spread of the sort of ultra-sociality that makes our species so puzzling, because they create the conditions that allow altruists to surround themselves with other altruists. Then, something of a mutual aid society arises in which dispositions toward costly acts can more than pay their way because these other altruists are just as likely to help you, as you are to help them. As a result, everyone is better off than had they tried to go it alone, just as a solitary amoeba has little chance on its own.

What form might these dispositions take in our species? Humans seem to be equipped with emotions that encourage us to treat others in our societies as if they were "honorary relatives." This is more than just a metaphor: we seem to practice a special and limited form of nepotism in which—just like the amoebae—we target our aid toward others who might not be related to us save for the fact that they are members of the same cultural group. Our nepotism is "special" and "limited" because our helpful emotions on the one hand, and prejudicial ones on the other, might be based on being related to these other members of our societies at just a single locus among our thousands of other genes. That single locus is the helping gene, and it would have spread in our evolutionary past, just as it has in the amoebae, by identifying others who carry copies of that gene and then helping them. For instance, the gene might simply code for an emotion that disposes people to be friendly toward those they perceive to be members of their societies.

In fact, nearly everyone will have experienced the vivid and controlling emotions that get us to favor members of our own societies over members of other societies for the simple reason that they are members of our societies. Maybe it is the feeling you get at a sports event, or hearing of your country's troops in battle, or when your country is attacked by terrorists. We call these emotions "nationalism" or "patriotism," or when directed against people from other groups, "jingoism," "bigotry," "xenophobia," or "prejudice." Most of us probably have no idea where these feelings come from; they appear spontaneously, they are visceral in nature, and they are alarmingly easy to teach to the young. They can also direct some of our most poignant and some of our most repugnant actions, and this leads us to believe they have played significant roles in our history in promoting our cultural survival vehicles, and more to the point, ourselves.

It might seem incredible that some of our most profound actions as a species could be based on such a simple mechanism, but we must take seriously the possibility that the nature of our cultural survival vehicles evolved because it creates the conditions that make our peculiar and yet powerful kind of altruism possible, and this is why they have been such formidable vehicles of our success. As we saw with the amoebae, even a single locus is enough to motivate costly ultra-social actions when cooperation serves all of our genes' interests and not just the gene for cooperation. Next time you feel that warm nationalistic pride at the sound of your national anthem or the news of one of your country's soldiers' valor, think of the amoebae!

Still, maybe this is just a fanciful story in which our similarities to the amoebae are merely coincidental, and not of great significance. So before we are willing to accept that we are just jumped-up social amoebae with jingoistic tendencies, we want to find that this explanation somehow fits with other sides of our behavior. It is easy to say that the amoebae surround themselves with like-minded individuals because the mere presence of an amoeba in the tower means it carries the tower-building gene, and it is the act of building the tower that is altruistic and normally self-sacrificial. But it is not so easy for us. Someone's mere presence in your society might not tell you anything about whether they share your altruistic dispositions. So we are forced to look for more than mere presence, and this might be why as a species we are so sensitive to cues of a shared cultural history, and so eager to create them. All those shared beliefs, customs, religious systems, languages, accents, rituals, songs, styles of dress, and mannerisms are the cues we instinctively and subconsciously use to assess our cultural relatedness to others. Our societies' tendency throughout history to restrict the movement of people and ideas, and to develop strong identities around their languages and cultures, makes these cues more reliable.

The search for shared history can take powerful, dangerous, and even amusing turns, but it is never dull, and always revealing of our nature. I do not know whether it is true but I have heard it said of the "troubles" between Protestants and Catholics in Northern Ireland that Catholic youths upon encountering strangers would often ask them to recite the catechism as a way to determine if these strangers were fellow Catholics. Failure to get it right could mean a severe beating, or worse. I was once selling a house and it emerged that a person interested in it had attended the same university as me, although at a different time. I am sure this helped clinch the sale. In 1993, the then Welsh secretary (a position that at that time was a little like a governor general sent from London to oversee the running of a far-flung nation) was a man called John Redwood, and a member of the ruling Tory government. Redwood was prone to controversy, but his most famous gaffe came at a public meeting in Wales at which everyone sang the Welsh anthem. Redwood didn't know the words, but aware this would offend the Welsh, he attempted to mime them. His strangely animated attempt was caught on camera and Redwood was broadcast to the nation looking like a marionette on the end of its puppeteer's strings. At America's Super Bowl in 2011 Christina Aguilera lost her way singing "The Star-Spangled Banner" and suffered weeks of abuse on the Internet. Ask any Briton what "LBW" means and they are likely to know, but it is doubtful that a visiting American would. Someone from America is likely to know what "RBI" means, but it is doubtful a British person will (hint: they are both related to sports).

I use some of these examples whimsically, but that is not to say the emotions that accompany them are mild or indifferent. Our special and limited form of cultural nepotism can even produce a disposition toward self-sacrifice that is eerily like that of the social amoebae. The journalist Sebastian Junger in his book *War* describes the experiences of a small platoon of U.S. soldiers in a remote, brutal, and violent region of Afghanistan known as the Korangal Valley. Junger spent time embedded with these men, who were on their own, with few amenities, ramshackle shelters, almost completely cut off at any given time from outside aid, and almost constantly under attack from deadly guerrilla forces. The men knew their fates rested on how well they performed together as a fighting force, and Junger came to realize that at the core of the platoon was a commitment on the part of each of the men to sacrifice his life for the others in battle.

It is tempting to romanticize such a commitment as the expression of noble heroes willing to die for their comrades, and this is the popular image of war films and national propaganda. But we have to ask ourselves how such dispositions could ever become widespread when for every noble hero there are many others of less noble disposition whose genes survive at the hero's expense. We saw that the solution to this problem lies in recognizing that a disposition toward self-sacrifice can evolve so long as those who share it can identify each other. Then the deaths of some copies of the genes carrying this disposition help other copies of those genes to survive. Weirdly, a tendency to give your life can be adaptive to the gene that causes that disposition by virtue of promoting copies of itself that reside in others. Put more bluntly, your disposition toward self-sacrifice can be beneficial to you so long as you can surround yourself with like-minded people, because one of those people could save *your* life.

For many soldiers, one of the most disabling fears is the thought of being wounded and left to die, or of being captured and tortured. It can cause soldiers to hold back and not take chances. And yet, it is one of the tenets of small-scale battle that the group that escalates to violence most quickly often comes out on top. Junger describes how the men's commitments to each other reduced their fears, and allowed them to fight more confidently. This meant the platoon was more likely to be triumphant in battle. And so, remarkably, the men *were individually more likely to survive when they were all prepared to die for each other.* This is not to say that anyone had a good chance of surviving, just a better one when they all pulled together. Holding back is not an option because anyone who shows signs of wavering in their commitment knows they will be left alone on the battlefield, where they would be almost certain to die. Generals know that groups of men who fight together are more likely to survive, and this is why armies spend so much time drilling these instincts into combat troops. But it might just be true that the dispositions to behave this way are already a deep part of our psychology.

A willingness to die in battle—or at least some sort of disposition for it—might reside to a greater or lesser degree in all of us as an emotion that draws on the same kinds of motivations that get us to protect our own families, although here the group assumes the role of the family. It might indeed be this part of our makeup that responds to real threats, but that can also be exploited by propagandists to produce Kamikaze-like or other suicidal behaviors. It is certainly relevant to this discussion that the political narrative that is offered up to justify such acts is normally one of intergroup conflict and great honor bestowed on one's family. The act of suicide then becomes a way that one set of genes or ideas promotes others like itself by killing off ones that compete with it. Imagine what havoc a suicide bomber amoeba could do that broke away from its tower and somehow joined and then destroyed a competing tower, perhaps by releasing a poison. Its death would promote large numbers of copies of its altruism gene in the others back in its tower. The psychology of suicide bombers could be just this simple and cold-blooded, derived ultimately from our special and limited form of cultural nepotism.

Taking a step back, our more general tendencies to help others in our groups at some expense to ourselves should not be taken as evidence that we are "nice" or robotically dedicated to our groups. We are nice, but we should recognize this as an emotion that encourages us to cooperate, because cooperation pays individual—not just group—dividends. Indeed, our social nepotism, being based on just a single locus, is an emotion that is easily overrun by the cacophony of our other genes when their interests might be better served by selfishness. We can be cooperative and collaborative on the one hand, opportunistic, calculating, and selfish on the other, even toward members of our own groups. This is the unavoidable tension of group living: even with shared fates and shared purpose, what is best for your group can conflict with what is best for you. Soldiers in combat platoons, for instance, know that their colleagues will become less helpful near the end of their tours. When soldiers have fewer than thirty days remaining on their tours, they are called "short timers" or just "short," as if to acknowledge that they have shifted out of the mutual aid society.

In a less deadly situation, the Tour de France bicycle race covers around 2,000 miles in a series of long stages spread out over several weeks. During a race a small group of riders might try to lead what is called a "breakaway," riding off to the front of the main pack of riders. The riders in this small group must then work together, taking turns to ride in the lead, the rest saving energy by riding in the slipstream behind the lead rider. Every time a rider takes the lead, he is sacrificing some of his endurance, and this benefits all of the other riders. But it is a sacrifice that each of them must make to have any chance of staying ahead of the larger pack. On the other hand, cyclists in the breakaway pack will often try to do less than their share, and the pack will sometimes swerve and veer to try to shake them off. But it is when the race nears the finish line that the riders are clearly revealed not as selfless altruists but as self-interested competitors. If this breakaway group has managed to stay ahead of the larger pack behind them, the mutual altruism now unravels, the riders break ranks and sprint to the line, doing everything they can to beat those they have been helping.

Even the social amoebae's sociality turns out to be shrewd and calculating. The outwardly serene towers of cooperating altruists conceal a society of competing strains for which making the tower is an act draped with suspicion and the potential for duplicity and manipulation. The tower provides only one route into the future, but it is open to whichever amoebae can get to the top first. Some strains of related amoebae within the stalk cheat, advancing more of their members into this privileged collection of spores. Even before the tower is constructed, a carpet made up of many different strains moves more slowly across the forest floor than those composed of a single strain or of just a few. The delays arise from conflicts and jostling over desirable positions in the moving mass, just as the selfish riders in the breakaway packs ultimately slow their escape from the larger pack behind them.

Before leaving this discussion of how our altruism toward members of our societies might have evolved, I want to point out how the principle of identifying like-minded others invades another part of our lives: it is the same principle that governs our altruistic dispositions to favor our actual relatives, not just our honorary ones. Our nationalism really is a special case of a disposition to protect "our own." Relatives, by definition, share many of their genes, so they are also likely to share any disposition that someone might have to "help relatives." Surrounding yourself with relatives makes it likely you will receive as well as dispense benefits. There is nothing special about relatives, then, at least not from an evolutionary point of view, except that our relatives share many of their genes with us.

This weight of shared genes is why our familial nepotistic emotions are so strong—many genes are pulling in the same direction. In fact, the arbitrariness of the disposition we have toward relatives is revealed when we realize that we don't really know who our relatives are—we merely assume they are the people around us in early life. This rule can go wrong, as when infants are mistakenly assigned to parents other than their own in hospital, but then reared normally with no one knowing. But this just proves that as we don't really know who our relatives are, we use a rule of thumb. And that rule of thumb has worked well throughout our evolutionary past as a means of identifying people who are likely to share genes and hence our altruistic dispositions. Similar rules of thumb govern our dispositions to help other members of our societies.

Some readers might think the idea of a mutual aid society falls short of explaining cases in which our nepotistic help feels one-sided, as in the aid we altruistically pass on to our children. Many parents live in the probably vain hope that their children will return their favors one day. But if it is vain, why do we bother to help them so much? Thoughtful as it might be for children to help their parents, the altruistic tendencies we think of as our love for our children do not depend for their evolution on getting children to reciprocate. In fact, to expect this reciprocity is to misunderstand the nature of the altruism we are attempting to explain. That altruism is not based on getting something back from the person you helped; instead, it is based on attracting benefits from others who share your disposition to behave altruistically. The principle of attracting benefits is still fulfilled in the parent-offspring relationship because most of us will have received the same sort of help from our own parents, with whom we probably share the genes for helping relatives. The familial mutual aid society is a generational one.

the co-evolution of war, parochialism, and moralistic aggression

WHATEVER THE social capabilities of the little jellylike amoebae, we can expect ours will be far more developed, and not just because of our vastly greater intelligence. The amoebae in their fleeting moments of building the tower are like Macbeth's poor player, who "struts and frets his hour upon the stage, / And then is heard no more." By comparison, all of human life is played out on the social stage, and so we are entitled to expect that selection will have acted strongly on the emotions we carry to make it work. Not only would genes, or culturally transmitted ideas, have spread that give us that warm glow from cooperating with other members of our group, but elements of an entire psychology of cooperation would have sprung up to encourage it. Having a conscience keeps us from straying into selfish territory; feeling guilt puts a brake on our appetites; empathy, by getting us to feel what others feel, helps us to be helpful to them, and also reduces any tendency to harm others because we can feel their pain; shame motivates us to put things right; and dispositions to be kind and generous build reputations and attract allies.

At the same time, if we accept the view of our cultures as survival vehicles, the nature of our altruism can also help us to understand some jarring facts about our social behavior. It is a melancholy feature of our species that there are two situations in our everyday lives in which humans can act with such explosive violence toward others as to make us question if it is ever safe to trust another of our kind, and both of these are linked to protecting our societies. One is when trust unravels between two tribal or ethnic groups, and uncontrolled slaughter breaks out. In a period of one hundred days in 1994, Hutu tribesmen in Rwanda massacred somewhere around 800,000 Tutsi tribal people in a terrible spasm of violence. There were no weapons of mass destruction and few modern tools of warfare. Most of the killers were armed with clubs and machetes. People who had lived near to each other for generations descended into a maelstrom of genocide, chasing down, cornering, and then hacking and chopping each other to death.

The other situation arises when we turn on each other within our own group or society over what is taken to be a defection by one person against the social norms that glue society together. When this happens, humans are capable of abandoning the restraints that normally govern their actions toward members of their own group. Some of the people the Hutu killed were other Hutu, their crime being to be committed to peace with the Tutsis. In 2008, a woman in a South London grocery shop summoned her boyfriend to confront a man she claimed had jumped the line in front of her. The boyfriend smashed the other man in the face, knocking him down and killing him. Life can be cheap in human societies: the dead man hadn't in fact jumped the line, and it was a line to buy a packet of cigarettes. Two young boys were beaten to death with wooden bats in Pakistan in late 2010 for carrying a cricket bag. The bag resembled one some thieves had stolen earlier that day. But the boys were simply on their way to play cricket.

These are not isolated incidents. If the anthropological accounts of our history are correct, there are good reasons to believe that natural selection has not only favored in us a willingness to kill other members of our own species, it has equipped us with a set of dispositions that make it alarmingly easy for us to do so. Just consider that the action our societies reserve their greatest moral sanctions against—murder—can earn someone that society's highest honor, if directed at the right kind of person in war. In the absence of evidence to the contrary, it would seem that humans are capable of throwing a switch in their minds that allows them, even with little or no provocation, to treat members of other societies—and even in some circumstances their own—as something considerably less than human in moral terms. The supreme and unhappy irony of our species is that these behaviors are connected. Both arise out of the fragility of our "special" and "limited" form of altruism toward people who are not related to us.

The first—our tendency toward *parochialism*, or a disposition to be hostile toward people outside of our groups—has a simple, if disturbing, explanation. As our societies occupied the world after leaving Africa beginning around 60,000 years ago, they would not just have replaced the premodern populations of *Homo erectus* and the Neanderthals in their paths, they would have increasingly encountered other modern human populations like themselves, looking to occupy the same territories. To survive, our cultures, like the vehicles of any other organism, would have had to evolve the means to compete against other cultures. Other human cultures would be our most natural competitors, and so we can expect that our history has been characterized by a to-and-fro arms race of societies meeting each other's improvements at warfare with countermeasures of their own. If holding a parochial or hostile view toward outsiders makes it easier for you to kill them in battle, parochial dispositions might have spread as good ways to end any particular arms race. But if so, dispositions to commit oneself in battle—even including a disposition to give your own life—might have been necessary in return.

The second—our ability to turn on members of our own society when we think they threaten its integrity—is sometimes called *moralistic aggression*, and is more subtle. When we rebuke people for deviating from norms, shout at or honk our horns at people who jump lines, or worse kill them, these are ways of punishing someone who has shown from their actions that they do not share the altruistic dispositions on which our shaky but valuable form of cooperation rests. They lose the special and limited sense in which they are honorary relatives by demonstrating that they are not related to us at the *altruism locus* that underpins our joint cooperation. This might tell us that our societies are worth protecting, but doesn't explain the ease with which we can so violently turn on someone from within our own group. But, of course, moralistic aggression is as easy to evolve in our societies as parochialism is toward those outside—in both cases, our aggression is directed at someone not related to us. Once someone has shown by their actions that they do not share your disposition at your altruism locus, it doesn't matter whether they are in your tribe or not—their relatedness has fallen to zero. This is why it was easy for Hutu to kill other Hutu who had sided with Tutsi.

A quote often attributed to the Roman orator and lawyer Cicero tells us that moralistic aggression is conceivably an even stronger and harsher emotion than parochialism:

> A nation can survive its fools, and even the ambitious. But it cannot survive treason from within. An enemy at the gates is less formidable, for he is known and he carries his banners openly. But the traitor moves among those within the gate freely, his sly whispers rustling through all the alleys, heard in the very hall of government itself. For the traitor appears not a traitor—he speaks in the accents familiar to his victims, and wears their face and their garment, and he appeals to the baseness that lies deep in the hearts of all men. He rots the soul of a nation—he works secretly and unknown in the night to undermine the pillars of a city—he infects the body politic so that it can no longer resist. A murderer is less to be feared. . . .

The physicist Steven Weinberg has said, "With or without religion, good people can behave well and bad people can do evil; but for good people to do evil—that takes religion." Weinberg's remark is something of a rallying war cry among those who decry the evils of religion, but it is sophomoric and naive. As the previous examples show, it hardly takes religion for good people to do evil things, unless Weinberg thinks Hutus are all innately evil, or that Hutu is a religion. The causes of the massacre are still debated, but there is no suggestion that this was a religious war. Instead, Hutus are agriculturalists or farmers and Tutsis are pastoralists who herd cows. Both need land, and where pastoralists and agriculturalists overlap, there is often conflict over who gets it. All it takes, then, is for something to spark violence into action and our culturally defined parochial tendencies can take over.

Another physicist, Stephen Hawking, seems to get closer to the truth. Hawking warns that we should expect the very human tendency to exploit new lands and the people and animals that inhabit them to be true of any aliens who might visit us, noting: "If aliens visit us, the outcome would be much as when Columbus landed in America, which didn't turn out well for the Native Americans. . . .

We only have to look at ourselves to see how intelligent life might develop into something we wouldn't want to meet." It is easy to be amused at Hawking's alarm, but we may have been something we would not want to meet for at least the last 60,000 years, maybe much longer. In *War Before Civilization,* Lawrence Keeley documents how battles, raids, and massacres have occurred in our past at far higher rates that in modern nation-states, and mortality rates were far higher than in modern combat, sometimes exceeding 10 percent of the local population in a single massacre. That would be like a country the size of the United States losing 30 million people in one battle. Rates of homicide among the San Bushmen—long fabled as a peaceful society and sometimes called "the harmless people"—exceed those of industrialized democracies. Keeley's conclusions are remarkably similar to those reported by Steven LeBlanc and K. E. Register in their book *Constant Battles*, whose title is intended as a description of our hunter-gatherer history.

Humans have routinely throughout our history treated members of other racial or ethnic groups with hostility and wanton violence, torture, massacres, and disfigurement. Keeley recounts how in Tahiti, a victorious warrior would pound his dead foe's corpse flat using a heavy war club, then slit open the flattened body and wear it as a trophy. Some think this level of aggression is an anomaly caused by malign Western influence; or that it is only true of modern agriculturalists, whose possession of crops means they have something to defend and something worth stealing; or that it is limited to a few psychopaths. But such views are simply incompatible with what we know of our species. The evidence is now undeniable that conflict has been a routine and debilitating feature of our history, with high levels of violence documented for many hunter-gatherer societies. One often quoted survey from 1978 by the anthropologist Carol Ember, entitled "Myths About Hunter-Gatherers," revealed that somewhere around two thirds of them waged some form of war at least every two years. The study is technically flawed, but even accounting for those flaws, war has never been very far from our doorstep.

We should not be surprised. We were the species that discovered the division of labor and the cooperative dispositions that allowed it to work. Together, these developments allowed us to assemble and deploy what was by anyone's reckoning a formidable fighting machine of specialized and cooperative warriors. But our tendency to aggression goes even deeper than this, right back to our breaking of what we called the "rule of two." It is likely that wherever we have gone, we have filled up our environments to their carrying capacities. In earlier times as hunter-gatherers this might not have meant very many people per square mile of land, but they were still at their carrying capacity because hunting-and-gathering is an inefficient way to use land. So once we had spread around the word, there never would have been a time when great tracts of usable land lay unused. Neighboring groups, each at their respective carrying capacities, will therefore always wish to have the land next door, and killing off those neighbors would have been a good way to get it.

This might have been the fate of Ötzi, the Ice Man, the 5,300-year-old man whose body was discovered in 1991 sticking out of the end of a glacier on Tisenjoch Pass in the Alps spanning the Italian-Austrian border. Ötzi (the name given him by researchers who studied his remarkably well-preserved body) was fully clothed and carrying a bow and arrow. He was wearing shoes made of brown bearskin, with deerskin side panels that were drawn up tight around his feet using a bark-string net inside the shoe. Anthropologists speculated that he was a shepherd who might have fallen into a crevasse and frozen to death. But the more Ötzi was studied, the more this story seemed incomplete. Ötzi turned out to be armed to the teeth. In addition to his bow and arrow, he carried a dagger and a hafted ax. But it was only when the body was finally X-rayed that the cause of his death became clear. Ötzi had an arrow embedded deep in his chest. He probably froze to death fleeing his attackers.

Measured in terms of their cost to society, the Trident nuclear submarines of their day might have been the Maori war canoes. They could be 100 feet long and hewn from a single tree felled in the forests of New Zealand (or Aotearoa, as it is known to Maori), then hollowed out to produce the hull. The hull would have been filled with seats, riggings for oars, and space for up to fifty-five paddlers and seventy warriors. New Zealand's Auckland Museum contains the last one of these to be built, known as *Te Toki a Tapiri*, carved in the middle of the nineteenth century from a single totara tree (*Podocarpus totara*). It is 82 feet long, painted, ornamented, and decorated, and would have required hundreds of hours of work from scores of laborers. And yet this canoe would not have contributed food or shelter or help to rear young or look after the elderly, or treat disease. It would have consumed a considerable portion of a tribe's "gross domestic product," but it was built exclusively to mount raids on other tribes, and to defend the tribe against others wishing to do the same to them.

The practice of slavery is not simply a moral stain on the fabric of seventeenth- to nineteenth-century British and American societies. Nearly every human group that has ever been studied has at one time or another kept people against their will and used them in a slave capacity. Arab slave traders bought people from East Africa perhaps as early as from the eighth century onward, and in fact one can still see the slave auction blocks in the modern tourist destination of Zanzibar, just off the Tanzanian coast. Pacific Northwest Coast Native American tribes captured slaves from each other. In Oceania, some Samoan and Maori groups practiced slavery, as was true of western regions of New Guinea. Ancient Greece and Rome imported slaves from Africa, and some African slaves were sold as far away as China. Slavery was common in India, and modern-day Pygmy communities often live in an economically subservient capacity to nearby farmers. What we regard in modern society as morally reprehensible acts are all too predictable from the nature of our societies and the nature of our cooperation. Slaves are not members of our groups, and as such lose that precious degree of cultural relatedness that only barely knits our own societies together. As we have seen, once that fleeting bit of relatedness is removed, humans seem to have the capacity to lose their normal moral restraints, using others in whatever ways serve their interests.

Aggression toward members of our own societies is not even limited to people whom we think have violated the cooperative pacts they rest on. For the same reason that moralistic aggression is easily evolved, so too is it easy for our societies simply to rid themselves of people who can no longer contribute, even though this is now something we regard as obscene in modern societies. The Aché (ah-CHAY) are a tribe of hunter-gatherers living deep inside a remote part of the Amazonian rain forest. They did not make contact with the outside world until late into the twentieth century, and were living what most of us would think of as a Stone Age existence. The anthropologists Kim Hill and Magdalena Hurtado lived among the Aché for years. They report that killing of members of their own group is common, especially children whose parents have died, and adults who are injured or potentially fatally diseased.

In fact, this killing is so routine for the Aché that they frequently bury people alive, and then the group walks off leaving them to die. One account describes a middle-aged man whom the group abandoned because he was too ill to keep up. He was so weak that he lay on the ground under a tree. Vultures came and sat in branches right above his head and defecated on him while waiting for him to die; he was too frail to wave them away. Unusually, this man eventually managed to recover enough from his illness to catch up to the group. But upon his return the members of his hunter-gatherer band called him "vulture droppings" because of the feces on his head. Others, especially the children, were less fortunate.

The Aché's ruthlessness toward other human beings must be understood in the context of a roving hunter-gatherer band that depends for its survival on all members of the band contributing. The Aché acquire everything they need for their survival from the forest, and each day they are on the move, hunting and gathering. Life is hard and most days everyone is hungry. Hunger is almost certainly the normal state for most hunter-gatherers for the simple reason that were there more food to eat, the environment would support a larger number of people: we expect life to push up to the edge of carrying capacity. In the cold arithmetic of a group living on the edge of survival, orphans, the diseased, the frail, and even just old people are a burden on the group. They can no longer contribute and diseases can spread to others. Our life of plenty is an anomaly in the history of the world and one that not even all of us enjoy.

the target of our altruistic dispositions

THIS ACCOUNT of how altruism arises can help us to understand one of the most contentious ideas about the nature of our behavior and morality. It concerns the question of whether we perform our altruistic acts—especially our acts of self-sacrifice—for the "good of our group" or for more self-interested reasons. Proponents of what is known as *group selection* believe that Darwinian evolution can choose among groups, thereby selecting for sets of behaviors that make it likely that one group will outcompete another. Group selection theorists believe that the most successful groups of people in our past—our hunter-gatherer or early tribal ancestors—were those whose individuals submerged their own selfish interests to the interests of the group, because this would have created highly cohesive and formidable opponents in battle.

The idea is that over time, this process molded our psychology and social behavior so that we became—as one of the proponents of group selection, David Sloan Wilson, puts it—like cells in a body, or bees in a hive, devoted to the well-being of our group, even willing to sacrifice our health or survival for it. And indeed, suicidal charges in battle or falling on grenades will surely promote your group's success. Proponents of group selection interpret music, dance, religion, and even laughter as aids to promoting the sense of group membership and mutual well-being that gives rise to these self-sacrificial emotions. It is an account of our nature and individual psychology which has us content to accept that we are part of a larger organism that looks out for its interests, even if this means sacrificing some of its individuals. It is a view that chimes with our sense of duty as taught to us in patriotic songs and national anthems, and is most vividly manifested in our acts of courage and bravery in war. This sense of submerging ourselves in the group is also thought to be why we do such things as help the elderly, give money to charities, put on identical silly shirts to attend football matches, obediently wait in line, and why we positively ripple and snort with righteousness and indignation when we think others don't do some of these things.

But the account we have given is clear. Even dispositions that can predispose someone to an act of suicidal self-sacrifice can nevertheless evolve out of self-interest, by attracting benefits from others who share that disposition. Sebastian Junger emphasizes that the men fighting in the Korangal Valley were not there to die for their country, to defend their land, or even necessarily to like each other, and many did not. In talking to the men, Junger came to realize that abstract and symbolic commitments like doing your best for your country were of little relevance in battle when it is group effort that saves lives. The men wanted to survive and realized this was more likely the more they all pulled together. It is difficult to overemphasize the importance of this peculiar and even shocking form of altruism for understanding our evolved psychology. But we must bear in mind that we are a species that has lived for nearly all of its history in cultural survival vehicles that compete with other such vehicles for the same territories, resources, and potential mates.

We owe the heroes of war enormous gratitude; but few young men join armies willingly except when their home country is directly threatened, and historically those who do have often had few alternatives. Drafts, press-ganging, economic hardship, and ultimatums frequently serve as the most efficient recruiting offices, and desertion in war is common. The trench warfare of World War I is better remembered for callous commanders willing to sacrifice men in their charge rather than men eager to be sacrificed. The least eager were often persuaded by the threat of being shot on sight for desertion. The spontaneous truces that sometimes broke out between the warring sides had to be forcefully put down by these same commanders. Erich Remarque's *All Quiet on the Western Front* is a story of individual heroism, but that heroism is a desperate struggle for mental self-preservation in the face of unspeakable horrors, not selfless military sacrifice. Despite what seems like daily reports of suicide bombings, the numbers of people who do this are negligible compared to those who could if our motivations really were to promote our groups at our own expense.

It is the view of war films, patriotic songs, and propaganda campaigns that we do things for our group; but if self-sacrifice were a part of our nature, why do we need to go to such lengths to sell the idea? No one asks us to eat, talk, sleep, or have sex, things that are very much wired into our nature. If we really have evolved to do things for the good of our groups, what are we to make of our tendencies to cheat, deceive, manipulate, and coerce? Why do we need so many laws, police forces, jails, surveillance cameras, and tax offices? Why do we gossip incessantly about others' behaviors and reputations; why do we compete so strenuously to get ahead and pay so much to get our children educated? All of this would mystify a sentinel bee or an immune cell in our body, neither of which would blink at giving its life and certainly not expect nor receive reward. Indeed, both would race to the scene of invasion clamoring to die for their hive or body.

What seems far more likely than that we are somehow nobly disposed to the idea of self-sacrifice is that natural selection has duped us with an emotion that encourages *group thinking*. It is an emotion that makes us act as if for the good of the group; an emotion that brings pleasure, pride, or even thrills from coordinated group activities. It is the emotion we feel on the sports field, when singing together, and probably when going into battle. It is the emotion that national anthems, flags, war recruiting posters, patriotic songs, and military commanders exploit. And it is an emotion that by encouraging coordinated group behavior has brought our ancestors and us direct benefits.

Warriors do die, but when the choice is to fight or be killed, fighting with coordinated gusto is often the best option for staying alive, and the spoils of victory can be great. Cooperative altruism of the style we can find in our species has paid handsome dividends in our past—dividends that arise from assembling a powerful and cohesive social vehicle made up of individuals committed to cooperating with each other. It is this that makes human culture the survival vehicle that it is, and we have evolved an entire psychology around it, from our acts of kindness and self-sacrifice to our xenophobia, parochialism, and predilection to war.

CHAPTER 3

The Domestication of Our Talents

That culture might have domesticated us in a manner not so different from the ways we have domesticated dogs

butchers, bakers, and candlestick makers

GO OUT INTO the wild sometime and observe a group of animals. Maybe it is a flock of birds, or a herd of cows, or if you are lucky you might travel to Africa and watch giraffes or monkeys. One of the things you will realize after you have been watching for a while is that, apart from the usual division of labor between males and females, all of the animals in these groups will be doing more or less the same things. If it happens to be cows you are watching, they will all have their heads lowered to graze, they will be twitching their tails, and lowing and mooing. If you are watching a group of monkeys, they will be feeding, grooming each other, and occasionally grimacing or shrieking. If it is a flock of birds wheeling around the sky, there won't be leaders and followers, at least not for any length of time; the birds' positions in the flock will be in a constant state of flux. In all of these groups, most of the individuals will routinely do a little bit of everything.

Now imagine yourself up in the air—perhaps having climbed

high up a tree—looking down on a human settlement maybe 40,000 years ago. The scene will be different from watching the cows, giraffes, birds, or monkeys. Yes, like them there will probably be a division of labor between the sexes, but even by that time in our history someone might be making a musical instrument, carving a figurine, or crafting jewelry. Someone else might be flaking a stone blade or making an arrow or a spear. Someone else might be building a shelter, making a net or bow, and someone who spent the day foraging might trade some of his or her food for one of these efforts. These humans are doing different things and exchanging what they produce or acquire for things others have built or acquired.

This kind of task sharing and exchange among unrelated people does not occur elsewhere in the animal kingdom. It is something we take for granted, but it might just have had profound effects on our makeup and be yet another way that culture has sculpted our species. The question we want to ask in this chapter is whether, as a result of thousands or even tens of thousands of years of being able to specialize or "do different things," people have come to differ from each other in innate abilities and dispositions that make them suited to alternative ways of prospering in society. Why, for example, are some of us so good at music, or art or architecture, and from such early ages, and others not? Why do others have exceptional ability at mathematics, singing, sculpture, or design? Why do others have remarkable spatial abilities or eye-hand coordination? And what of social skills such as charm, leadership, and persuasion? Where does this variety come from? Is it all a product of upbringing or chance, or something more hard-wired?

These differences are a puzzle because if natural selection is the process by which some combinations of genes survive at the expense of others, we normally expect differences among us to get "used up." Natural selection favors speed in the antelope fleeing from a lion, as well as in the lion. Slow antelopes get caught, and slow lions go hungry. When male songbirds try to attract the attentions of females, natural selection favors the melodic singers. Poor singers remain lovelorn, and more important, childless. But we will see evidence in this chapter that human populations carry what appear to be wide genetic differences related to performance, skills, and personality. Does this variety exist because life in the presence of human society has, throughout history, provided a range of opportunities for advancing our interests, opportunities that have cultivated genetic differences among us? Have we, in short, been domesticated by culture?

specialization and self-interest

IN RAISING this question I do not suggest, much less advocate, any sort of moral judgment or natural order built on the possibility of differences among us in our innate skills or abilities. In fact, we will see that if such differences do exist, they are likely to lead to equally good outcomes in those who have them—indeed, we will see that there is no other way the differences could be maintained. I also do not want to give the impression that I am ignoring our brains' prodigious abilities to learn and adapt, or the disabling effects of social deprivation. Rather, I want to call attention to the possibility that we have been subjected to forces of natural selection that won't have arisen in other species. These forces are what evolutionary biologists call "diversifying selection," or selection favoring more than one outcome, and they arise in our species because of the opportunities culture provides.

In fact, the possibility of domestication by culture is a scenario we should recognize, having inflicted it on countless animals. For example, in just a few thousand years humans created breeds of dog ranging from Chihuahuas to Great Danes. They are all the same species—they can all interbreed, if with care in some cases—and the differences among the breeds are genetic. There are genes for longer legs, shorter ears, fluffier fur, or wider snouts. But our domesticating efforts didn't stop at appearances. We have also selected for particular behaviors, intellectual abilities, and temperaments. Alsatians are aggressive, sheepdogs and collies are intelligent, having been bred to be good at the complex task of herding animals (try it yourself), spaniels are gentle, and bloodhounds are exceptionally good at tracking scents. Other dog breeds are good at guarding, or at sports; some are good with children, and others, like the sled dogs, have extraordinary physical stamina.

The environment that cultivates the various dog breeds is the environment of human preferences. Calling them breeds is just a shorthand way of saying that among the species we call the dogs, there are genetic varieties that have been selected for and maintained by humans. From the standpoint of the dogs' genes, humans constitute a social environment that favors many different solutions to the problem of how to satisfy their whims. Dog genes have been only too happy to comply, advancing as this does their survival and reproduction.

From the standpoint of our own genes, human culture also constitutes a social environment that presents many different ways to solve the problems of surviving and prospering. The opportunities for task sharing and specializing might have been limited in our hunter-gatherer past, but there seems little doubt that a tendency toward specialization has always lurked inside our societies. Even among the hunter-gatherers, several hundred different products might have been available—including foodstuffs and local technologies—and it is conceivable that some of these required specialized skills to produce. By about 10,000 years ago, humans had invented agriculture and animal domestication. These practices can produce surpluses of food, and so for the first time in our history some people were freed from having to hunt and forage. Almost immediately after—by 7000 to 8000 years BC—the first stirrings of "urban" life arose as towns were built at places like Jericho, in what is now modern-day Israel, and Çatal Hüyük in Turkey, both with populations of several thousand people. The settlements at these sites supported people who specialized at pottery making, metalworking, and jewelry, and a merchant class arose. There was even enough surplus food with the invention of agriculture to support armies and a religious and political elite. These were people who made no other contribution to society than to protect it, pray for it, or attempt to run it.

By the Middle Ages the pace of cultural change had produced a range of professions that would have left a skilled cave dweller of

40,000 years ago, or even an early farmer, baffled and incredulous. The medieval Italian city of Siena is best known for the *Palio*, a horse race of three laps around the Piazza del Campo in the city center. The race has been held, with few exceptions, annually since at least 1238. Riders race their horses bareback, circling the piazza three times. It can be an untidy and even dangerous race as riders are allowed not only to whip their steeds on the steeply banked and slippery track but other horses and riders as well. Most races see at least some jockeys thrown or dislodged from their mounts and horses running freely around the course.

But it is not the horses or their riders that interest us here. Siena is also a city of around seventeen small social and political groupings called *contrade*, and it is the *contrade* that contribute the horses that run in the Palio. Imagine an irregular shape, like a poorly rolled out pizza crust. This could be a map of Siena. Now divide that shape up into seventeen further irregular shapes that fit together like the pieces of a jigsaw puzzle. This could be a map of Siena showing the outlines of the areas given over to the *contrade*. Each *contrada* is associated with a particular trade: there are notaries, silk workers, tanners, cobblers, bankers, painters, smiths and goldsmiths, bakers, potters, dyers, carpenters, apothecaries, weavers, stonemasons, wool carders, and silk merchants. The same set of *contrade* has been in place since at least the seventeenth century and some sort of *contrade* structure has existed since the Middle Ages. They are different ways of making a living in Siena, and the differences among them are maintained from one generation to the next as their survival in their present form over at least four centuries—and maybe eight—shows.

In our modern world we see the trajectory toward doing different things, or "specializing," having reached its endpoint. Some of us are bakers, some of us are lawyers or engineers, butchers, medical doctors, hedge fund managers, mechanics or accountants. Once you become one of these, that is about *all* you do, acquiring everything else you need from others. Even so, unlike the domesticated dogs, we have a choice as to what we do, so why do we think there has been a cultural current pushing us toward greater specialization—why can't everyone do a little bit of everything? The simple answer is that once a species works out the rules of cooperation that allow individuals to exchange their goods and services, it no longer pays even to try to be good at everything. We know this from the work of the nineteenth-century economist David Ricardo, whose book *On the Principles of Political Economy and Taxation* explained an idea that would come to be known as Ricardo's law of comparative advantage. It is a vexing idea, simple and yet deeply counterintuitive, but one that, without our even knowing it, has penetrated our lives. Our interest here is that it is a law about the virtues of specialization, which can be summed up as "do what you are best at and do only that."

Ricardo asks us to imagine two countries (the example could just as easily be two people). One is good at producing food and very good at producing clothing. The second is bad at producing food and very bad at producing clothing. Our instinct is that the first country—being better at both tasks—should make food and clothing and ignore the second country. But Ricardo pointed out that this would be folly. Assuming both countries want food and clothing, Ricardo realized that the first country should specialize in making clothes and the second in producing food. The reason is this: the first country can make more clothing than it needs and then trade its excess clothing to the second country for food. Because it is better at making clothing than it is at producing food, when it trades for food, it is acquiring it with an efficiency that rivals that of its clothes making. For the second country, there is little point in making clothes. It should focus on producing or acquiring food and use that food to buy clothes. In doing so, it will be acquiring clothes with an efficiency that rivals that of its food production, which it is better at than making clothes.

Another way to think about the problem is that the first country could raise the most cash per unit of work (if transactions were in cash) by producing only clothes. The same argument applies to the second country. If both countries follow Ricardo's advice, both will be better off. Ricardo's law is regarded as one of the most important rules of economics, lying behind the push toward free trade agreements and eradication of tariffs. It is why economically developed countries have had to all but abandon coal mining and manufacturing as countries that pay lower wages can do these things more efficiently. It is why most call centers are located in these same lower-wage-paying societies. Countries need to focus on what they are good at, not protectionism.

What Ricardo probably didn't realize is that he was describing something that modern humans had stumbled upon within their own societies perhaps more than 100,000 years before him when they discovered social learning and cooperation. Ricardo's law tells us that when people naturally gravitate toward doing what they are best at, and then exchange their goods and services, everyone is better off. Everyone is better off because they will each be "purchasing" what they need from the returns they get from doing what they are best at. If I am good at making stone tools, I can trade them to you in exchange for food you have found. If you are good at shooting poisoned arrows through a blowpipe, you might trade dead birds to someone good at climbing trees for honey. Someone else good at tracking big game might lead hunting parties in exchange for someone making a shelter or a boat. These opportunities are simply not available to any other animal.

Specialization and cooperative exchange are revealed as the routes of self-interest. If you stubbornly refuse to follow this rule but others do, they will be better off and you will be left behind. Still, Ricardo's law has a "too good to be true" feel about it, an almost glib account of why our societies produce such riches. And yet there are striking precedents for it in nature. In fact, you inhabit one of them. Multicellular organisms such as ourselves, or elephants, or even clams, had discovered Ricardo's law of task specialization perhaps 500 million years before humans discovered culture or Ricardo discovered his law. The society of cells that is our bodies ticks over smoothly and efficiently because it is a vast citadel of specialization, resembling a city or town in miniature. Some of our cells build hearts, others make livers, or muscle, eyes, kidneys, or brain cells. Skin cells erect a defensive wall or protective barrier around the body, and other cells form aggressive armies of the immune system. Natural selection had to coordinate exchanges among these parts, and so it built communication networks in the form of nerve cells carrying electrical signals from one part of the body to another, and it produced liquid chemicals called hormones to contact many other cells simultaneously. Tubes carrying blood act as road networks to ferry oxygen and food to the tireless specialized workers, extending all the way out to the fringe territories of fingers and toes.

This would all be repeated millions of years later when the first social insects—the ants, bees, wasps, and termites—evolved. Their teeming communities of brothers and sisters came to act like a single coordinated body, only more loosely organized. Now, instead of millions of cells packed inside a protective skin and laboring to transmit their genes, millions of brothers and sisters came to labor inside a protective hive or nest on behalf of their mother, the queen. Workers would get assigned to separate castes of foragers, nest repairers, and nannies that looked after the brood. Others would become immune system commandos, capable of defensive but also of offensive actions to protect the nests against invaders or to attack other nests. They acquired airborne communication systems of chemicals called *pheromones*, and workers acted as blood, carrying food to developing embryos.

Ricardo's law can tell us that specialization is a good thing and a simple rule of thumb can tell us how we might make it happen without even knowing it. *Win-stay, lose-shift* is a strategy that follows the rule "If you find yourself winning, stay put, otherwise move on." You play this strategy when the gambling machine you are pouring money into comes up all cherries and you try again, or doesn't for some time and you move to another. You do it when the bar you frequent looking to attract a mate turns out to be a literary bar but you are a sporting type. You do it when the people you are cooperating with turn around and steal from you. Bees follow *win-stay, lose-shift* when they search for nectar in flowers. In some cichlid fish species, males produce one of two color patterns—red or blue patches on their scales—and females have a preference for one or the other of these. Female cichlids with the preference for red-colored males find it easier to detect them in deeper waters, and vice versa for the females with the blue color preference. If males of both colors are initially randomly distributed at different depths, a policy of *win-*

stay, lose-shift will see the blue ones moving toward shallower waters where they will be more successful, and red ones will move toward deeper waters for the same reason.

Human couples might unwittingly play *win-stay, lose-shift* when they sort themselves by height. In most couples, females are shorter than males but not by much, so that if you make a plot of women's height against men's height among couples, you get a positive relationship beginning with the short couples in the lower left of the plot and moving up to the tallest couples in the upper right. There are of course exceptions, but the relationship is surprisingly consistent. It is also surprisingly easy to get it to emerge. Imagine women have a preference to find mates taller than themselves. The tallest women will seek to pair up with the tallest men (the tallest women have no choice). Once the tallest males are taken, the tallest of the remaining women will tend to pair up with the tallest of the remaining men, and then the next tallest men and women will pair, and so on, right down to the shortest men and women. The pattern can be helped if men also prefer shorter women, even if they don't mind too much how much shorter. Next time you are at a large social gathering, look around to see how closely this pairing rule is followed. And it needn't be limited to height.

Now, suppose among a group of people there is a range of talents and skills but people find themselves randomly scattered among a set of tasks to be done. Suppose further that those who just happen to have the most suitable skills or temperaments for their task are more successful. These *winners* stay put while those who are less successful—*losers*—move on to some new task where they can get better returns for their efforts. Over time, if everyone plays a strategy of *win-stay, lose-shift*, people will come to be highly sorted such that everyone is doing something they are good at. If we imagine that something like this has been going on for thousands of years of our history, and we are prepared to believe that different tasks really do require different skills, then we would expect a variety of different talents and skills to emerge and be maintained in our societies. If we find this suggestion surprising or even alarming, we need look no further than just about any sports field. Games such as football, soccer, and basketball attract people of different body shapes and sizes. Height is far more important to a basketball player than it is to a soccer player, and bulk and strength are more important to a football player than to a basketball player. Jockeys and rowing coxes are small.

unmasking our latent abilities

IT IS true, tall basketball players and short jockeys are not the same as saying that notaries are notaries because they are genetically good at being notaries and only at being notaries, and vice versa for silk workers, or that those who climb trees for honey do so because they are better at it than those who are best at throwing spears. But what if there were some natural number of different dimensions of human existence? Maybe, for example, our societies have always needed people with good analytical skills, particular physical abilities, manual dexterity or hand-eye coordination, musical, artistic or linguistic talents, social skills or features of personality. If so, the mere possibility of sorting combined with Ricardo's law tells us that societies that encourage specialization will be the most efficient and will therefore return the greatest benefits to their inhabitants. But this also means human societies will have placed the greatest demands on their inhabitants as competition to be good at *something* will be intense. The demands arise from the fact that Ricardo's law makes it inevitable that those who are best at what they do have the most to exchange with others. In these circumstances, it will pay you to find what you are best at and do it. If you don't but others do, you will inevitably compete against people who are better than you at what you *want* to do.

A simple but true feature of cultural evolution also tells us to expect that our cultures will have a built-in tendency to evolve in the direction of producing innovations that unmask and then cultivate the talents that naturally exist among a group of people. The variety of things around us, and the variety of different things we do, exist because they are the things that, with few exceptions, we are capable of exploiting for our own gains or pleasure: the products of cultural evolution will tend to be adapted to us. For example, until someone produced the first guitar, there would not have been any guitarists, but there would have been soon after as those with the ability to play them did so. If this example seems frivolous, it tells us something important. It tells us that when human culture produces innovations such as guitars, those innovations will tend to attract those who possess the skills best suited to them. There will be exceptions of course, but Ricardo's law and a policy of *win-stay, lose-shift* make this almost inevitable. Modern educational and social systems do their best to offer equal opportunities to children, but everyone—and not least schoolchildren themselves—knows that some of us are just better at some things than others (playing guitars among these) and are so right from the beginning. There seems little doubt that technological innovations—and not just limited to musical instruments—reveal latent abilities lying hidden amongst us. Indeed, school music teachers often say the instrument finds its player.

Among our near relatives the chimpanzees, there are no guitarists and there never will be, at least in nature, because chimpanzee society will never produce guitars. If there are latent abilities among chimpanzees for strumming guitars, we will never see them emerge. But we could imagine a chimpanzee strumming a guitar if one were made available. We know this because chimpanzees don't naturally paint either, but they can if they are handed a paintbrush. Picasso even owned a painting by a chimpanzee called Congo. Joan Miró is said to have swapped two of his sketches for a Congo painting, and Dalí was reportedly smitten with Congo's works. Several of Congo's paintings were auctioned in 2005 alongside a Warhol painting and a small Renoir sculpture, both of which had to be withdrawn for failing to meet their reserve prices. Congo's work sold for around $25,000. This is either a delightful story or a sign of human frailty, that the art emperors may indeed have no clothes.

We may not be able to agree on whether Congo's paintings are art, but this doesn't matter. Congo produced paintings that others found attractive, despite the fact that chimpanzees do not have any history of using paintbrushes, paints, and easels. It doesn't matter that Congo might not have had any notion of producing a painting, that he might have been quite happy to knuckle-walk away from his painting at any time, or that if left in front of it long enough he might continue to add paint, indifferent to the folly of covering over work he had previously done. We don't require of Congo that he deliberately set out to produce a work of art. What matters is that when the right cultural innovation came along, Congo could use it. And this is where we differ so from the other animals, because human culture has been producing the "right" innovations now for tens of thousands—maybe more—of years. Where artists have no particular role to play among the chimpanzees, they do among humans, and this might have meant that genes for artistic abilities have prospered just that little bit more in our past.

When a complex behavior such as the ability to paint emerges abruptly rather than over long periods of time, we are entitled to expect that it already existed in some form but that the conditions capable of unmasking it had never before arisen. It is precisely this scenario that seems to describe human artistic abilities. In the middle of December in 1994, three friends out spelunking (exploring caves) in the Ardèche region of France felt a waft of air emanating from a part of a cave they had visited before. They dug a passage through some fallen rocks and came upon a spectacular system of limestone caverns, now known as the Chauvet-Pont d'Arc cave after one of the three friends.

At first the explorers didn't notice anything except the beauty of the cave's spectacular stalagmites and stalactites. But then deeper in the cave, they saw something that hadn't been seen by human eyes for over 30,000 years. The walls contained startlingly accomplished paintings of rhinoceroses, mammoths, panthers, bears, horses, and owls. A drawing of a bison depicts it with eight instead of four legs as if the artist was trying to convey its motion. A rhinoceros image is repeated several times, each successive image shifted a small amount to the left, almost like a cartoon flip book. The shape of the cave walls is used to give some of the paintings a sense of having three dimensions. There is a negative image of what appears to be a male's hand, created when one of these prehistoric painters filled his mouth with paint and then blew it at his hand resting against the wall, leaving an unmistakable symbol of robust self-awareness.

All of these images had remained in utter darkness in the sealed cave since they had been created at least 32,000 and possibly as much as 36,000 years earlier, making them the oldest paintings yet discovered. Nothing would have hinted that they would burst onto the scene like this, and cave paintings are found around the world. It is as if our ability to produce them had lain dormant in our species, merely awaiting an outlet. That outlet might have been little more than the caves and perhaps charcoal to draw with taken from fires, or the availability of some local colorful mineral to mix into paint. There is simply no evidence the Chauvet paintings and others like them are the product of generations of earlier and less talented humans learning step by step how to paint. Instead, around 13,000 years after the Chauvet painters—a period of time about three times longer than all of recorded history—another set of cave artists in the much more famous nearby Lascaux Cave produced paintings that are not substantially different. They are more detailed and colorful, but these differences probably owe more to technological achievements that produced better paints, paintbrushes, and the fire lamps to venture into caves than to any changes to our underlying abilities.

In fact, the paintings in Lascaux, Chauvet, and other caves are so remarkable that for many years they were thought to be fakes, put there by mischievous modern graffiti artists. But someone noticed that the mammoths painted in one of these caves, the Grotte de Rouffignac, had a small bump on their derrières. No one knew about this bump until a mammoth was found somewhere else in the world virtually intact frozen in ice. It too had one of these small bumps and in the same place. That bump turns out to be a gland called the *operculum*, and it was probably used for scent marking. Modern elephants don't have opercula, and so it is unlikely that a modern artist would have known of them. Carbon dating of the paintings now confirms the conclusion of this fortuitous discovery of anatomy. Upon seeing the paintings at Lascaux, Picasso reportedly said, "We have invented nothing." It had not escaped Picasso's attention that the paintings were not mere illustrations but often fanciful representations that would anticipate elements of the Cubist style he would introduce to wide acclaim 17,000 years later.

Some of them depict horses with large distended bellies and short skinny legs, or aurochs with sweeping horns. One of these fanciful creatures at the Chauvet Cave has the head and body of an auroch and the hips and legs of a human female. Not far away at the Pèche Merle Cave there is a horse covered in polka dots and the figure of a woman with auroch features.

These works demonstrate not just imagination and symbolism but the fluidity of our minds that Steven Mithen, in *The Prehistory of the Mind*, suggests underpinned our leap of symbolic thinking. They show that not only could we think symbolically; we could combine different symbols in new and creative ways. Another 17,000 or so years after Lascaux, and again not far away, fourteenth-century fresco painters at the cliffside village of Rocamadour would produce the now familiar, and to modern eyes somewhat stiff and wooden, two-dimensional representations that characterize Byzantine religious art. It is not difficult to imagine that one of the Chauvet or Lascaux painters transposed seventeen to thirty millennia forward to this time could easily have done the same, and would have been tempted to do more. Indeed, when the film director Werner Herzog was allowed to view the Chauvet paintings, he was astonished at their sophistication, saying, "Painting never got any better through the ages, not in ancient Greek and Roman antiquity, nor during the Renaissance. It's not like the Flintstones—the work of crude men carrying clubs. This is the modern human soul emerging vigorously, almost in an explosive event."

The abrupt appearance of technical skills is not unique to painting. There are now examples of musical instruments, flutes and percussion devices, dating back to around 36,000 years. An ivory carved figurine of a startlingly sexually exaggerated female figure and dated to nearly 40,000 years ago has been found in Germany. It is not some primitive object, but rather a compelling example of what could pass as high figurative art in the modern world. As with cave art, we cannot know if there are far earlier and primitive sculptural forms, but it does seem that sculpting ability in some sense precedes sculpture as a regular or conventional art form; this figure, as with the cave paintings at Chauvet, seems almost to appear out of nowhere. Or not quite out of nowhere: analyses of the sculpted figurine reveal that the artist must have had access to fine scraping and carving tools. But that is of course the point: it was these technologies that unveiled the sculptor's abilities. And a recent find suggests those abilities might have existed at least as long ago as 75,000 years. Human hands are responsible for a prepared ochre stone about 1.5 inches wide and 3 inches long, found in the Blombos Cave in South Africa and dating to this time. It is geometrically engraved with a cross-hatched design on a face of the stone that had been ground to make it flat. It is a breathtaking view of our species' symbolic thinking and artistic abilities in a dim and distant past.

made or born?

THERE IS no reason to restrict our thinking to artistic, sculptural, and musical abilities; it is just that they have left a remarkable trail of artifacts. Now, whether we believe that human culture has cultivated differences among us, we really have only two choices to understand our obvious variety of talents: are we, in a word, "made" or "born"? The latter view says that genes are the dominant force in our lives, and it is our genetic inheritance that influences what we become, and why we differ from each other. Stacked against this notion is the point of view that says our environments make us what we are. Our differences might have little if anything to do with our genes, but much to do with upbringing and opportunities that arise by chance. People come to their various stations in life by good or bad luck, social immobility, long-lived cultural influences, and teaching and training.

No one doubts that children's early learning and social environments and the opportunities these provide strongly influence their eventual educational and economic attainments. But the truth, as with so many questions that pit nature against nurture, almost certainly lies somewhere in between the two extremes. In fact, modern genetic and behavioral studies are revealing that we are neither made nor born. Rather, these studies show that what we are born with—

our genetic endowment—can exert a powerful influence on what environments we are attracted to: someone good at kicking balls is attracted to sports, someone good at music is attracted to music, someone good at mathematics is attracted to working with numbers. And this brings us full circle to the theme of this chapter. We have seen how human culture uniquely provides a range of opportunities that might encourage genetic variety to be maintained in our ranks (not least in sports and music)—genetic varieties that would otherwise lie dormant—by providing the outlets that attract those with differing talents and skills.

Animal breeders are quietly aware there are wide differences among animals in temperament and behavior that they can encourage or suppress through careful breeding. These are genetic differences, but there is reluctance bordering on zealotry to acknowledge the possibility of genetic differences related to disposition or performance among humans. And yet, it turns out that a substantial portion—often 50 percent or more—of the differences among people in just about anything that can be reliably measured seem to be attributable to differences in their genes, at least insofar as these can be measured by looking for similarities between parents and offspring. The list includes physical traits, reaction times, spatial reasoning, verbal and mathematical intelligence, attitudes, personality, and even occupational, social, and religious preferences. No one suggests there is a gene for your religious preference, or for your particular occupation, for that matter; but temperamental, intellectual, and personality differences that do have a genetic origin might influence your attraction to religion or to a particular occupation.

Fifty percent is a high figure, but one that also of course allows for a large—50 percent—input from the environment. Still, we need to interpret this statement carefully. When newspapers and televisions quote scientists who say it is meaningless to ask which is more important, genes or environment, because "all genes require an environment in which to express themselves," these scientists are right but saying something misleading. To suggest that genes influence a behavior or characteristic is not the same as saying that feature is *determined* by genes. Rather, it says that two people, one with a particular gene or set of genes and one without, and each brought up in typically human environments, will on average differ. The proviso "typically human environments" is important. If it were possible to rear people in identical environments right down to the smallest details, then any and all differences among them would be attributable to their genes. On the other hand, if we were to rear people in extreme environments that overwhelmed any influence of their genes, then the genetic contribution to how they turned out would appear negligible. For instance, our genes have not evolved in the environment of being a toad or mushroom. Were we to attempt to raise a group of us as mushrooms (or even toads), any genetic differences among us would be swamped by the utter unsuitability of this unfamiliar upbringing; indeed, most of us, if not all, regardless of our genes, would simply die.

The skeptical stance that it is meaningless to ask about genes versus environment conveniently ignores that our genes have evolved over very long stretches of time in what has been a fairly predictable and typical human environment, beginning with life *in utero*. Yes, some of us trace our ancestry back to people who lived in hot environments and others cold, or some of us have enjoyed enriched environments during our lifetimes while others have suffered deprivation; but all of us throughout history have faced a similar set of challenges of staying alive and prospering *as a human*, and not as toads or mushrooms. These are the challenges that our genes, having been shuffled around among the bodies of our ancestors, will have had to adapt to. And so when we study large numbers of people, we hope these environmental differences get averaged out, leaving something we can interpret as the effect of a gene in a typical human environment.

Bearing these points in mind, we can try tentatively to interpret whether the variety we see around us might bear the signature of genes. The psychologist Thomas Bouchard undertook beginning in the 1970s a now famous series of studies of identical twins who had been reared apart owing to adoption. Identical twins reared apart share all of their genes, but not their environments, and so can help to reveal the relative importance of one over the other. If genes play only a small role, then we would not expect twins reared apart to be similar. But study after study shows that they are—and often strikingly so—on measures of intelligence, aptitudes, spatial reasoning, and features of personality. Sometimes the similarities among identical twins are unnerving. Bouchard reported a pair of identical twin sisters reared apart who were both afraid of water. Even though they had never met, they both recounted tales of being taken to the seaside, and when approaching the water, both recalled turning around and backing in; twin brothers reared apart both grew up to be firefighters, and both were captains of their firefighting teams; two sisters reared apart both had a habit of wearing seven bracelets.

Bouchard's and others' twin studies are criticized because twins who are reared apart are frequently adopted by families of similar backgrounds, even if unwittingly. But we can compare identical and non-identical twins who have been reared apart. Non-identical twins are no more similar genetically than ordinary brothers and sisters. When these two cohorts are compared, the identical twins reared apart routinely prove to be more similar than the non-identical twins reared apart, implying a strong role for genes. It also seems that the influence of genes becomes stronger in environments to which they are suited, as if finding the right environment reveals the effects of genes. An adopted child with musical talent who just happened to be brought up in a musical environment would achieve more than someone with a similar set of genes in a non-musical environment, and more than a non-musical person brought up in a musical environment. If you think this is just common sense, it is also a statement of culture (a particular home environment) unmasking innate abilities. And it is something that would not have been lost on prehistoric cave painters handed the latest new ochre pigmentation or a sculptor handed new carving tools.

Still, we must be careful. Bouchard's studies do not specify in advance what sorts of personality similarities and differences are expected, and so stories of how two people back into water, or how two brothers both became fire captains might just be quirky or unusual things that emerged by chance from trawling through hundreds of answers given in interviews. Indeed, years ago a colleague of mine undertook a study in response to Bouchard's work that he sardonically called "similarities among perfect strangers reared apart." He got people to fill in questionnaires asking them about likes and dislikes, habits, personality traits, educational background, and so on. Then he randomly sorted the respondents into pairs and compared their responses. As might be expected, if you ask any two people enough questions, there will be something unusual they agree upon, and he found them. Two "perfect strangers" might have shared the same favorite opera, the same favorite song, or the same hobby (maybe perfect strangers reared apart are nevertheless reared in similar environments!). But this should not be taken as a refutation of the main body of Bouchard's work. Identical twins reared apart are more similar to each other on a wide range of measures of intelligence, personality, temperaments, and morphology than non-identical twins reared apart, and they in turn are more similar than random pairs of strangers.

For an evolutionist, the intrigue in Bouchard's work is that we are able to see any real genetic variety at all in these measures. Remember that natural selection's propensity is to use up differences among individuals, as our example of antelopes, songbirds, and lions shows. And, indeed, in many of our traits, such as how many fingers we have, or how many arms and legs, there is no variation, at least beyond occasional genetic abnormalities. Even among all the 5,000 or so different mammal species, there is no variety in the number of neck vertebrae: we all have seven, yes, even the giraffes. They build their startlingly long necks not by adding vertebrae but by merely elongating those they, and we, have. Equally, one won't lose many bets by saying a person has ten fingers or ten toes. Natural selection has all but erased differences among us in the ways that our genes give rise to the numbers of arms and legs or fingers and toes we have. Why, then, hasn't natural selection erased differences in height, or running speed, musical talent, personality, or mathematical and spatial ability?

Normal numbers of fingers and toes illustrate an important and often misunderstood feature of natural selection. The term *heritability* has a technical meaning, referring to how much of the *variety* among a group of people on some trait—such as height or hair color or mathematical ability—can be attributed to differences in their genes. Heritability is a property of a group of people, not of an individual. Traits that are highly heritable are those that differ among individuals, and those differences are preserved in offspring from one generation to the next owing to the genes they inherit: height is heritable because tall parents tend to produce taller than average children, and vice versa. Intelligence is heritable because more intelligent parents tend to produce offspring of higher than average intelligence, and vice versa. If there is no variety among the people in the group, then there are no differences for genes to explain, and so we say there is no genetic heritability of that trait. Number of fingers and toes is a trait that does not vary among individuals, and so we say it is a trait that lacks heritability. Even so, numbers of fingers and toes is *inherited* because it is caused by genes—it is just that those genes don't differ among us.

But here is the surprise. Standard genetic theory tells us that genetic traits with high heritability (lots of differences among individuals) are *not* strongly linked to our ability to survive and reproduce. An amusing pastime in national art galleries is to identify the long noses, thin lips, or so-called aristocratic chins that crop up over and over in the Old Masters' paintings of the generations of a nation's ruling class. These are heritable traits and they indeed differ among individuals. But we don't think they tell us much about someone's chances of surviving. On the other hand, the same genetic theory tells us that the traits we think *are* most important to our ability to survive and reproduce, what is sometimes called our *fitness*, are least likely to vary among individuals. This is because natural selection weeds out the differences among us on the traits that are vital to our survival and reproduction. Try grasping a heavy round object such as a metal pipe or a glass filled with water while not using one of your fingers and you will discover that your five fingers are well suited to grasping (although one could counter this by saying that these objects have evolved culturally to be shapes we can grasp, and that if we had just three fingers these objects would have different shapes).

This conclusion leaves us in an unexpected place. Do we really believe that the heritable differences in language or mathematical ability, or differences in musical or artistic ability, have somehow not been important for promoting our survival and reproduction? Perhaps they are not, and our genetic variety does not confer any meaningful differences among us. But if we believe that these heritable differences have conferred meaningful differences in our abilities, then only two plausible answers can be given for the existence of our genetic variety. One is that the purposes the genes serve have arisen only so recently in our history that they are still sweeping through human populations, and eventually we will all have them just as we all have ten fingers and ten toes. Thus, perhaps certain mathematical, musical, and spatial abilities and certain aspects of our personalities only became important once human societies grew and became complex. Some of us have, for example, inherited genes for mathematical ability already, and if we wait long enough, all of us will eventually have the genes that confer these abilities.

A second answer emerges from thinking of the variety we observe in our traits as arising from the different roles or strategies the social environment of culture has made available. If there is no single best way of being a human, then natural selection would not "use up" genetic differences among us, the way it uses up, for example, differences in running speed among gazelles. Instead, our genetic variety in things like mathematical, architectural, analytical, linguistic, mechanical, engineering, creative, oratorical, physical, and artistic abilities would be maintained because people with differing genetic makeups have filled the different roles that our many and diverse cultural innovations have created throughout our history.

alternative but equally good strategies

IT IS important to emphasize that this is not an argument for the superiority of some combinations of genes. It says that different combinations of genes make people suited to tasks that have been approximately *equally successful* throughout our history in promoting our survival and reproduction. Were they not, the less successful combinations of genes would have disappeared. One of the more subtle and powerful ideas to emerge from evolutionary biology in the past fifty years or so is the notion of *evolutionarily stable strategies*. We can think of a strategy as a behavior you adopt in your encounters with others. An evolutionarily stable strategy can be roughly defined as a strategy that when enough individuals adopt it, it cannot be bettered by any other strategy. A simple example might be your behavior in standing in line to board a train or bus. If you are polite, others will crowd in front of you and you won't get a seat. So, being polite is not a stable strategy for getting a seat because others will outcompete you. To counter this, you could become pushy and now you probably will get a seat. But is being pushy a stable strategy? Well, if everyone adopts this strategy, fights might break out and you could get injured. Now it might be better to be polite—you won't get a seat but at least you won't get hurt.

In fact, this made-up scenario describes a situation that evolutionary biologists call the *Hawk-Dove* game, and it turns out that neither the Hawk nor the Dove strategy is the "best," because whenever one of them becomes popular, the other can either take advantage of it or obtain higher rewards. So, maybe there is some intermediate "self-interested but not too pushy" strategy—we could call it prudent—that gets you a seat much of the time, but allows you to avoid fights. Maybe you position yourself near to where the train door will open, or you crowd in just a little bit to get in front of someone, but not so obviously that they become indignant and accost you. This strategy of prudence could have a higher payoff than either of the others, and if so it would become the one that most people would adopt. It could even be the evolutionarily stable strategy if it avoided the injuries the Hawks get and simply stepped in front of the Doves.

The simple parlor game of *rock-scissors-paper* is a game of stable and alternative strategies in which all three exist simultaneously. Each strategy (*rock*, *scissors*, or *paper*) can win against just one other, and each strategy beats a different other strategy (*rock* beats *scissors*, *scissors* beat *paper*, but *paper* beats *rock*). Knowing this, one way to approach the game is to adopt a mixed strategy in which you randomly play each strategy one third of the time. This will produce a win on average once in every three encounters. If you deviate from this and play, say, *scissors* over and over, someone who switches to playing *rock* will beat you, and your payoffs or "fitness" will fall. Alternatively, players could adopt *pure* strategies: we could imagine thirty people playing the game, with ten assigned to play each strategy and only that strategy. If they mix amongst each other at random, each person will win, on average, one third of the time. If someone deviated from their strategy, say a *rock* became a *scissors*, the people playing *rock* would begin to do better because they would meet people playing *scissors* just that little bit more often.

Males of the common side-blotched lizard (*Uta stansburiana*) play a *pure strategy* version of the *rock-scissors-paper* game in their competition to mate with females. Each male has one of three different genetically determined mating strategies. *Polygynous* males are large in size and can therefore control large territories. By controlling these large territories, the polygynous males can guard and exclusively mate with a number of females. The polygynous males can even take females away from smaller *monogamous* males which, having smaller territories, attempt to guard just a single female. But the polygynous males don't ever take over completely. A small, *sneaky* male can take advantage of the polygynous males by snatching a liaison with one of his females when the polygynous male is not watching. On the other hand, the sneaky tactic does not work against the monogamous males because they can guard their single female. As with the parlor game, each one of these strategies can "beat" a different one of the others, and whenever one becomes too numerous, one of the others will take advantage of it.

The idea of stable alternative strategies tells us that genetic variety can exist and be maintained for good adaptive reasons because no one tactic, strategy, or role is always best. Combined with Ricardo's ideas on the benefits of specialization, it provides a way to understand how societies can maintain variety in their ranks. Think of the Sienese *contrade*. It is reasonable to expect that they have lasted in a relatively stable form for so long because they feed off of each other in much the same way that *rock, scissors*, and *paper* do in the parlor game. When there is an abundance of carpenters, they cannot find enough work, so the number declines; too few carpenters and their numbers can increase; too many or too few bakers and the same thing happens. In this context, we call the force regulating the numbers who can "play" each of these strategies "supply and demand," and it means that over the long run the variety of strategies can persist so long as there is some demand for their different services.

Variety alone does not say people differ in their innate abilities. People might end up in a particular profession merely by chance or perhaps because their parents introduced them to it, and then they get good at it from practice. It would be like the example we gave of a group of people ten of whom choose to play *rock*, another ten play *scissors*, and the final ten play *paper.* There are no initial differences among them, and they will all have equal returns over the long run. Alternatively, at the other extreme, one could imagine something like Aldous Huxley described in his futuristic novel *Brave New World* in which people were mass-produced to have different predispositions. In that world we are more like the side-blotched lizards, endowed with dispositions to adopt different behaviors or to perform different tasks. This is anathema to the modern liberal view, and no one suggests there are specific sets of genes for each of the many different roles in society. But those many different roles might rely on some smaller set of different talents and skills.

Look around you. Do people tend to settle into tasks they are predisposed to by dint of their genes or because of upbringing and hard work? Is being good at languages helpful for a job in journalism or technical writing? Is being good at spatial reasoning a boost to a career in architecture or design? Does being good at mathematics grant an advantage to being a notary or financier? Do scientists, inventors, and entrepreneurs have a psychological disposition to focus single-mindedly on a goal or to wonder if the world could be different? Do musicians have *innate* musical talent? Do children often follow their parents into similar walks of life because of the opportunities their parents provide, or conversely because of a lack of opportunities for social mobility? Or, might children often do what their parents do because they have inherited their parents' temperaments and abilities? These might be uncomfortable questions and the truth will often be "some of each," but they are not questions we can dismiss outright. Some combination of a slight genetic push combined with a *win-stay, lose-shift* strategy over long periods of time can sort people by their abilities, even if the contribution of genes is small.

The possibility of equally good but alternative strategies might even explain aspects of personality. For example, when people are asked what aspects of their cultures they are drawn to, they tend to divide into five broad categories ranging from *aesthetic* (drawn to creative culture) and *cerebral* (drawn to information) to *communal* (relationships and emotions), *dark* (intense and hedonistic pursuits), and *thrill seekers*. These categorizations must be treated with caution, but biologists are coming to realize that alternative and equally successful personality strategies can even arise in animals. For instance, people differ in how likely they are to take risks, and a hypothetical evolutionary scenario can give one reason why. Imagine you and I differ in our perceptions of what the future holds. You occupy a position of prestige and great reputation in society and consequently you have an expectation for a long and productive life. I am not so highly regarded and this makes me pessimistic about my future. If each of us is right in our assessments, then it might pay you to be averse to taking risks in order to ensure many future returns. Perhaps you will bypass the big mammoth we are hunting out of fear that one swipe from its large tusks could do us in, and wait for a smaller mammoth to kill. But the opposite will be true of me—I will be motivated to adopt a short-term view, to cash in on what is at hand, even if it means taking a risk of being gored. If I don't take a risk, I might be injured or dead before I get another chance.

By some estimates, 5 percent of us might be *sociopaths*. Such terms are notoriously difficult to define and harder to judge, but it is said that to understand what it is like to be a sociopath, you need to imagine yourself without a conscience—no sense of guilt or shame, no remorse for your actions. In *The Sociopath Next Door*, Martha Stout says that lacking these emotions sociopaths are unscrupulous and manipulative, and capable of emotional and physical cruelty. They are often able to conceal this from others, and maybe even themselves. It is difficult to get them to change their behavior because they don't respond to appeals to morality, disgrace, or shame—these are precisely what they lack. In extreme form, this sociopathy can tip over into the malevolent and often deadly violence of a psychotic killer. But most sociopaths are not like that and indeed might wander among us in society. It is suggested they are likely to be chief executives and other people in positions of power in organizations, people not troubled by the nagging voice of their conscience, because sociopaths don't have this voice, or if they do it is easily overridden. In fact, as the joke goes, being a sociopath means "never having to say you're sorry."

Could sociopathy be an alternative strategy, a personality style that can exist on the margins of society, given that most people are not sociopathic? The cooperation on which human society rests depends upon exchanges among people, trust, and a sense of fairness, and so we expect those dispositions to be widespread. But of course the more widespread they are, the more they present a target for others to exploit. Lacking a conscience might make someone the perfect impostor, able to deceive others as to their intentions because they don't struggle with—and therefore have no need to conceal—the normal feelings of remorse a conscience brings. And what if getting a difficult job done sometimes requires someone who doesn't worry too much about the consequences to others? Maybe it is even the unpleasant job of attacking that tribe in the next valley that is competing with yours for resources. Societies might only ever have room for a few sociopaths; but those opportunities will always exist, and that might be why we always seem to find them lurking in boardrooms, running companies, or shouting from a dictator's pulpit. On the other hand, they will limit each other's numbers because when two sociopaths meet, there is bound to be trouble—like *rock* meeting *rock* in a game of *rock-scissors-paper*.

It must be emphasized that there is no good evidence one way or the other that cultures have sorted us according to genetic predispositions, at least beyond the commonplace observations we all have made. Even so, when we look at how natural selection has molded differences among people who inhabit different parts of the world—differences in skin color, eye shape, stature, hair color—we can be sure it has had the time and ability to differentiate us within societies. Modern genetic technologies make it increasingly easy to collect evidence relevant to this question, but it is an issue we approach as a society the way we would approach Pandora's box, and for the same reasons. Even among those prepared to ask such questions, the hunt for specific genes related to differences in lifetime performance is still in its infancy. The simple reason is that no one knows in advance which genes to examine. It will only be with the collection of large numbers of human genomes that cohorts of people with differing outcomes can be compared. This is not different from the approach that attempts to find genes for various medical risks, and the technology is becoming available to sequence human genomes cheaply and in large numbers.

If our societies have for millennia been sorting us by our talents, even if weakly so, this is something society might benefit from knowing about. Most of us would agree that a society that promotes "equality of opportunity" is a desirable one. But we must also recognize that if there are inherent differences among individuals that make them more or less suited to a particular role or job in society, an inevitable consequence of equality of opportunity is to produce a society differentiated by innate predispositions, a *genetic meritocracy*. It will produce this meritocracy because equality of opportunity merely ensures that everyone has a fair chance of being delivered to the doorstep of a job or role in society, but does not ensure that everyone has an equal chance of being good at those roles. Inevitably, then, and the more so the greater the equality of opportunity, competition with others will sort people by their genetic predispositions. Or, as the sociologist Peter Saunders has written, the "essence of a meritocratic society is that it offers individuals equal opportunities to become unequal. There is open competition for the most desirable, responsible and well-rewarded positions, and the most able and committed people generally succeed in attaining these positions."

culture and the selection of our genes

WHAT *is* staring us in the face is that there is now striking evidence that, coinciding with the advent of culture, our genes appear to have undergone an exceptionally rapid rate of adaptive evolution. This could, as mentioned, be evidence that recently evolved genes are still moving through our populations, but it could also mean that culture has been sorting us—the genetic phenomena themselves do not as yet clearly distinguish between these two alternatives. The evidence pointing to an increase in rates of evolution exploits the fact that our genes are arrayed on long strings of DNA called *chromosomes*. Humans have twenty-three pairs of chromosomes, one of each pair inherited from the mother and one from the father. This means that each of us has two copies of most genes, called *alleles*. The alleles might be the same or slightly different versions of the same genes. You might, for example, inherit a gene for brown eyes from your mother and blue eyes from your father. At another place along the same chromosome, you might have inherited a gene for long fingers from your mother but a gene for short fingers from your father. Now, if one allele of a gene—say blue eyes—is always found in combination with some other allele for a different gene—say short fingers—this tells us natural selection has acted strongly and quickly on that region of the chromosome to cause what is called a *linkage* between the two alleles. It is merely a statistical association, not necessarily having anything to do with what these genes are for.

The reason that linkage identifies strong natural selection is that our pairs of chromosomes (the maternal and paternal copies) sometimes exchange some genes between them when we produce egg or sperm cells. Eggs and sperm carry just half our genes, and they are a random assortment of our maternal and paternal alleles. As a baby, you might have inherited the blue eyes and short fingers alleles from your father, but later in life one of these alleles might get exchanged with the alleles you inherited from your mother when you produce sperm or eggs. The process, known as *recombination*, occurs infrequently, so the probability of it splitting up a given pair of genes is low. This means that if something can propel a gene through a population rapidly it may come to be inherited by everyone before recombination has had a chance to separate it from other genes on that chromosome. That force is of course natural selection, and when it strongly favors an allele—because the allele confers a large advantage to survival and reproduction—the genes on this chromosome will become linked: they become more likely to be found together than expected for genes not undergoing strong selection.

When researchers study blocks of human DNA, they find thousands of linked genes that appear to have been subject to strong forces of natural selection. It is possible to attach estimates of the timings of these events of evolution, and the conclusion is that humans began to experience unusually rapid evolutionary changes to many of their genes beginning sometime around 40,000 years ago. One possibility is that this represents the limit of how far back in time this kind of genetic analysis can go. But the period beginning 40,000 years ago is also the period coinciding with the rapid expansion of human cultural groups around the world and with a flowering of human cultural innovations. It is possible, then, that the blocks of linked regions could be the signature of culture ushering us into different roles within our societies that favored particular combinations of genes. On the other hand, these strongly linked regions could just be the signature of widespread adaptation of entire groups of people to new environments. Forty thousand years ago was a time when humans were spreading out around the world, adapting to hot and cold climates, to new diets, and fossils show it was a time of rapid changes to our body size and shape. There were also new diseases such as smallpox, malaria, and yellow fever. Genes for brain size, pigmentation, immune responses, olfaction, nervous system regulation, and body size and shape all show signs of changing around this time. One intriguing finding is that genes associated with hearing seem also to have been subject to strong effects of natural selection. This could reflect the increasing importance in human societies of language and communication.

So, with current knowledge, we can't say much more than that our genes have been undergoing strong selection in recent times, but another feature of our societies also leads us to believe that latent differences among us might still be being revealed. Human population sizes have been increasing over the last 40,000 years. Natural selection operates more efficiently in larger populations because when populations are small, chance or random events from one generation buffet the population of genes, making it hard for natural selection to pick out the varieties it prefers. This is the phenomenon of random drift discussed in Chapter 1. Equally it will be less effective at removing other varieties that are less beneficial. This means that the small early human populations might have carried many different genetic varieties owing to random genetic drift. These varieties—sometimes referred to as evolutionary *debris*—would have been increasingly revealed to the honing effects of natural selection as populations grew. In the presence of cultural opportunities and larger groups, what was once debris might have become the raw materials of our differences.

Some of this debris may surface in the modern world in unexpected ways. Dyslexia is a heritable condition often associated with difficulties in reading and writing. It might have gone unnoticed before the emergence of writing because it did not affect any capabilities that mattered, and it is even possible that it granted some benefit in times past. For instance, we know that in modern society many dyslexics are good at mathematics, spatial reasoning, and computer programming. Autistic people often have profound deficits in understanding others' motives and feelings, can be infuriatingly literal-minded, and are often socially withdrawn and isolated from others. Autism is also heritable and about four times more common in boys than girls. The psychologist Simon Baron-Cohen suggests that autism might represent the extreme end of a normal continuum of abilities to focus on and single-mindedly persevere at a task to the exclusion of others—he calls it the *extreme male brain*. The dispositions that produce the extreme male brain might have been useful in our evolutionary past, for tasks such as warfare or hunting large game that might demand a concentrated and unempathetic commitment to tracking down and then executing one's prey. Baron-Cohen also suggests that boys might be more prone to what he terms *systemizing*, and this is seen in hobbies such as collecting things, bird-spotting, repairing cars and motorbikes, becoming a pilot, or sailing, but also in interest in mathematics and computer games and computer programming.

In our current environment, autism is an awkward condition and at its worst it can be disabling. The brilliant early twentieth-century physicist Paul Dirac is sometimes compared to Einstein for the intellectual depth of his discoveries. Dirac was also autistic, and some biographers speculate that his single-mindedness and ability to shut out or not even be aware of those around him might have benefitted his scientific work. Dirac was awkward and he would exasperate colleagues by not speaking for days. He would often go on long country bicycle rides with his daughter and not utter a word. One time after giving a lecture he agreed to take questions and a man raised his hand to say he didn't understand the equation in one of Dirac's slides. Dirac stood motionless for about a minute and a half, at which point the moderator asked him if he was going to answer the question. Dirac replied that it wasn't a question.

the push of our genes and the gentle pull of the future

TO SOME, the idea that cultural innovations unlock differences among us that quietly await their destinies in finding a task or role they are good at might suggest that we have been on a preordained trajectory, an unfurling, or inevitable march to modernity. Some social anthropologists see in this a tendency to believe there is an inherent superiority to modern society, it being further along the path of inevitable progress. But this is a view that confuses different senses of the idea of progress. Cultural evolution has produced increased technological complexity, and most would probably agree it has raised our standards of living, even if unequally so around the world. If improved health and well-being makes modern societies superior, then these social anthropologists are correct, but then again one would not wish to see this as a bad thing. Biological evolution has also been progressive in this sense of producing greater complexity. The first life, starting billions of years ago, comprised simple, single-celled organisms; only later did natural selection discover how to build big, complex things like elephants.

But neither cultural nor biological evolution has been progressive in the sense of working toward some predetermined or preordained goal. Both merely lumber along producing new varieties of things, some of which catch on and some of which don't. Nevertheless, we might have to accept that there is at least some broad inevitability to the outcomes of both cultural and natural selection. Given the physical nature of our planet, were we to rerun the tape of biological evolution it is highly likely the plants would evolve again, and if they did, it is highly likely we would see things like birds, fish, and land animals evolving to make use of the oxygen they released. These new forms might not be identical to the ones we have now, but there is good reason to expect that the broad outlines would reemerge. We know this because evolution has already played the tape independently many times on the plants and animals we do have. Penguins, seals, otters, porpoises, and fish have all independently evolved similar streamlined shapes for coursing through water. Birds, insects, bats, and some fish have all evolved wings for flight. Consider that in any world that produced insects that flew in the air, it is highly likely something like a bat or insect-eating bird would evolve to catch them. And if trees evolved, things would evolve to climb them, or giraffelike, acquire long legs and necks to eat their upper leaves.

Whether another sentient being like ourselves would reappear is impossible to know, but we do know that there is no reason to rule it out. The sentient being that walks this planet is upright, bipedal, naked, has superb eye-hand coordination, good color eyesight, highly dexterous hands, and is capable of social learning. I suspect that were we able to replay our cultural evolution, we would get many of the same things we have now, because our cultures will always have had a tendency to come up with and retain innovations that we are good at using or good at exploiting. This is an obvious point, but it is easy to forget how much it determines the cultural innovations we see around us today and that we developed throughout our history. Wheels were invented several times independently in different parts of the world, as were stone tools, clothes, spears, bows and arrows, atlatls (spear throwers), boats, shelters, agriculture, and fishing technologies.

The nature of our physical world and of our bodies makes not just these inventions but even the forms they took nearly inevitable.

For example, bicycles were invented sometime in the nineteenth century, and their principles have remained unchanged since. They were probably destined to be part of our societies from the moment we first emerged as modern humans, having only to await the equally inevitable accumulation of the metallurgical and other technologies that made them possible. So, too, perhaps were hammers, steam engines, airplanes, toasters, and computers. For the same reason there are probably also some things that, if we thought hard enough, we could rule out will ever be a part of our societies. Strap-on wings, telepathy machines, and color television programs in the ultraviolet spectrum have never caught on and probably never will because none exploits our innate abilities.

Uniquely in the case of our species, this interplay between our genes pushing us toward particular outcomes and our cultures producing technologies that respond to our talents means that the goalposts of our existence are constantly being shifted. But even the shifting is somewhat predictable because it must fit in with what we are disposed to want to do and are capable of doing, and so cultural developments merely reduce the demand for old skills, while raising it for others. Increasingly, there is little need for the backbreaking physical labor that might have built our modern societies, and so little need for the physiques that did this work. Instead, the modern world will call for a more domesticated set of abilities, among them mental agility, concentration, and communication skills. Culture has not yet finished sorting us by our talents.

CHAPTER 4

Religion and Other Cultural "Enhancers"

That the arts and religion evolved to enhance the expression of our social behaviors

THE ARTS, MUSIC, and religion are what many people have in mind when they refer to "culture." This is so-called high culture, there to move, uplift, entertain, or console us. But what really seems to distinguish these cultural forms is that they can lodge in our brains and exert a grip on us that is sometimes beyond our control. They can wrench our emotions, make us believe things that are false, cause us to steal, fight, or even die for them, bring us to rapture, get us to construct monuments and halls to them, beguile and intrigue us, and get us to lay down large sums of money. Unlike other aspects of our cultures, the arts, music, and religion can affect us these ways without feeding or clothing us; they do not provide shelter, and they do not change the weather or smite our enemies. They can even get us to act on their behalf when they provide nothing. Crowds flock to see Kazimir Malevich's 1918 painting *White on White*, a monochrome board of white. During John Cage's composition for piano entitled *4′33″*, a pianist sits at a keyboard for four minutes thirty-three seconds without playing. Audiences can find it difficult to know when the piece has begun or ended (see for yourself; *4′33″* can be downloaded from the Internet). In the Gospel of John, 20:29, Jesus says to Thomas, who has doubted him, "blessed are they that have not seen, and yet have believed."

Why do we have the arts, music, and religion, and why can they seemingly command our attention and allegiance? These are not questions we can easily ignore, or write off as obvious, saying, "we just like them." We do like them, but this just raises the further question of why, from an evolutionary perspective, our minds have evolved to be attracted to these cultural forms. Something about them is of such significance to us that human societies have for thousands of years supported entire classes of people—the priesthood or knowledge elite, artists and musicians, but also poets and magicians, storytellers and shamans—who do not make any direct product that enhances our chances of survival and reproduction. Rather, their existence derives solely from appealing to our appetites of mind, whims, and aesthetic tastes.

But this should give us pause. Natural selection is a stern and severe master that generally avoids profligacy—in a competitive and evolving world, the things we do, in general, must pay their way. If I spend my days in pursuit of religious truth or artistic beauty while you are out bringing home game, I may be content, but not survive the night. If you are amassing an army across the valley while we have a good night's singing, you might overrun us by morning. So we are entitled to ask what the arts and religion have been good for throughout our evolution, not just now in modern societies that—awash with money and free time—allow us to pursue aesthetic, psychological, and hedonistic pleasures. We want to know why our early ancestors welcomed them into their lives, and how they could have afforded to do so.

In trying to answer these questions I want to develop the idea that the arts, music, and religion evolved in our distant past as what I will call *cultural enhancers*. We can think of an enhancer as something that exists solely to promote the interests of a replicator inside its vehicle. Here is an example from genetics. Short strings of

DNA in our genomes known as *genetic enhancers* exist solely to help genes function more effectively within their bodies or vehicles. For instance, genes for growth need to be active early in life but less so or even not at all later in life. Genes for eyes need to be switched off in the backs of our heads, but switched on in the front. Genes solve this problem by recruiting genetic enhancers that tell them when and where to be switched on or off. This serves the gene's interests by helping it to make a better vehicle, and this means the gene is more likely to survive and be passed on. What makes genetic enhancers fascinating is that they do not make anything of their own that we use to build our bodies (for instance, they do not make proteins). Like poets and magicians, genetic enhancers also do not make any direct product that enhances our chances of survival and reproduction. Rather, it appears that genes have recruited them solely to influence how much they are expressed in different parts of our bodies or at different times.

I will use the term *cultural enhancer* to refer to cultural forms we have recruited because they help us to express ourselves more effectively as replicators inside our cultural survival vehicles. Their existence might be yet another example of remarkable convergence between the ways that biological and cultural evolution have solved similar problems. For instance, prehistoric people might have discovered that they could recruit something like what we now might call the arts, music, and religion to elicit or motivate emotions, to strengthen beliefs or resolve, to transmit information, to increase cohesion or remind people of shared history and interests. Like a gene in a body, we might have used these enhancers to alter our "expression" at different times and places in our societal vehicles, and in ways that served us. Maybe it is to gain courage before battle or seek hope in the face of uncertainty. To prehistoric people, the arts and religion might have been like having a class of performance-enhancing drugs (another class of enhancers we have built, if only in more modern times) at their fingertips. Like genetic enhancers they needn't produce any direct product, such as a better blade or spear, that directly affects our chances of survival and reproduction. Instead, they will have evolved to "alter our expression" by having access to our emotions and pleasure centers because these are the parts of our makeup that motivate us to behave. And this could explain our attraction to them—and even our weaknesses for them—in the modern world.

brain candy

I WILL have far more to say about religion in this chapter than about art and music, and some might even see it as a folly or confusion to combine them. After all, art and music are aesthetic forms, whereas religion is doctrinal. But I beg the reader's indulgence because in spite of the differences among these cultural forms, I suggest they are linked by having been recruited and shaped by us to play the role of cultural enhancers.

Still, the suggestion that we groomed the arts and religion to serve us must confront the simpler idea that these cultural elements might exist for no other reason than that they have evolved to be good at manipulating, exploiting, or taking advantage of *us* to aid *their* transmission—not to help us. Culturally transmitted ideas, behaviors, and objects, or memes do not share the same route into the future as our genes do. Unlike our genes, whose survival depends upon keeping our bodies alive, culturally transmitted elements can jump directly from mind to mind. This means they will have a tendency to evolve toward forms that get us to talk about them, sing them, perform them, build models of them, or anything else that gets us to transmit them into someone else's mind, even sometimes at our expense. Of course, not all cultural transmitted elements evolve to take advantage of us. Most of what we think of as our accumulated knowledge and technology takes the form of culturally transmitted ideas and instructions that help us to survive and prosper. These elements serve a useful purpose to us, and that makes us want to retain them, and share them with others. But the feature of memes we have to bear in mind is that there is no necessary reason they have to help us: whatever form makes them likely to be transmitted, they will be likely to adopt.

Perhaps, then, the arts and religion are little more than hedonistic or exploitative *mind drugs* or *brain candy*. We like them, or suffer from them, because we can't help it! Thus, we make paintings, songs, and stories, and we invent religions. We pass these elements of culture one to another by word of mouth or by producing objects. At each stage of the retelling, singing, or drawing or sculpting of some object, changes creep in—some by accident, others by design. Most of these changes work against the cultural form, but others will make the items more visually attractive, memorable, psychologically compelling, and even perhaps irresistible. Elements that fail to compete for our attention will be cast off, but the survivors will be those whose allure outpaces our minds' defenses, allowing them to bypass our normal filters, and grab our attention or acquire unnerving abilities to control our emotions—like crack cocaine, only ingested differently. Karl Marx said almost exactly this about religion when he described it as "the opium of the people."

The nature of cultural evolution means that some memes evolve as parasites that live at our expense. Indeed, religion has been described as a culturally transmitted virus of the mind, and if this is true, it makes no sense to ask what we get from them. Religions would owe their existence simply to the fact that they are good at exploiting us to aid their transmission; we cannot shake them off, and they compel us to teach them to others, especially our children. If religion is such a virus and has gripped you already, you won't believe this, but the charge that religions are parasites cannot just be dismissed. For example, everyone has received chain letters promising good luck if sent on, but warning or even threatening bad luck if not. Chain letters are not religions, of course, but like them, they have a tendency to play on our hopes and fears. It is an inevitable truth of cultural evolution that you are likely to receive the most compelling versions of these chain letters—the ineffective ones will have been ignored and simply faded away, but the best will have mutated to forms that cause anxiety and then control your actions, getting you to send them on to someone else.

Other aspects of culture get us to aid their survival and reproduction, often without our even being aware of them or costing us anything. In the biological world, such things are called *commensals*. Cattle egrets are small birds that spend their time around cattle whose hooves churn the soil, exposing insects and worms the egrets like to eat. The egrets' presence costs the cattle nothing; the cattle might not even be aware of them, but neither do the cattle obviously benefit from having them around. A cultural commensal might take the form of an advertising jingle we cannot get out of our minds. It is difficult to see how singing one either benefits or harms us unless we do it uncontrollably or perhaps too absentmindedly in public, but this singing surely aids the transmission of the jingle. This is what it has evolved to do, and if it lifted its head too high above the parapet of your awareness, you would probably jettison it. It is a melancholy thought that our brains are reservoirs of the leftover detritus of thousands of these commensals; most of us can remember advertising jingles from our youth, and they have a knack of popping back into our minds unexpectedly. But this is just what we expect of a system that will inevitably evolve in the direction of taking advantage of us. They will multiply and we will be largely unaware of them.

Cutthroat competition among cultural forms to attract our attention is why the best art galleries can take our breath away, the Old Master paintings are so good, why the classics are such good literature, why the best films are the old ones, why we so frequently return to styles of times past—so-called retro fads in dress, music, and design—and why the best songs grab hold of our emotions. It is not that everything was better long ago, just that the survivors we see today were the best of their time. Thus, there might have been many Homers alive at the same time writing their own *Iliads*, but whose stories could not compete with Homer's. Then *the Iliad* went on to see off all competitors for the next 3,000 years. Equally, we can be sure there was plenty of other art, music, and fashion created at the same time as other classic works that has long since vanished, been pulped, or painted over. So good are the survivors at influencing our minds, people have over the centuries shown themselves willing to travel great distances to see them, lay down huge sums or extort and murder to own them, or risk their well-being or lives to steal them—indeed, that is why these cultural elements have survived.

Because the arts, music, and religion are not bound by the usual constraints of physics and nature that constrain objects with some overt function—such as a toaster or a bicycle—the range of forms can expand without limit. As we grow inured to the forms that have been tried, this leaves a more and more extreme or even bizarre set of untried forms to be explored. Many rejected the first Impressionist paintings as vague, Matisse's colors as garish, Mark Rothko's panels of color and hues, Jackson Pollock's drip paintings, or Andy Warhol's soup cans as lacking substance and talent. These now seem tame by comparison to the London artist Tracey Emin, whose art installation of an unkempt bed made headlines, or Damien Hirst, who displayed a shark cut lengthwise and pickled in formaldehyde. In *The Guardian* on January 17, 2002, the director of London's Institute of Contemporary Art at the time, Ivan Massow, declaimed "the British art world is in danger of disappearing up its own arse. . . . Most concept art I see now is pretentious, self-indulgent, craftless tat that I wouldn't accept even as a gift."

In 2004, at the Tate Britain gallery in London, a cleaner threw out a bag of rubbish. Awkwardly, this bag of rubbish was part of an art installation, this one by Gustav Metzger. In 2006, the artist David Hensel contributed a carved head to London's Royal Academy of Art. The head reclined on a bone-shaped wooden rest set on top of a plinth, but it got separated during shipping and was returned to Hensel. The empty plinth, with wooden rest, was exhibited and later sold at auction. The lesson of cultural evolution of art forms is that the progression to minimalism such as this, and eventually to nothingness, is as inevitable as the progression to greater garishness is in the other direction. Maybe this is what the historian E. H. Gombrich meant when he said: "There really is no such thing as Art. There are only artists." Jules Renard, the French writer (1864–1910), adopted a more economic tone: "The beauty of literature. I lose a cow. I write about her death, and this brings me in enough to buy another cow."

a psychological predisposition to religion?

THIS ACCOUNT of the evolution of cultural forms treats our minds as passive receptacles. But of course they are not. We have biases in the ways our minds work, and we have likes and dislikes. These biases constrain the forms that cultural evolution is most likely to take because competing forms must evolve to take advantage of the peculiar environment of our minds. Thus, most musicians play a limited range of instruments, and most singers sing from a limited range of song styles. It is no accident that we find the tones our musical instruments make so pleasing and emotionally resonant: it was the varieties of these instruments throughout our history which best conformed to our tastes that we retained. Even when we have freed ourselves of instruments by creating music electronically on computers, we often get the computers to imitate musical instruments. Most artists paint the same range of subjects, and most fiction follows one of a small number of basic plots—some say as few as seven—and even most of these include sex, love, money, betrayal, or death.

Our biases extend to religions, where all or nearly all of them draw on a restricted range of typically human forms with magical powers, such as the ability to be everywhere at once; they can subvert causation, pass through physical barriers, or create something out of nothing. All or nearly all promise things that we can never attain on our own, such as salvation, redemption, or immortality, and for which demand is unquenchable. But why do religions so often take these particular forms, usually headed up by a God who has a purpose? Ironically, the answer might lie in the nature of our minds as organs designed by natural selection to understand our world.

The psychologist Paul Bloom says that children are naturally prone to a *dualist* view of minds as something distinct from ordinary matter such as the matter our brains are made of. This is a philosophical view that finds its way back to the seventeenth-century philosopher René Descartes, who famously wondered how the "incorporeal" substance of our minds interacted with our decidedly corporeal brains, and it is a question to which we still do not have good answers. The significance of dualism according to Bloom is that it predisposes us to allow that other things, like rocks, trees, the sky, waterfalls, or even clouds, can have minds. After all, if minds can exist independently of physical objects, but can nevertheless somehow communicate through them, why not? But of course if a mind can exist independently of a body, then it can also wander alone as a disembodied spirit.

We are also, Bloom suggests, psychologically predisposed to see purpose in things. We have a taste for *teleology*, or the expectation that things happen and exist for a reason. Thus, children might tell you that clouds are "for raining" or that lions are so that we will "go to the zoo." Bloom goes further and suggests that this predisposes us to be creationists at heart, because if things have a purpose, our naturally dualistic minds consider that something—a *creator* perhaps—gave them that purpose. In an adult mind, these tendencies turn into an appetite for religious explanations of what can otherwise be an inscrutable world. In *Nothing to Be Frightened Of*, the English novelist and essayist Julian Barnes describes the novel as telling "beautiful, shapely lies which enclose hard and exact truths," saying, "We talk of the suspension of disbelief as the mental prerequisite for enjoying fiction, theatre, film, representational painting." The scenes and stories they depict "never happened, could never have happened, but we believe that they did. . . ." Religions, says Barnes, "were the first great inventions of the fiction writers. A convincing representation and a plausible explanation of the world for understandably confused minds. A beautiful, shapely story containing hard exact lies." Or, as Voltaire put it, "If God did not exist, it would be necessary to invent him."

Voltaire might be right. In *The Natural History of Religion*, the philosopher David Hume said, "We find human faces in the moon, armies in clouds . . . and ascribe malice and goodwill to every thing that hurts or pleases us." Daniel Dennett calls this tendency to attribute agency to things the "intentional stance." Most things in our environment that move and behave—especially other humans—can be dangerous, or can at least make a difference to our lives. If we adopt the stance that they have intentions—and, in effect, minds—like our own, it allows us to make predictions about how they might behave. We can use the intentional stance to simulate what an agent might experience and to make guesses about what its goals might be or what it is thinking. Any particular stance we adopt need not be correct; it merely needs to provide a useful shorthand for engaging with the world. If it does so, the conclusions and predictions we derive from adopting the stance then become embedded into our understanding of a particular agent, and allow us to make quick decisions that could, in the extreme, save our lives.

Like our dualism and teleology, the intentional stance might set us up to embrace religion as we seek to understand and even predict what our imagined creator might do next. Still, how can we so easily and uncritically accept this sort of religious stance when it asks us to believe things that are palpably false—that there are beings that can pass through walls, have no weight, live forever, and otherwise break the laws of physics? An unexpected answer is that religious beliefs might have flourished throughout our history precisely because they do not appear to be palpably false, at least to our minds. One of the more compelling demonstrations of experimental psychology of the mid-twentieth century was to show how easy it is for animals to acquire what we might call superstitions or magical thoughts that connect some action to some outcome. In the late 1940s, the provocative and influential learning theorist B. F. Skinner studied pigeons put on what students of learning call *reinforcement schedules*. A reinforcement schedule describes how often and in response to what behavior or behaviors an animal receives a reward or punishment. Not surprisingly, animals will generally do more of the things they come to associate with rewards, and less of the things they have been punished for in their past. Indeed, this is the implicit and, to some, sinister message of the Jesuit motto that goes: "Give me the child for his first seven years, and I'll give you the man."

In a typical learning experiment, animals will be rewarded either on some *fixed* schedule, in which they receive a reward perhaps every time or every second or third time they perform a particular behavior, or a *variable* schedule, in which rewards come at variable intervals, perhaps sometimes after one bout of behavior, other times after three, or maybe other times two, the precise interval constantly being varied. Skinner's insight was to ask what would happen if animals were put on reinforcement schedules that provided food rewards not in response to what the bird was doing, but at *random* intervals of time. Skinner's random reinforcement schedule meant that a food pellet would drop into a food hopper independently of what the bird happened to be doing in the moments before, and the provision of food was not dependent upon any particular behavior.

To his surprise, some of the pigeons on this regime began to twirl in circles, or raise and lower their heads, others would swing their bodies from side to side, or prance around the cage. Skinner realized that pigeons produce some of these behaviors spontaneously anyway, so if they produced one just by accident shortly before a piece of food arrived, they would somehow associate the behavior with getting food and be more likely to perform it again. It was as if the pigeon had come to *believe* its behavior would make more food appear (although Skinner, who treated the mind as an unknown "black box," would never have used such language). Because rewarded behaviors are more likely to be performed, they were also more likely, just by chance, to get rewarded again. It wasn't even necessary for the reward to be presented every time the pigeon produced the behavior (remember it was a random reinforcement schedule). The occasional twirl or prance might be missed, but just so long as some proportion of the behaviors was rewarded, they would continue. In fact, one of the most robust findings of learning experiments is that behaviors rewarded every now and then rather than consistently are far more resistant to fading away when rewards are eventually withdrawn. It goes by the term *partial reinforcement* and explains why children nag so much.

We posed the question of how people can come to believe things that are false, and Skinner's work provides an answer. In fact, the irony of Skinner's work is that it shows how our hard-wired tendency to look for causes and to attribute agency to things means it is precisely where no real cause exists, or the true causes are beyond our grasp, that we are vulnerable to finding a false one. Thus, to a hungry pigeon unable to predict when the next food will come, twirling in a circle, or prancing around the cage, or raising or lowering its head or wings would have been as good a bet as any for obtaining a reward, because, after all, the food appeared at random. To a thirsty person, unable to predict when the next rain will fall on their parched savannah landscape, whatever they happen to be doing just before it does finally come might come to be associated with *making* it happen, and in the blink of an eye a religious or superstitious belief would be born.

But this raises the question of why Skinner's pigeons pranced around or raised and lowered their heads rather than doing precisely nothing. It is an important question because a pigeon that did nothing would have received just as much food as those that had behaved, or done something. In fact, we could even speculate that Skinner's work provides us with a framework for understanding the origin of quiet and reclusive monk- or nunlike contemplation. Imagine that one of his pigeons had happened to be doing precisely nothing just before the food was presented! The answer to why Skinner's pigeons behaved is that animals probably have a bias hard-wired into their brains to do so—to try things out, and to poke and prod at their environment. For most of the things that matter in our lives we have to behave to get them, or to make them happen. We might suspect then that genes for behaving—that is, genes that encourage us to act on our environments—will have spread as those who carried them would have been more successful, on average, than their more retiring counterparts. And, if it is generally true that doing *something* is better than doing nothing, this bias is probably stronger in humans than in any other animal. We have a great ability to use our behaviors to change our world, and in ways that suit us, and so natural selection will have strongly favored dispositions to have hunches and to try things out. Indeed, this might be one reason why we are so prone to finding causes of things in the first place.

If you are feeling smug and superior to the pigeons, remember we throw salt over our shoulders, touch wood, avoid black cats, and become anxious on Friday the 13th. Next time you are in a tall building, check to see if it has a thirteenth floor. Many don't, even though of course one of the floors *is* the thirteenth. Each time we perform one of our superstitious acts and nothing bad happens, the superstitious behavior is rewarded and we breathe a sigh of relief. It is a logic our abstract minds can engage in, and they do. If allowed to get out of hand, people can enter the beckoning corridors of obsessive-compulsive behaviors, including hand-wringing, repeatedly checking that doors are locked, and obsessive cleaning—these are just the human versions of Skinner's prancing pigeons. Some religions get people to hug trees or to worship animals, to hit themselves with sticks, recite mantras, starve themselves, prance around, and move their bodies from side to side. They also get us to perform strange rituals of bowing, genuflecting, burning incense, chanting and singing in special buildings we call churches, all in the hope of bringing about things we want to happen, but which are utterly out of our control—Skinner's pigeons again? Our capacity for language puts us at even greater risk for developing false beliefs because, unlike Skinner's pigeons, we don't even have to witness an event to know about it. Churches take advantage of this by widely publicizing miracles, and then beatifying or even granting sainthood to people, all as ways of advertising a connection between beliefs and outcomes.

On top of our predilection to find causes even where they don't exist, a simple and yet true law of nature means that in just those extreme circumstances when all other natural explanations have been exhausted, the supernatural, magical, or superstitious explanation is likely to appear to work. The phenomenon of *regression to the mean* tells us extreme circumstances are likely to return spontaneously to less extreme circumstances over time. Thus, when you roll two dice and get two sixes, your next roll will normally be less fruitful. This is because many improbable chance factors come into play to produce two sixes, and it is unlikely they will happen twice in a row. For two dice, it is easy to calculate the probabilities exactly. There are 36 possible outcomes to the roll of two dice, ranging from (1,1) to (6,6). For each of the possible 36 outcomes of the first roll of two dice, there are 36 possible outcomes of the second roll, or 1,296 possible outcomes. Only one of these corresponds to (6,6),(6,6). By comparison, if you get two sixes on the first roll, 35 of the 36 outcomes of the second roll are not (6,6), and so (6,6) followed by something else is far more probable.

The same principles are true of other extreme phenomena, even if it is difficult to calculate their probabiities exactly. Regression is why you shouldn't count on a long run of good weather to plan a picnic for tomorrow, why the tallest parents tend to have children shorter than them and vice versa, why exceptional school reports one term often disappoint the next, and it is why luck does run out. In each case, a set of improbable events has to combine to produce an outcome, and it is less likely that all of them will come together again than that they won't. So, regression is also why praying for a hurricane to end, a flood to recede or for rains to end a drought is more likely to work the longer your desired outcome hasn't occurred. If you happen to think up a supernatural explanation just before some extreme event spontaneously returns to "baseline," you might just attribute it to your belief rather than to regression.

In other walks of life, the stock analyst's advice to buy stocks when they are low and sell when they are high capitalizes (in this case, literally) on regression effects, but we pay stock analysts to give us this advice. All doctors know most common illnesses spontaneously get better within about two weeks. If a shaman-doctor proposes some sort of ritual or ceremonial treatment in that time, even if ineffective, it can come in our minds to be connected to the improvement, just as Skinner's pigeons associated their twirls with getting food. We often consult homeopaths as a last resort only after other medical treatments have failed. Homeopaths might owe to regression a greater proportion of what success they do have than they care to entertain, but if you have been "cured" by one you will not believe this. Finally, regression is also why, as the psychologist Daniel Kahneman has pointed out, praising your children seldom works nearly as well as a rebuke.

three reasons to bet against exploitation

THE FOREGOING gives us plenty of reasons to expect that cultural evolution can sculpt the arts, music, and religion to fit into the particular crevices of our brains. But I want to suggest that there are at least three good reasons to doubt that they are exploitative or hedonistic brain candy. One is they ask and receive far too much of us. Religions can command our loyalty even sometimes to the point of death; music can take control of our emotions; and art can consume our time and money. The "life-dinner" test of the Prologue tells us it is precisely those things that are costly in time and effort, or capable of causing us harm, that we would be expected to evolve defenses to—unless of course they somehow pay their way. A second reason is that—far from being frivolous or purely hedonistic pursuits—it should at least intrigue us that the arts and religions everywhere concern themselves with ideas, shapes, and emotions directly relevant to our lives. In all societies most art is about other people, romance, landscapes, animals, and social relations, except where this has been prohibited by religious beliefs. Music in all cultures expresses or elicits the range of our emotions, and our emotions exist to motivate our actions. Religions are about understanding an inscrutable and even terrifying world. They respond, as Hume in *The Natural History of Religion* says, to "the incessant hopes and fears that actuate the human mind."

The third reason is that it would be surprising indeed had humans missed an opportunity to hitch our survival and well-being to cultural traits that can get our minds so hungrily to pursue aesthetic and psychological rewards. Precisely because the arts and religion can get us to spend the night in rapturous song, believe things that are false, and even put our money, well-being, or lives on the line, they become forces we could put to our own use, somehow to *enhance* our performance. Any human group that failed to acquire these cultural forms could find themselves in competition with others that had. Religion, art, and music become part of the environment of being human that others have to compete against. If they can somehow promote our survival and reproduction, it will behoove us to adopt them, whether or not they are true, frivolous, or hedonistic!

consolation, hope, and optimism in religious belief

EVEN IF we are prepared to accept one or all of these reasons for believing there is more to the arts and religion than mere exploitation and fun, we are left in the case of religion with the nagging worry that it is all a delusion or spell. Religions don't really work as they say they do because nothing really is exchanged between you and the gods: miracles don't happen, your prayers don't bring immortality, the winds don't change direction and blow the enemy's ships away, and you are not granted protection against arrows or bullets. Even so, it might be important to take a step back from the perspective many of us have in our modern societies of religion as an irrational and non-scientific way of seeing the world. Today, we separate science from religion, but our ancestors would not have had sophisticated scientific knowledge, and what we think of today as a supernatural explanation might not have appeared to them all that different to a natural one.

It is difficult to overstate the importance of trying to imagine the perspective of someone alive early in our history as a species. Religion at that time, in its earliest forms, might have been something closer to the received causal understanding of the world rather than "religion" in the carefully articulated and restricted modern sense. It would not have been the rather awkward poor relation to "real" scientific understanding that it is today. It would not have been this poor relation because there wasn't any richer relation. Instead, to prehistoric people, religion provided an explanatory framework for why things happened, and would have acted like a sophisticated model of the cosmos, giving a rationale for people to behave one way as opposed to another in an arbitrary, dangerous, capricious, and unpredictable world.

David Hume, again in *The Natural History of Religion*, put it this way:

> The first ideas of religion arose, not from a contemplation of the work of nature, but from a concern with regard to the events of life. . . . Accordingly we find that all idolaters, having separated the provinces of their deities, have recourse to that invisible agent to whose authority they are immediately subjected, and whose province it is to superintend that course of actions in which they are at any time engaged. Juno is invoked at marriages; Lucina at births. Neptune receives the prayers of seamen; and Mars of warriors. The husbandman cultivates his field under the protection of Ceres; and the merchant acknowledges the authority of

> Mercury. Each natural event is supposed to be governed by some intelligent agent; and nothing prosperous or adverse can happen in life, which may not be the subject of peculiar prayers or thanksgivings.

Religion in prehistoric times would not even have had to be very good at providing solutions. We have seen, for instance, how little it takes for us to acquire false beliefs. Even today, the natural environment can brutally remind us of the impotence of our best medical treatments, engineering solutions, or resources for staying alive and prospering in the world. No matter how good one's scientific knowledge, there is little we can do in the face of epidemics, tsunamis and earthquakes, floods, many cancers and inherited disorders, or even tomorrow's weather. Nature taunts us to appeal to something stronger than our rational human best, and for animals with our brains this has often meant looking to supernatural powers.

The economist Rodney Stark maintains that when we are driven to such desperate circumstances, we often enter into straightforward exchanges with the gods, seeking to purchase the commodities only they can offer—such as better weather, more plentiful game, and immortality—in return for prayer, ritual, offerings, ceremonies, and sacrifices. There is often little or nothing to lose in exchanges with the gods and much potentially to be gained. This was of course Pascal's famous wager for belief in God: "If you gain, you gain all; if you lose, you lose nothing. Wager, then, without hesitation that He is." Because regression effects will often cause extreme situations to improve anyway, the wager will frequently appear to have worked. If you pray long enough for rain, it will eventually come, and if your god predicts an earthquake is coming, it eventually will. And the psychologist Nicholas Humphrey reminds us that "in a dangerous world there will always be more people around whose prayers for their own safety have been answered than those whose prayers have not."

Now, imagine further that confronted with this harsh world, merely holding the view that things can be made better by belief, effort, or hard work—whether or not they can—improves your hope, motivation, or confidence, and eventually your performance or well-being. Thus, whether or not some god can actually control an outcome that matters to you, a belief that it can affects your motivation to make something happen. Robert Trivers has called this tendency to look out for reasons to be optimistic *perceptual defense* and *perceptual vigilance*. In his *Social Evolution*, Trivers points to the tendency for humans to "consciously see what they want to see. They literally have difficulty seeing things with negative connotations while seeing with increasing ease items that are positive."

The destructive Haitian earthquake of 2010 flattened the town of Port-au-Prince in Haiti, killing thousands and making more homeless. I watched on television as a journalist interviewed a man whose wife and children had been killed in the earthquake and his home destroyed. Not only had he survived that earthquake, he had also survived the ferocious hurricanes that swept the island in 2008 and 2009. From this he concluded that God had chosen to spare him: a delusion perhaps, but a useful one. He was full of hope and confidence for the future even while he stood among the ruins of his life. His belief was, as Julian Barnes might have put it, a "convincing representation and a plausible explanation of the world for understandably confused minds."

This gives us an answer to a question that bedevils the subject of religion. People often say that religions somehow satisfy our longings to understand the universe, or how we got here. But this merely raises the question of why we have minds that want to know the answers to such questions. Our minds, as the above example shows, might want answers because they give us hope or direction, and illusory or not, that hope is in itself useful. Once we have come up with a belief like that of the Haitian man, we can set it running like a piece of computer software in our subconscious minds, where it can intercept and disarm our worries and anxieties. This does not say why it is religious rather than some other form of belief that we acquire and make use of; but we have seen how our minds might be partial to constructing gods, and those gods might provide as useful an explanation for what happens in the world as anything else.

It is easy to adopt a supercilious tone about the Haitian man's views, but people like him in our past will probably have produced more children than those of us who disconsolately withdrew from life, and genes that granted a sunny disposition would have spread. The alternative of facing the stark truth head-on can, for many people, be debilitating. Psychologists have discovered that people susceptible to depression often have more accurate perceptions of the world than non-depressives. When they say they have no friends, nobody likes them, they are hopeless at their job, or have no future, they are often more right than not. No wonder they are depressed!

religion and group conflict

THE EVOLUTIONARY psychologist David Sloan Wilson gives a vivid example of false beliefs conferring benefits when people act together in groups. He uses the language of genes but it is just as readily thought of as a cultural example in which an idea rather than a gene spreads through a population by people copying one another for what they perceive to be some advantages. In his article on "Species of Thought," Sloan Wilson imagines a tribal world in which

> A mutant gene arises that causes its bearer sincerely to believe certain distorted versions of reality. For example, the mutant might believe that his enemies are by nature despicable people when in fact they are by nature just like him [and thus think of him as despicable] and are enemies merely because they compete for limiting resources. Nevertheless, fear and hatred of despicable people is more motivating than accurate perception that one's enemies are the same as oneself. The mutant is a more successful competitor than his truthful rivals and the mutant gene spreads through the population.

Sloan Wilson is, of course, describing the familiar beliefs that power xenophobia, racism, bigotry, and parochialism, and the violence that often attends those views. In his example, the beliefs take hold because they promote survival, not because they are true. Odious as we might find them, once again we have to imagine a world in which this kind of belief becomes part of the environment that others have to adapt to. Up against a group in battle who consider you despicable, it might be useful for you to acquire your own brand of motivational bigotry.

Many people think that religions promote group conflict by means similar to those that Sloan Wilson imagines here. Indeed they might, and it is important to see that in Sloan Wilson's example, any tendency to adopt religious or other precepts that make you and your group more formidable foes can bring real Darwinian advantages to those who hold them. Recall that even dispositions that put your life at risk can nevertheless bring you benefits if enough people around you share that disposition. It is then easy for a fledgling tradition of false beliefs to grow as it acquires a collection of different beliefs that get woven together, and all of which motivate people to action. Tribe A on this side of the valley says that Tribe B treats their women badly, and that their greed is a threat to Tribe A's territories. The first belief acts as a justification to steal their women in battle (indeed, it would be to do them a favor). The second is a justification to eradicate the others owing to the threat of their greed.

Once such a story catches on, it can become self-promoting, and people will follow it without even knowing why. There is an anecdote about group beliefs in monkeys that is probably apocryphal, but so instructive it bears repeating. A group of monkeys is in a room with a banana tethered from the ceiling. They can reach the banana by hopping up on a box. But whenever one of them does this, they are all sprayed with water. Monkeys don't like water because in their natural environment, ponds, lakes, and rivers often hide crocodiles and other predators. So, after a while, they all avoid hopping up on the box and even restrain each other from doing so. Then a monkey is removed and replaced by a new monkey. It is naive so it climbs up on the box to get the banana. The others quickly pull it down and eventually it too stops trying to get the banana. One by one the monkeys get replaced this way, and one by one the naive ones are trained by the others not to jump up on the box, until none of the original monkeys remains. At this point not one of the monkeys knows why, but they all avoid climbing the box to get the banana. As far as they know, they have always behaved that way.

The success of religion at promoting coordinated action might be one of the chief reasons why we welcome it into our ranks rather than try to chase it away. The uncomfortable truth of natural selection and cultural evolution is that dispositions—including religious ones—that serve individuals will be favored. On the other hand, this does not mean that all religious beliefs serve us. A religiously motivated suicide bomber might indeed be a case of religion taking over someone's mind for the purpose of *meme-meme* warfare. The suicide bomber's religious belief is only too happy to commit suicide—along with the suicide bomber's genes—if it can kill copies of competing ideas in other people. Suicide can advance memes as well as genes. But examples like this shouldn't make us think that all of religion is a mind virus. The life-dinner principle proposes we will evolve to evade memes that bring us harm, but it is not perfect: some of us will get infected despite our desperate attempts to evade these brain parasites.

In this light there is a suggestion that many suicide bombers come from the ranks of disaffected young men with little future, and this might make them more malleable or vulnerable. But on the other hand, as the life-dinner princple would expect, suicide bombers are rare, given the numbers we might expect if we had a psychology like that of a colony of ants or bees in a hive. They happily and enthusiastically stream out in their thousands in suicidal charges to save their queen. We should also not rule out the possibility that by their actions suicide bombers bring great honor to their families, and so in an obscene sense their actions can be seen through the lens of kin selection, or nepotism. Saddam Hussein is reported to have paid widows and mothers of successful Palestinian suicide bombers up to $25,000 each. The *CBS Evening News* quoted Mahmoud Safi, leader of a pro-Iraqi Palestinian group, the Arab Liberation Front, as acknowledging that the support payments for relatives make it easier for some potential bombers to make up their minds: self-sacrifice can pay in more ways than one, especially if it promotes copies of your genes in your relatives.

Here is a passage from the Gospel according to Matthew (Matthew 12:47–50) that draws on the psychology of kin selection. Someone in a crowd gathered around Jesus says to him:

> "Behold your mother and your brethren stand without, seeking you." Jesus replied, "Who is my mother? Who are my brothers?" Then he pointed to his disciples and said, "Look, these are my mother and brothers. Anyone who does the will of my Father in heaven is my brother, and sister, and mother."

Imagine if you were a believer the excitement these words would cause in you—you could become Jesus' mother, brother, or sister simply by acting out God's will. In Chapter 2, we saw how an altruistic disposition toward other members of your group can arise as a special and limited form of nepotism, in which we recognize we are related to someone at the altruism locus, if not on our other genes. Well, here Jesus is telling you that your relatedness is about to get much higher. If you believe his words, you might experience the same emotions toward Jesus and his religion as you do toward a member of your own family. It is a ploy that religions use widely, often referring to God, and even his priestly representatives on Earth, as "Father." We in turn are often referred to as his children, making us all brothers and sisters.

religion as a way to advertise commitment

THE IMPORTANCE of the cultural survival vehicle in our history means that groups will want to know whom they can and cannot trust within their ranks, and it turns out that religious beliefs might be one of the best ways to advertise your commitment to your group. To understand why, we need to take a detour into a different arena: the arena of animal sexual displays. The connection might not be immediately apparent, but it will emerge once we have understood this puzzling topic. Darwin was troubled by the bizarre and seemingly useless but costly displays and behaviors of many animals in their attempts to acquire mates. Peacocks, for instance, produce spectacularly ornate and beautiful tails, but such tails are so big they almost prevent the peacocks from flying, they make them more vulnerable to being attacked by a predator, and they provide useful homes for disease-carrying flies and ticks. What bothered Darwin was that his theory of natural selection was an explanation for the survival of the fittest. How could his theory be true if such a thing as a peacock's tail can evolve?

Darwin's theory of *sexual selection* was an attempt to answer this question. Like the arts and religion, traits such as a peacock's tail seem to lack any function other than to please our senses or occupy our minds. And indeed, it was by pleasing our senses that Darwin thought these costly traits could evolve. Darwin's insight was that females would find males with long tails, bright colors, or melodic songs more attractive even if these traits made them less fit. In this way, the costly or sexually selected trait would more than pay its way because males with the best displays would get the most matings. You might die younger because your bright colors give you away to some predator, but you will still leave more offspring than someone who lacks your extravagant display, and it is leaving offspring that counts in the sweepstakes of natural selection.

Darwin was right; these gaudy ornaments, songs, and other displays do attract females. But there was a problem. Why should a peahen have these particular aesthetic tastes? Why should she prefer a peacock with a trait that slows him down and can even harm him? Why not just go for the more ordinary fellow? He might be dull, but at least he will survive, and probably stay at home and help you look after the children. A convincing answer to that question wouldn't be proposed until the 1970s, when the evolutionary biologist Amotz Zahavi brilliantly conceived of a peacock's elaborate tail and other ornaments or displays as a form of showing off, or what Zahavi called *handicaps*. Zahavi said that the peacock uses his large tail to advertise to the peahens that his genes are of such quality he can afford to drag his long, ungainly, and costly tail around behind him and still survive. Like a runner carrying extra weight in a foot race and yet still winning, handicapping yourself is the key to the message of your genetic worth. Less genetically fit peacocks could perhaps make a large tail, but could not withstand such a drain on their energy and resources. Natural selection favors males that can make ever more extreme ornaments because this is a way of showing the females who is wheat and who is mere chaff.

The females' aesthetic preferences now evolve for a good reason: the "medium is the message" in Marshall McLuhan's memorable phrase. Here the medium is a wasteful display whose message is, "you can believe me." It is a symbol of something greater lurking underneath, and this is why the peahens prefer it. The sociologist Thorstein Veblen anticipated this idea seventy-five years before in his *Theory of the Leisure Class*. Veblen made the bold claim that rich people throughout history have advertised their wealth through acts of what he called *conspicuous consumption*—ostentatious and extravagant displays that genuinely reveal how much they have by showing how much they can afford to throw away on otherwise useless objects. As with the peacock's tail, the waste allows us a way of glimpsing what must lie behind the flamboyance. One of Veblen's favorite examples was the bizarre behavior of the so-called potlatch Native American tribes of the American Pacific Northwest. These tribes would invite a neighboring tribe—potential adversaries—to a lavish feast. When their guests had finished and were preparing to return to their village, the hosts would erect a bonfire and throw blankets, food, and even canoes on to it. Was this superstition, an offering to a god, or perhaps disgust at the thought their guests carried lice or some infectious disease that needed to be eradicated? No. To Veblen it was their way of showing, or perhaps warning, their neighbors just how much they had in reserve by showing how much they could afford to lose.

Veblen's and Zahavi's ideas tell us something startling: some traits and behaviors evolve precisely because they are reckless and wasteful, and the more reckless and wasteful they are, the more they tell us something believable about the owner. This is why a $25,000 wristwatch is better than a $10 one, despite both keeping equally good time, or a Ferrari better than a Ford even if you only need to get to the nearest shop. It is why larger diamonds make better engagement rings, why silver is better than silverplating, why turning your back on your adversary is so compelling, and why fifty pairs of shoes are better than the one pair you need to protect your feet. Famously, in Jane Austen's *Pride and Prejudice*, it is the size of Mr. Darcy's house and not his looks or personality that finally takes Elizabeth's breath away, helping to convince her to marry him. No one *needs* a home with hundreds of rooms, but that is not the point. It is precisely because Mr. Darcy does not need the rooms that his house advertises he has wealth to burn.

We are now in a position to see why religious belief can be a powerful indicator of someone's commitment. Your religion is not just a marker of group membership, such as your language might be. Faith is about believing things that by all known rules cannot possibly be true or verified, and could even get you killed. It is about acting without evidence, participating in its rituals, fasting (a form of starvation), memorizing scripture, scarification, crucifixion, and paying of tithes. Veblen and Zahavi's insights tell us that it is the utter recklessness and costliness of adhering to religious beliefs that makes them a believable way of advertising your commitment to a group, and thereby of attracting altruism from others (you could try to demonstrate your commitment to your group by, for example, helping to build a boat, but its usefulness means your effort might be seen as partly for your own gain). Now, if you are the sort of person who can hold false beliefs, or have an ability to act on blind faith, you are probably also the sort of person who could be persuaded of the moral superiority of your group over the one next door. When group conflict is never very far away, religious believers become the kind of people others like to have around.

The fourteenth-century Chapel of Our Lady at Rocamadour in southern France is partly perched and partly cut out of a rocky cliff high above the Alzou Canyon, not far from the Lot River. The chapel is approached via the Great Stairway, 216 steps chiseled out of the limestone rock, giving breathtaking views out across the canyon valley. The laity and ordinary believers who would have daily trudged up these steps might have noticed embedded in them a burdensome irony to the modern mind: they contain fossils of ammonites and other shellfish millions of years old that stand in direct contradiction to a literal reading of scripture. Someone who demonstrates religious faith in the light of this sort of contrary evidence can be counted on as the sort of person disposed to make a commitment, not an evaluation that could change as the evidence or understanding of it changes. Or as Jesus tells Thomas in the quote from the Gospel of John that opened this chapter, "blessed are they that have not seen, and yet have believed."

(Having said this, I recently visited the magnificent eleventh-century Durham Cathedral in the north of England. A kind man who showed me around told me that as part of their training, guides are taught to expect people to ask them what the oldest part of the cathedral building is. Like most such cathedrals, Durham's has been added to and altered over the centuries. For instance, its lower levels are built on Romanesque arches while its upper levels carry the distinctive pointed arches of the later Gothic period. But rather than pointing out this or that arch or column, my guide looked down at the floor toward a black marble tile known locally as Frosterley marble, after the village where this tile had been quarried centuries before. He told me to look at it closely; upon inspection the tile turned out to contain small fossil corals that had been exquisitely preserved in the marble for hundreds of millions of years. Quietly, and I thought just a little subversively, he confided in me that *they* were the oldest part of the cathedral.)

The existential philosopher Søren Kierkegaard recognized what he saw as the virtues of faith without evidence in the story of Abraham. Genesis tells the story that "God tempted Abraham and said unto him, Take Isaac, thine only son, whom thou lovest, and get thee into the land of Moriah, and offer him there for a burnt offering upon the mountain which I will show thee." Kierkegaard's philosophical work *Fear and Trembling*, written in 1843 under the pseudonym Johannes de Silentio, examines Abraham's story as the ideal of commitment in a world in which our presence seems arbitrary, our purpose is unfathomable, and our existence a matter of chance. Abraham decides to kill Isaac and binds him to an altar. But all of a sudden an angel appears and intervenes to stop him. How could Abraham have known this was a test of faith, and that his son would be spared? He could not, and this is why in the biblical account God showed him compassion. Taken to Abrahamic extremes, we think of this sort of commitment as insane. But this kind of faith is often just what a group is looking for.

religious exploitation revisited

IN *The God Delusion*, Richard Dawkins writes, "The God of the Old Testament is arguably the most unpleasant character in all fiction: jealous and proud of it; a petty, unjust, unforgiving control-freak; a vindictive, bloodthirsty ethnic cleanser; a misogynistic, homophobic, racist, infanticidal, genocidal, filicidal, pestilential, megalomaniacal, sadomasochistic, capriciously malevolent bully." Why, unless religion is just a manipulative brain parasite, would we embrace a god who endorses such behavior? In trying to answer that question, we must be clear that the reasons I have given here for considering how religion might act as a cultural enhancer do not suppose that religion is true, or that it is harmless, just that throughout our history it has, on balance, acted to promote individual survival and reproduction. This statement can be true even if, for example, large numbers of people might have died in its name. To expect that only "good" things will evolve is to miss an important point about Darwinian evolution. Natural selection does not wear moral glasses; it promotes collections of genes and ideas that triumph in competition with other collections of genes and ideas. No one would ever ask if a snake's deadly poison has been a force for good or why the snake embraces it. No one would ask why our armies embraced better longbows, or later on, better guns. If religion has been an enhancer in our past—a bit of social technology—we shouldn't look to understand its grip on us by expecting it to do good.

Instead of religions manipulating us, we might just have to consider the proposition that we concocted and groomed religions to motivate and give justification to behaviors that have served our groups and us individually throughout our history. Their particular forms might have evolved because they have proven to be good at giving us courage and hope, at coordinating our actions, uniting us against common foes, controlling weaker people, or suppressing those we think challenge the norms that glue society together, even if these norms are arbitrary. Then, those individuals and the societies that adopted them would have been at an advantage over those who did not. It is just possible, then, that we created religions and their gods in *our* image, not the other way around. To think otherwise is to ignore—depressing thought that it is—that violence and hatred are all too human characteristics that transcend any one religion or culture. It would be dangerous for us to blame all this on religion, and it would represent us as feckless in its presence. Indeed, to blame religions would do little more than shoot the messenger, albeit an often very efficient one.

This is in no way to justify our brutality and violence as a species, and it is in no way an apology for those who would use religion for these purposes. But it is to remind us that there is no reason to expect our species would stop killing, raping, and pillaging were we able somehow to expunge religions from our consciousness. The analogy to performance-enhancing drugs might be worth revisiting here. Those drugs might improve your performance in a foot race, but they are not the reason you are running. Equally, if you think religions cause us to fight wars, just consider for a moment that, throughout our history and even today, the people we are most likely to fight or have a war with live next door. It would be a coincidence indeed if we were fighting them for no other reason than that our two different religions had told us to. A far simpler explanation is that it is our neighbors that we will often be in competition with for the same territories and resources. Aware that they will be thinking the same about us, we could say that the fuse of conflict is always smoldering. Indeed, the most terrible violence of the twentieth century in terms of lives lost—World Wars I and II, the Korean War, the Vietnam War, and the many millions who died in Stalin's 1930s purges in Russia, Pol Pot's Year Zero slaughter in Cambodia, or Mao's Great Leap Forward, to name just a few—have little or nothing to do with religion per se (I am taking the Nazi atrocities toward Jews, Gypsies, homosexuals, and others as something other than a religious agenda).

I have based some of the logic of this chapter on the life-dinner principle. That argument says that we will desperately evolve to avoid things that bring us harm. In this context it says that if religion is a mind virus, natural or cultural selection will have acted on us more strongly to evade its pursuit of us (because it can take our "life") than it will have acted on religion to infect us (each of us is merely one of its "dinners"). It is a principle, not a law, so we shouldn't expect too much of it. But it does give us a way of understanding how selection acts on competing sets of replicators, in this case, our genes versus religion's memes. Biological replicators that take over our minds and bring us harm, such as biological brain flukes and other biological brain parasites, are really rather rare, and this is as the life-dinner principle would expect. So are things that we could imagine are culturally transmitted and could bring us harm, such as reckless drug taking, obsessive-compulsive behaviors, or simply an annoying (to others) inability to stop singing some song. If this is what we expect of the life-dinner principle, then we are granted at least some permission to turn the argument around and wonder if the things that are common in our lives are less likely to be things that have brought us great harm. If so, we might wish to seek adaptive or functional explanations for their presence.

It is often countered that colds and flu are common, in fact ubiquitous, and yet we don't seek adaptive explanations for them. This is true, but misses an important point. Colds and flu might be ubiquitous but they are not omnipresent. A person infected with a cold or the flu might be ill for two or three weeks out of the typical year, and even then in most cases not very ill. But a person infected with a mind virus—as is suggested of religion—is infected twenty-four hours per day from early in life. The life-dinner principle would therefore predict that if religions really do exploit us in harmful ways, selection will have acted more strongly on us to avoid them than it has acted on us to avoid the common cold (or, it could be that in our past colds and flu did routinely kill us and our immune systems have responded). If we stack up all of the different diseases we get, they probably do approach being omnipresent in us. But this is not a relevant comparison. Each of these diseases is a separate "gene pool," and it is their individual success that we can predict from the life-dinner principle.

If we accept this argument, then religion's omnipresence might suggest we are not trying very hard to avoid it, and that could be because it is not harmful to us, at least on average. A comparison from the biological world might be helpful. There is a biological infection we have that is omnipresent, and we could wager it is the exception that proves the life-dinner rule. The billions of bacteria that reside in our gut are with us from life, beginning shortly after birth. They are an infection, but as we might expect from the life-dinner principle, they are often described as *symbiotic*. Still, the religions-as-mind-viruses argument is at its most persuasive in drawing attention to how this set of culturally transmitted ideas uses parents and others to infect vulnerable children, who then go on to do the same to their children. Human children are almost certainly unique among animals in the extent to which they have evolved to look to their parents and others in positions of authority to learn the rules of their society. More than any other species, we rely for our successes on the accumulated knowledge of past generations, and so this has wired our brains to have a certain docility and openness about learning per se—as youngsters we don't discriminate too much what we are taught. Our rule is to believe elders, and it is difficult to imagine it being any other way in our species.

This is all true, but the religion-as-mind-virus account requires the further assumption that the religion memes are so incomparably good at their task that, even when they don't return any rewards (as the mind-virus argument presumes) and can be downright costly, parents somehow will still wish to teach them to their children because they can't help themselves. But this seems peculiar because we know that adults can easily reject many of the mental infections that fool children. It is also peculiar because if our children's success is, uniquely among animals, dependent on learning, it will also be dependent on parents teaching their children useful things.

Given this, it seems that human parents will have been equally as strongly selected to exercise careful judgment about what they teach their children as their offspring have been to learn from them. Put it this way: imagine parents who could see through a mind virus and not teach it to their children, or better yet, inoculate their children against religious memes. If the religion meme really is harmful, these lucky children would outcompete children of less discriminating parents. The awful chain would have been broken (although memeticists might counter that the religion memes would fight back by getting others in the community to ostracize the lucky but wretched nonbeliever!).

I offer the thoughts in this chapter in a speculative way and not as proof or refutation of others' views. My purpose is to raise the possibility that the forms of religion we see around us today are indeed collections of highly evolved memes, but ones that we have shaped to suit our needs. This is not to say we should necessarily cling to them, teach them to our children, or have established (i.e., state) religions. We have cast off many old practices that might have once served a purpose but have outlived their usefulness: for example, we no longer build defensive walls around our cities. Equally, no one should think that even if religion has historically played the role of a cultural enhancer—and that it is not merely an exploitative mind virus—its uses are limited solely to consolation, engendering cohesion in groups, or demonstrating commitment to them. An entire topic of religion that we have not even touched upon concerns how *purveyors* of religion might use it for their own ends. I do not include here the many people who use or rely on religion in genuine and selfless attempts to help other people. Rather, I have in mind people who use religion to control other people's behavior, to exploit the vulnerable, to marshal armies of "crusaders" to attack foreign lands or gather wealth. The list of these purveyors runs from the pious to the ludicrously crass and rich televangelists, from prehistoric shamans to modern-day popes, and from child abusers to the charismatic leaders of mass suicide cults.

Perhaps nowhere more effectively do the purveyors of religion excel than in the great cathedrals of Christianity. Visit one sometime, such as the immense Durham Cathedral whose subversive tour guide I mentioned earlier. Imagine yourself as a typical eleventh-century peasant standing in front of this imposing edifice 80 feet tall at the roof, 200 feet at the tower, and over 400 feet long, comparing it to your home—probably a one-room hovel with mud floors, no windows, and possibly shared with your animals. Then, you enter one of these palaces to God. The space inside takes your breath away. There are gently curving Romanesque and daintily pointed Gothic arches. The spectacular vaulted ceilings are so high above you, you couldn't hit them with a stone you cast. They are supported on rows of enormous round columns, far bigger in diameter than the giant oak that sits on your village green. They are engraved with geometric patterns whose perfect symmetry and alignment trumpets the skills of the stonemasons. The cathedral is ornately decorated with carved objects, precious paintings that tell stories, and it has vivid stained-glass windows that dazzle your senses, normally accustomed to the dull drab gray tones of your life. And a group of choristers is singing in celestial harmony.

This is not so much a palace to God as a peacock's tail of an advertisement of the wealth and power of the religious social clique that built it. Its construction would have required the quarrying, transportation, and then carving of unthinkable quantities of materials. Still, you can't eat it or live in it; it doesn't sow your corn, feed your children, or look after your animals. Nevertheless, it commands your loyalty because its unmistakable and probably irresistible message is "Join us and we (or He) can do for you what we did here."

But this is another topic and one that is not necessarily about religion per se. Rather, it merely reminds us that we shouldn't necessarily expect people to look out for others' well-being. These purveyors of religion to a greater or lesser extent are behaving in ways that do not differ from anyone else who might try to manipulate social systems for their own gain. Religion might just be a particularly effective tool for them to use, and it is the purveyor's skills that we should marvel at—and often denounce—rather than the particular way he or she does it. Thus, religious indoctrination or exploitation might be one way to take advantage of others, but Maoist style political reeducation is another. In the hands of a skilled politician, nationalist rhetoric devoid of religious connotations can also be a powerfully motivating tool; just think of Churchill's many rallying speeches during World War II. The point is that to denounce religion because of the ways others might exploit it is potentially to blame the wrong agent, even though we should be aware that this dangerous agent lurks in our midst. We surely wouldn't denounce morphine because some use a different form of this opiate to make enormous profits from the sale of heroin, even killing people along the way to do so.

some possible darwinian roles for music and visual art

WE SHOULD remind ourselves that, just as with religions, our early ancestors' art and music might have started out as little more than simple acts, nothing like the sophisticated forms we have today. Some of these would have caught on, then grew, filling what was before them an empty space, not one occupied by some other artistic or musical traditions. The first art might have been a map or an image of an animal, scratched into the earth as it is known Australia's Aborigines and some Native American tribes do, or something cut into a tree, or maybe a sketch on the wall of a cave using a piece of charcoal. It is easy to imagine the curiosity at an image being able to resemble something real, almost as if the image itself captured some essential quality of the object, as some traditional societies still believe is true of photographs.

A similar story could be told for the origins of music. We might be primed for music by our evolutionary history with the other animals. Songs or calls or vocalizations have an ancient evolutionary foundation in animals as ways of attracting prospective mates, warning off competitors, and marking out territories. We retain in our modern brains an equally ancient structure known as the *limbic system* that processes many of these signals. It is sometimes called our "lizard brain" because it acts as a powerful emotion-charged center for responding to sights, sounds, and odors. Emotions are, of course, some of our most powerfully motivating forces. Appropriately, the word "limbic" derives from the Latin *limbus*, meaning "edge," and it is the limbic system that can so quickly put us on edge.

Stephen Mithen reminds us that "if music is about anything it is about inducing emotions," and, as we might expect of something processed by an ancient part of our brain, the emotional message of most songs can often be understood without even knowing their words. This helps us to understand how it is that singing seems so naturally to enhance the cohesion of groups: when people sing the same song, they recognize that they are feeling the same emotions. This shared feeling will enhance the sense of cultural relatedness that we have seen can drive our cooperation within groups. Layered on top of the pure emotion of music, the words to hymns often have a military quality to them, as in "Onward, Christian soldiers, marching as to war," or extol the benefits of cooperation, as in "Bringing in the sheaves, bringing in the sheaves, / We shall come rejoicing, bringing in the sheaves." A national anthem is an emotive and historically charged symbol of an entire nation's collective cultural nepotism. Music reminds us of our shared destinies and our common histories—how else could we all know the same songs?

At the same time we should not grant special status to music as a means to enhance the emotions and dispositions that encourage group cohesion. Forming into a circle and praying, or listening to a rousing speech, can have the same effect. Listen to Shakespeare's St. Crispin's Day speech spoken by Henry V just before the Battle of Agincourt:

> *From this day to the ending of the world,*
> *But we in it shall be remembered—*
> *We few, we happy few, we band of brothers;*
> *For he to-day that sheds his blood with me*
> *Shall be my brother; be he ne'er so vile*
> *This day shall gentle his condition:*
> *And gentlemen in England now-a-bed*
> *Shall think themselves accurs'd they were not here,*
> *And hold their manhoods cheap whiles any speaks*
> *That fought with us upon Saint Crispin's day.*

Not only does Shakespeare evoke the emotional power of "relatedness" to get the men to throw themselves into battle as they might if battling alongside their actual brothers; he hints at the potential reproductive benefits of victory. In a different arena of battle, the New Zealand rugby team chants the *haka* before their matches, and no one would accuse them of breaking into song. It is an intimidating display of harmony and unity performed in full view of the opposing side. Players slap their thighs, stick out their tongues, and grimace together while reciting the words of the *haka*. It sends the unmistakable message that the All Blacks are a unified team that has spent hours and hours practicing together. It also tells us that the All Blacks are so good they can afford to waste time practicing their exuberant ritual and still be ready for the opposing side—it is also a peacock's tail of an advertisement if there ever was one, and opposing sides find it unsettling to watch.

Dispositions toward art and music can grant individual benefits outside of a religious or group setting. Both can enhance our memories. We all know that it is easier to learn the words of a song than merely to learn a set of words. For the same reason, long messages or oral traditions, such as Homer's *Iliad*, were often written in verse and sung rather than merely recited. Music and language might even draw on similar capabilities in our brains, as there is a suggestion that the intervals between the sounds we make in speech match those of the musical scale. Early in our evolution as we were developing our language abilities, musical ability might then have been of great value as a means of promoting communication and remembering, and genes for musical ability might have spread. Just think of all the time you have wasted trying to follow a sequence of instructions for how to get to someone's house, and then think how dangerous this might have been 40,000 or more years ago when you lost your way trying to navigate through hostile territory—you might have lost your life. Now that we have writing, we take for granted that we can know and benefit from the thoughts of our ancestors. But for our ancestors to know the thoughts of their ancestors—or even just some directions—might have meant singing a long story. In fact, this is just what Australia's Aboriginal people did when they told their "dreamtime" stories of the creation of their world.

The invention of writing would gradually remove the need for this link between music and memory. The earliest known writing and literature was not used for comedy or pure entertainment, as we might expect if it was taking over for a cultural form that was devoted to hedonism, but rather to keep track of things. By 8000 BC, early Mesopotamian civilizations were producing clay tablets that seem to have been used for some form of accounting or for keeping records. The cuneiform tablets of the Sumerians are regarded as the first documented writing system and they also were used for keeping track of transactions and amounts. Homer's *Iliad*, although perhaps not originally written down, is often regarded as the earliest surviving literature, and it tells a great historical story. These examples show, perhaps not surprisingly, that writing seems to have been used to enhance memory, and its invention would have been revolutionary in offloading a burden on our minds. At the same time it would have democratized information by casting it in a stable form and making it available to everyone.

Indeed, in Plato's dialogue *Phaedrus*, Socrates voices his suspicion of writing as an invention that will harm thinking and memory. Socrates is speaking to Phaedrus about the art of rhetoric. They are standing under a plane tree by the banks of the Ilissus River that ran just outside the defensive wall of ancient Athens. Socrates is agitated because he believes that writing things down means that speakers need no longer understand what they are reading. He says to Phaedrus:

> this discovery of yours [writing] will create forgetfulness in the learners' souls, because they will not use their memories; they will trust to the external written characters and not remember of themselves. The specific [thing] which you have discovered is an aid not to memory, but to reminiscence, and you give your disciples not truth, but only the semblance of truth; they will be hearers of many things and will have learned nothing; they will appear to be omniscient and will generally know nothing; they will be tiresome company, having the show of wisdom without the reality.

If writing might have interfered with rhetoric, visual art might have been an even earlier form of written communication with the same function of being a way of storing and transmitting information, acting as an external memory, or *aide-mémoire*. The artist Orde Levinson suggests that just as an ability to read written work is important in modern society, a capacity to read visual art for its information and emotions might have been so in our ancestors. Three forms dominate painting: religious scenes, portraits, and landscapes. The information these images carry and its relevance to our lives should not be underestimated, especially in the context of prehistorical people who lacked writing. Depictions of scenes and landscapes could improve cooperation, understanding, planning, route finding, and teaching. Australian Aboriginal paintings tell stories of the dreamtime and of everyday life. They are now popular among tourists as art, but the paintings are a modern invention, barely more than a few decades old. The images we see on canvas were originally scratched into the ground, in caves or on rock walls. They might have been used in religious or other ways to teach or instruct. They might have been left as descriptions to others of what went on at a particular place and how many were involved. They could also have been like medieval fingerposts, giving directions not to the next village but to oases, or meeting points, or sources of food.

The early cave art from the South of France consists predominantly of depictions of the game animals that would have been important in the diets of the early hunter-gatherer modern humans, or of predatory animals whose dispositions one might benefit from understanding. Visual images painted on walls might have been used for teaching or in parallel with religious ceremonies. These images would have made their effects available to a wide audience, and simultaneously. They could be consulted and re-consulted. We only have to ask ourselves why we still keep images in the form of photo albums, and remember how they make us feel. They motivate our behaviors by reminding us of our attachments to friends and family, calling up and making more salient and vivid our emotional memories. Emotions are of course instruments of motivation, and so here is another enhancing role for art. Stand and look at a statue of Caesar or Athena in a museum, and the person becomes at once more real, more intimidating, and more formidable.

In his great *Essay Concerning Human Understanding*, John Locke makes the case for visual imagery as a means to enhance our memories, saying,

> Thus the ideas, as well as the children, of our youth often die before us; and our minds represent to us those tombs which we are fast approaching; where though the brass and marble remain, yet the inscriptions are effaced by time, and the imagery moulders away. The pictures drawn in our minds are laid in fading colors; and, if not sometimes refreshed, vanish and disappear.

Indeed, so powerfully can images enhance our perception and memory, they can do so even when the image resides in our mind. The great Russian psychologist Alexander Luria showed how mnemonic devices for remembering events or lists could be remarkably effective when based around simple concrete imagery. Luria would get people to imagine scenes they knew well, such as taking a walk from their house and then along the adjacent street. He would get them to place objects from lists of things to be remembered at familiar landmarks along the route. For example, if it is a shopping list, you might place an apple on your car and then a bag of sugar at the end of your drive, some flour further down the street, and so on. Then Luria would get people to replay their walk in their minds, looking for the objects. Try this yourself. It is a surprisingly effective way to remember long lists. Imagery and visual art help us to think and remember more clearly.

PART II

COOPERATION AND OUR CULTURAL NATURE

Prologue

THE CHAPTERS OF PART I show us just how much the cooperative enterprise of human society can achieve, and how much it can influence our lives. Culture has produced riches beyond the imagination of any other species, and it has propelled us around the world in mobile survival vehicles in which individuals cooperate to defend each other and their jointly held assets—their technology and know-how, and their lands. And yet, it is one of the oldest themes of literature that the availability of riches, either actual or potential, is a source of treachery and betrayal. So, it is not enough to say that culture is a success because it has given us riches. We need to understand how we have contained our appetites to exploit these riches for our own personal gain, how we have avoided spiraling down a vortex of treachery matched by further and more ingenious countertreachery.

"Ingenious" is in fact closer to the truth than we might expect. It is a melancholy fact that natural selection will favor in me certain kinds of devious tendencies for acquiring more than my share of the wealth from a cooperative venture. I might deceive you as to how much there is, I might quietly steal some of your share and blame someone else, I might not work as hard as you, especially when your back is turned. My actions will in turn favor yet more nefarious and devious tendencies in you. Over long periods of time, a kind of arms race develops in which a shrewd, devious, and wily intelligence evolves, only to be matched by even more wily and shrewd intelligences that arise in response. But it is an arms race that is ultimately destructive. We all lose out as we get better and better at taking advantage of each other and spend an increasing amount of our time and energy defending ourselves against each other. In a televised debate on ABC's *Nightline* program in 1983, the cosmologist Carl Sagan described the arms race in nuclear weapons between the Soviet Union and the United States in the 1950s, 1960s, and 1970s by saying, "Imagine a room awash in gasoline, and there are two implacable enemies in that room. One of them has nine thousand matches, the other seven thousand matches."

We've avoided this explosive outcome by developing a set of rules and dispositions that allow us to cooperate with people who are otherwise our competitors. This kind of cooperation is a puzzle to evolutionists because a disposition to help someone else who doesn't share that disposition means they might prosper at your expense. So, how does cooperation ever win out against people who would take advantage of your kindhearted help, with no intention of ever returning the favor? Why, for example, when you offer assistance to me, do I not simply take it and run? Why when I walk past you on the street, late at night, do I not knock you over the head and steal your wallet? If cooperative systems can evolve in spite of these temptations—and they manifestly have in us—why haven't they evolved in the other animals?

Our style of cooperation is made all the more remarkable because it violates the rules that govern nearly all cooperation seen in the rest of the animal kingdom. That cooperation is explained by one of the great insights of evolutionary thinking: that by helping a relative, we help a little bit of ourselves, because our relatives are more likely to share copies of our genes. As we saw in Chapter 2, this is known as the theory of kin selection and it helps us to understand some of the most poignant of our behaviors, including risking our health and well-being, our opportunity to have children, or even our lives. A remarkable photograph from the December 26 tsunami of 2004 that struck Indonesia shows a group of people fleeing up the beach as the terrifying wave rises behind them. But one person is recklessly heading down the beach toward the wave. She was the mother of three of the children in the fleeing group, desperately trying to save her offspring.

If we greet this image with a mixture of wonder and horror, it may be because natural selection has not really prepared us for such selfless live-or-die decisions. Or it might be that—emotions aside—a parent sacrificing itself for its children only sometimes makes sense in the unsentimental arithmetic of kin selection: a very young parent with years of potential reproduction ahead might do better to let her children die and try to start over, an option that might not be available to an older parent. We know kin selection has created emotions to be discriminating because the success of kin selection rests on our degree of genetic relatedness. Parents and children share about one half of their genes on average. So do siblings. Cousins share about one eighth. We are more likely to help our own offspring or a brother or sister because doing so promotes copies of our genes more effectively than providing the same help to a cousin. Or, as J. B. S. Haldane is purported to have remarked, "I would jump into a river to save two brothers or eight cousins."

Who amongst us has escaped the Kamikaze charge of a bee defending its hive? These soldiers are only too willing to give their lives to protect the nest because it houses their brothers and sisters, and their mother the queen. Warfare and self-sacrifice is also common among the ants, which deploy sophisticated and well-drilled armies to intercept intruders or to invade other nests, giving up all hope of ever reproducing on their own. E. O. Wilson describes the soldier caste of one species of African termite (*Globitermes sulfureus*) as "walking bombs." The soldiers, in an act reminiscent of human soldiers carrying flamethrowers, have on their backs two glands full of a yellow corrosive liquid that when sprayed out entangles them and their enemy combatants. Sometimes the termite's body simply explodes, spraying the deadly fluid in all directions in an act chillingly equivalent to that of a suicide bomber. The ultimate expression of kin selection is the behavior of the cells in large multicellular bodies such as our own. These cells are genetic copies of one another, sharing all of their genes, and this makes them only too happy to exchange their lives for the good of the rest of the body. Our skin cells do this when they good-naturedly allow themselves to be burnt alive on our behalf protecting us from the sun. Some of our immune system cells do this when they find and attach themselves to a foreign invader. They then put up little chemical flags that summon others of our immune cells to come and eat them, thereby giving their lives to have the invader destroyed.

But human cooperation is not limited to helping relatives, and this makes it a risky proposition, not being bound by the usual constraints of family ties. Some anthropologists think our apparently helpful and altruistic nature is an error of judgment, a leftover from an earlier time in our history when we lived in small groups of closely related kin. Natural selection would have quite routinely favored high levels of altruism among these kin or extended kin groupings, and the argument is that now in our wider groupings we simply carry on behaving cooperatively because it is what we do. Humans probably did evolve in small groups around many of their kin, but so too do many animal species, and none of these shows the range of altruistic tendencies toward unrelated others that we do. We cannot easily imagine a chimpanzee or a fox giving its life for an unrelated member of its group, or even helping an elderly member. Animals do help each other—as in fabled stories of elephants assisting injured members of their groups—but this help is normally directed at kin.

Given all of the risks and complexities of cooperating with people outside of our immediate families, how did such behavior ever get off the ground, and what keeps it going? One answer is that we simply use our large brains to work out how and when to cooperate; but as we have seen, that can only be part of the answer because our large brains also grant us the ability to take advantage of others by being able to think up devious new ways of cheating, manipulating, or otherwise exploiting them. It is common for economists to appeal to trust and norms, but this is to replace one problem with another: where do norms and trust come from? In the absence of evidence to the contrary, far from being trustworthy, humans would seem to have an exquisitely evolved sense of when they can and cannot act in their own naked self-interest.

The chapters of Part II show how humans have solved these problems by evolving two kinds of cooperative behavior not seen elsewhere in nature. One is the topic of Chapter 5 and centers on everyday acts of reciprocity and exchange, governed by emotions of fairness and the expectation of future interactions. The other form of cooperation—the topic of Chapter 6—balloons in complexity to a set of behaviors, peculiar to our species, of helping others even when there is no possibility of return. It is as if we are predisposed to precisely those acts of kindness that others could take advantage of. What purpose do these behaviors serve in us? Chapter 7 finishes this section by asking the question, why do we humans have such exceptionally large brains?

CHAPTER 5

Reciprocity and the Shadow of the Future

That we owe human cooperation to conflicts of interests among people

conflict as a source of cooperation in society

In *On Human Nature* (1978), E. O. Wilson stated that "kin selection"—the helping of relatives—"is the enemy of civilization." What could he have meant by this? We saw in the Prologue to this section how that theory understands altruism toward our relatives as a mechanism for promoting copies of our own genes, and how this simple principle can lead to remarkable and poignant acts of helping and self-sacrifice. But Wilson is right; nepotism turns out to be a simple-minded, knee-jerk, sentimental distant relation to the more cerebral, shrewd, and imaginative forms of cooperation among non-relatives that characterize human society.

Two points that immediately follow from the theory of helping relatives tell us why. One is that we are less likely to take advantage of relatives, because to do so is to exploit a little bit of ourselves. Nepotism doesn't erase rivalries among relatives—after all, the first murder in the Bible is an act of fratricide—it just makes these rivalries less likely. The second is that we expect nepotism to fall away rapidly as our kin become more distantly related. A second cousin is sixteen times less likely than one of your siblings to share your genes. That is why we are far less likely to buy a second cousin a large birthday present, much less the person who lives down the street, than we are our own sister or brother.

Wilson's point, then, is that we don't expect kin selection on its own to produce the complex social arrangements we recognize as human culture—the alliances and coalitions, pacts and agreements—and we don't expect it to produce the psychological dispositions and emotions we use to control, forgive, chivvy, and take advantage of our rivals. There is just too much commonality of interest among close relatives and too little among distant relatives to expect kin selection to produce much cooperation. It is because most animal societies have not moved beyond nepotism that they have never had to evolve the cooperative arrangements and psychology we regard as distinctive of our societies.

The difference between human cooperation and most animal cooperation is even more fundamental. Fundamentally, our nepotism is "paid for" by promoting copies of our genes, and this is why, for example, parents don't expect something of equivalent value back every time they help their children. But our cooperation with other members of our societies is based on the idea of a social contract: when I do a favor for you, it is in the expectation that you will pay me back either immediately or at some future time. Simple as this sounds it requires a level of psychological sophistication that surpasses even the most poignant acts of nepotism. If I do something for you, how can I know you will pay me back, and what should I do if you don't? Equally of course if I do something for you, you will be tempted not to return my favor, or perhaps return it with something of less value. In an intelligent species, this situation leads quickly to two alternatives. One is to engage in an escalating spiral of exploitation and counterexploitation as increases in intelligence and cunning are met by similar increases in one's competitors. But the other is to accept that conflicts of interest can create opportunities for agreements that provide better returns than endless cycles of betrayal and revenge. If we choose this second course—and often we don't—something surprising emerges: our conflicts of interest become the source of our truces, pacts, and agreements, and over long periods of time, they are the source of our institutions, laws, and morality. If you and I both want the same thing, we can fight for it, or we can strike up an agreement to parcel it out equitably. The savings we make in not fighting or in not having to devote time and energy to defending the resource against each other's depredations often more than pay for the reduced income of dividing the riches between us.

god save the queen

THE THOUGHT that we owe our cooperative societies and other cooperative institutions to conflicts of interest might strike us as bizarre, but conflict has been the source of complexity in the biological world since the origins of life. The first life consisted of simple strings of chemicals that could make copies of themselves, or replicate, on their own. They were probably not even DNA (or deoxyribonucleic acid)—the molecule that supports our life—but a simpler molecule known as RNA or ribonucleic acid. These simple strings would have competed for the same basic chemicals in the early primordial soup, because for a string of RNA to reproduce, or copy itself, it had first to find enough nucleic acid building blocks to create a duplicate string. One course for this early life to follow was to get better and better at finding chemicals before your rivals did, or perhaps work out ways to kill competing strings. But there was a cooperative alternative even then.

The late John Maynard Smith gave an elegant example of how competing replicators might have learned to cooperate in a way that benefited them all. Maynard Smith imagined there were four of these simple strings, one called *God*, another called *Save*, a third called *The*, and a fourth called *Queen*. We can think of them as short segments of RNA or DNA, and Maynard Smith supposed that initially they all competed to make copies of themselves. Maynard Smith supposed that by virtue of being shorter, *God* and *The* could replicate more quickly and consequently we might expect them to dominate the longer and slower-replicating *Save* and *Queen* strings, and eventually even replace them. But this wouldn't mean *God* and *The* were safe from competition. Sometimes when they copy themselves, errors creep in and these change the word. For instance, *Got* and *She* strings might arise just by chance and they would be equally fast at replicating.

Life might have remained stuck on *Got* and *She* had natural selection not worked out how to exploit these strings' conflicts of interest. Rewinding back to the original four, imagine it just happens by chance that the strings can help each other to replicate. There is nothing special about this; perhaps they just incidentally act as chemical catalysts of each other (for instance, gunpowder is a combination of sulfur, carbon, or charcoal, and potassium nitrate or saltpeter. In that mixture sulfur catalyzes the explosive reaction by lowering the temperature at which the other two ignite). In particular, Maynard Smith imagined that the presence of *God* might have catalyzed or increased the rate at which *Save* copied itself. Similarly, he supposed *Save* increases the rate at which *The* replicates, and so on for *Queen*, which in turn promotes copies of *God*, completing the loop. Thus, think of the four strings as comprising a continuous circular chain in which each provides a benefit to the next one along. The result of this cooperation is that more copies of all of them will be made than when they were competing, because now their efforts are coordinated in such a way that helping others is a way of helping yourself.

If this all sounds too easy, it is, because there is a problem. On their own there is no incentive for any of the strings to get better at promoting the next one in the chain. In fact, the situation is even worse than that. Each time one of these strings acquired some new mutation that made it better at affecting the next one, it would have benefited that rival, but at its own expense. Any change to one of them that improved another's ability to copy itself would have been like digging its own grave because now there would be more copies of that other string, and they would have been in competition to find the same chemical building blocks. Once again, then, we would expect this set of replicators to come to be dominated by *God* and *The* (or any equally short variants of them) because they can replicate the fastest.

We learn from this that natural selection will not favor naive altruism. In this case it takes the form of handing over a useful piece of technology to a rival but without any guarantee of a return. As we might expect, natural selection does favor greed in the recipient of this altruism, but even this is shortsighted because the naive altruist will quickly be driven extinct and there will be no one for the greedy recipient to take advantage of. On the other hand, imagine that one of the strings, say *God*, all on its own just happens to get better at using the presence of the previous string in the chain, *Queen*. Nothing new is required of *Queen*; *God* has simply acquired a mutation that improves its performance in *Queen*'s presence. *God* has specialized. *God* has got better at making more *God*s. In contrast to naive altruism, natural selection will reward this specialization because by definition these new *God* strings will replace the older ones.

If *Save* now acquires a change that makes it better at using the presence of *God*, this too will be favored by natural selection, and for the same reason. *The* and *Queen* can also acquire changes that make them better at replicating themselves and the four molecules will have created a partnership that allows and even encourages certain kinds of specialization and a division of labor. We learn from this that natural selection will favor each of the strings getting better at using what is naturally available to it. In fact, specialization among these four replicators promotes rounds of competition, making them all get better at what they do. Each molecule benefits from making better use of what the others produce, and their dependency on each other makes it less likely one of them can "run away" and outcompete the others.

But we are not quite yet out of the woods. One of the strings might acquire a change that means it gets better at using the previous member of the chain, but it no longer affects the next one. It has become parasitic on the chain. Now, *The* might stop assisting *Queen* while receiving help from *Save*, and the world would fill up with these parasitic *The* strings that bleed the budding partnership to death. From our perspective, *The*'s behavior is pointless, but remember, natural selection, unlike you, does not look ahead. There is a way out of this dilemma, though. If somehow these four strings could have their fates linked such that they could only replicate when the others did, then cooperation would again be favored. Once fates are linked, the groups of four act as a single unit that must compete with other units like themselves. Now, if a string in one of these units received aid without passing it along, this would cause all of the strings, including itself, to fail.

Linking their fates might have been as simple as confining them all to one place, such as inside a cell. Natural selection will favor ever greater specialization and division of labor as cells compete against other cells, and it will reward changes that improve everyone's outcomes. The enterprise benefits the best cooperators because together as a cell they are able to produce more total "information" or more in the way of technology than had their individual strings competed against one another. So long as their combined technology means they can produce more copies of themselves than they could have on their own, this partnership will be a success. This indeed could be a description of the early single-celled bacteria that began to colonize the Earth perhaps more than 3 billion years ago.

We can see in this simple story two fundamental messages for groups of replicators, whether they are short segments of DNA, simple bacterial cells, or even groups of people. One is that when everyone's fate becomes linked, natural selection can powerfully favor cooperation and mutual acts of altruism over competition. The message for societies is clear: groups that can somehow come up with psychological and social mechanisms that strengthen the links among people will be at an advantage over less strongly linked groups. The second message is that conflict is an endlessly creative force of evolution because once one set of cooperative ventures arises, another might follow, and then these two might compete. There will always be a temptation to render your rivals impotent or even to destroy them, and the payoffs for doing so can be great. So, we don't expect conflicts to disappear. But we can expect the constant force of conflict to create increasingly sophisticated acts of cooperation, not just because it is nice to be nice, but because this cooperation can serve everyone's interests.

In our distant past, large groups of people formed into bands and then bands of bands formed into tribes and later villages or even nations. At each stage, other groups at a similar level of complexity are potential rivals, but a larger alliance can reduce conflict. The early nineteenth-century explorers of the American West, Meriwether Lewis and William Clark, whose observations on Native American tribes we saw in Chapter 1, were struck by the fluid, shifting, and opportunistic alliances among these tribes, often based on specializations one competitor could offer another. For example, the Cheyenne were at that time nomadic hunters, dependent on other tribes for corn, but they were good at making quillwork clothing and at breeding horses, and so would trade clothing and horses to other tribes for their corn.

One of the great mysteries of this daring expedition, known as the Corps of Discovery, is why the Native American tribes did not simply exterminate Lewis and Clark and their relatively small party of thirty or so permanent members (some members of the expedition made trips back to Washington with news of the expedition's progress). Remarkably, only one member of the Corps of Discovery and two Native Americans died, despite nearly continuous contact between the two sides for over two years. It seems that the promise of trade and commerce might have kept the Corps of Discovery from being slaughtered. For instance, the Teton Sioux or Lakota were the most powerful and feared tribe in the western plains. When Lewis and Clark arrived in their territory near the Missouri River, the Lakota were determined to maintain their control of trade up and down that river. They had a reputation for violence and had for years intimidated and bullied other tribes, along with Spanish, French, and other European fur trappers. A Yankton Indian chief had told Lewis and Clark that the Lakotas "will not open their ears, and you cannot, I fear open them."

Lewis and Clark met the Lakotas near what is now Pierre, South Dakota. They had with them an African-American slave called York who was part of the Corps of Discovery. The improbability of this meeting is difficult to overstate, bringing together as it did in this remote heartland of America in 1804 representatives of three of the most distantly related peoples on the planet, whose genetic and geographical paths had been diverging for most of the previous 50,000 years. York intrigued the Lakotas, but they resisted Lewis and Clark's attempts to pacify them with medals and other gifts. The Lakotas eventually sparked a tense stand-off with arrows notched into bows on one side and guns pointed on the other. The stand-off was only defused, and bloodshed avoided, when a Lakota chief stood down his side. This chief had realized that peace with the expedition meant they could retain their control of trade into the western interior, channeling the goods these highly technologically advanced whites would supply. Looking from that moment decades into the future, we can now see that the Lakota chief had sealed his people's fates by allowing Europeans in. But at that moment, his decision to cooperate rather than to fight ensured the Lakotas' prosperity for years to come.

Where raw competition often leads to survival of the simplest and fastest, or the nastiest, cooperation born of conflicting interests can give rise to complex forms (just think of our species' solution to the crisis of visual theft). Alliances, agreements, friendships, and coalitions can often pay their way by giving all of us more returns than we could have had by going down the path of outright competition. Eventually, institutions and even norms or expectations of behavior can emerge, all aimed at controlling the debilitating bouts of betrayal followed by counterbetrayal that inevitably occur without the guiding hand of cooperation to hold them in check. An optimist would say that the course of our recent history has been one of increasing interdependence, cooperation, and reduced violence around the world. Together, the institutions and the norms that emerge from interdependence can reach to the highest level of worldwide cooperation. In 2010, the nation of Greece was bailed out of a financial crisis by the alliance of nations known as the European Union, followed by a similar bailout of Ireland in 2011. These nations are all economic competitors but recognize that financial instability in Greece and Ireland is bad for everyone's economy, for the simple reason that all of the economies of the European Union trade with each other—their fates are interdependent. Looking into the twenty-first century and the rise of India and China as economic powers, it will be dependence of everyone's economies on trade with each other that will act as the best brake on conflict.

Conflicts of interest can often lead to uneasy allies. During President Lyndon Johnson's 1964–68 term in office, relations with his director of the FBI, J. Edgar Hoover, were tense. Hoover was increasingly using illegal wiretaps on telephones to gain precious evidence on other political leaders that could be used to coerce and blackmail them. Johnson worried that the wiretaps might even include his own phone and came close to sacking Hoover. But he decided against it, trenchantly concluding, "I would rather have him inside the tent pissing out than outside the tent pissing in."

four ways to be social and the shadow of the future

IN 1981, Robert Axelrod and William Hamilton chose to phrase the case for cooperation differently from President Johnson, proclaiming: "Many of the benefits sought by living things are disproportionately available to cooperating groups." They are right, but as we saw with the *God-Save-The-Queen* example, the trick of getting cooperation to work is somehow to contain the conflict before it consumes the riches that could otherwise be shared. This can prove surprisingly difficult to achieve and might be why, outside of our own species, cooperation beyond the family is comparatively rare in nature. Another problem is that there are really just four kinds of social behaviors or ways that two or more parties might behave toward one another, but only one of them benefits both parties: *altruism*, *selfishness*, *spite*, and *cooperation*.

If I rush out into the street to push you out of the way of an oncoming vehicle, I act *altruistically*. My altruism benefits you, but potentially at great cost to myself. We don't expect altruism to flourish on its own, because one party benefits at the expense of another, and it was just this asymmetry that undermined the four simple strings offering assistance to each other. At the other extreme, I behave *selfishly* when I take advantage of your altruism. For instance, maybe I surreptitiously eat more of my share of our stored food. Selfish behavior is tantalizing because of its immediate benefits; but of course if everyone behaves selfishly, the shared food will quickly evaporate and the selfish players will be thrown into unending conflict.

Sometimes people behave *spitefully*, as when they do something that might be costly to themselves, in an attempt to hurt someone else even more. If I see you standing near the edge of a cliff, why should I not just push you off? I will rid myself of a potential competitor and at little cost to myself, perhaps just some torn clothing as you grasp at me when you go over the edge. This is an example we must take seriously because we know the thought that it could happen to us or that we might do it to someone else crosses our minds. In fact, in the 1990s in New York, travelers on the subway were still routinely warned not to get too near the edge, and not just for the obvious reason that a train might hit them. City authorities had judged there were a sufficient number of unpredictable people about who might not be able to resist the temptation to push someone off the edge.

But even though spite crosses people's minds, it is unlikely to evolve because my spiteful actions toward you don't just benefit me in getting rid of you as a competitor, they help everyone else who competes with you. My spiteful actions toward you then become acts of altruism toward these others—and that spite might be costly for me to perform. For this reason, spite is thought to be rare in nature, and so it is something of a puzzle why for many of us the thought of spiteful revenge is often attractive, and spite itself so sweet. One possibility we will see later in this chapter is that spite might have evolved as a way people can advertise to others that they are not the sorts of people who can be taken for granted. Our spite then pays its way by bringing us better outcomes in future encounters.

This leaves the fourth social behavior—*cooperation*—as when two parties exchange favors. This sounds attractive, but even cooperation is difficult to get established, since it is easy for one of the parties to succumb to temptation and not return the favor. Economists and evolutionary biologists use the so-called prisoner's dilemma as a vivid metaphor to illustrate this point. The police round up two people suspected of being involved in a joint crime. They haven't enough evidence to convict either one without a confession. They put them into separate jail cells and tell them both that if they confess and implicate the other one, they will be treated leniently. They are also told that if they don't confess and their accomplice does, they will be treated harshly, with a long jail sentence. What would you do? You could be loyal (cooperative) and not say a word, and hope that your partner does the same, and you will both be released. On the other hand, you are worried that your partner will sell you down the road. So, you act first, implicating your partner in the crime. But of course your partner has done the same thing to you, and you both end up being convicted.

The prisoner's dilemma teaches us that if Axelrod and Hamilton are right that cooperators enjoy the benefits of life disproportionately, then cooperation has to overcome a big problem. It is that if two people are only going to meet once, it will pay them to act selfishly. And worse, evolution has created tendencies and dispositions in us to recognize when this is true and to act on it. If I am starving and see some apples out of reach in a tree, I might ask you, a stranger passing by, if I can climb up on your shoulders to get them. But when I climb down, I might then run off, not offering any to you. Or, if I do offer them to you, you might grab them all and run off, not giving me any. It might not be the nice or polite thing to do, but it will often pay to be greedy like this. That is, unless you are going to see the person again. Now it might pay to cooperate in hopes that your partner will remember this and cooperate with you.

Robert Trivers formalized this idea in the early 1970s, calling it "reciprocal altruism," and it involves a sort of promise of exchange between two unrelated parties. I help you now in exchange for help from you at a later time. If this exchange brings you both more than you could get on your own, then cooperation should flourish. Trivers also realized that even this simple act brings with it a truckload of possibilities for exploitation. The reason is that in every act of reciprocal exchange, initially only one of the two parties benefits, and does so at a cost to the other. The helper in these exchanges is taking a risk that the help will be returned. The dilemma this causes is nowhere better illustrated than in the scene from countless detective films when the good guys hand over money to bad guys in ransom for some wretched person who has been kidnapped. The good guys hold the suitcase containing the money just out of reach of the kidnappers, who in turn hold their hostage just out of reach of the good guys. They have to do this: how can the good guys know the bad guys won't just take the money and run? How do the bad guys know the good guys have put the right amount of money in the suitcase?

For reciprocal altruism to work means that the person you decide to cooperate with also wants genuinely to cooperate with you. Trivers recognized this would create an entire evolved psychology of traits and emotions surrounding every exchange. These include friendship, gratitude, sympathy, guilt, a heightened sensitivity to cheaters, generosity, withholding of help to people who do not reciprocate, a sense of justice or fairness, and even forgiveness. Each of these either encourages us to enter into cooperative relationships or protects us once we do. The fragility of reciprocal altruism and the psychological complexity even its simple acts require might be why it is surprisingly difficult to find in nature, outside of humans. The best-known example in the animal kingdom is that of vampire bats, and even that one is controversial. Vampire bats have long lives and they spend them living in colonies in which they see the same individuals repeatedly. They prey on mammals and birds at night, obtaining a blood meal from bites they inflict with their sharp teeth. Sometimes a bat will fail to obtain a meal and return to the colony hungry. A missed meal is not a problem for a large animal such as a human, but can cause a small animal to starve to death in one night because their higher metabolic rates mean they burn through their reserves of energy quickly. Nevertheless, a well-fed vampire bat can afford to share some of its meal, and will, in some instances, regurgitate some of it to a starving one, keeping it alive.

Why help each other this way? Why not let the other die the better to reduce competition for future meals? The reason might be that vampire bats repeatedly confront a risky environment in which, on any given night, through no fault of their own, a meal might be hard to come by. In these circumstances, if I save you from starvation now by sharing some of my meal, you may live to help me another day when I have been unlucky. The example is controversial because the bats in a colony are often relatives and so the behaviors might be little more than help directed at kin. Still, the argument tells us that when we might see someone again it can pay to be kind, even at a cost to ourselves, rather than to compete or even fight, especially if that person might repay our kindness at a later time.

The expectation that you will see someone again can restrain your tendencies to cheat them, but that alone is not enough to promote reciprocity. Let's imagine I know that after another ten rounds of exchanges in which I give you something and you give me something in return, we will never see each other again. After our ninth exchange, I quietly change my tactic. I will accept your offering but withhold mine. I have taken advantage of you, and unpleasant as my behavior is, it makes sense if I am trying to maximize my payoffs. But wait, if you also know that we are going to finish after ten rounds you will do the same to me, and on our tenth exchange both of us will betray the other's trust. It gets worse. Knowing you will betray me on the tenth exchange, I will act before you and defect on the ninth. Of course, you will have done the same. So, we step back to our eighth exchange and the same thing happens, and this backward spiral goes right back to our first encounter. A fixed end point means that I will want to cheat you and worry that if I don't, you might cheat me; so we will both cheat each other right back to the beginning.

Stable cooperation requires more than just the possibility of a future encounter. It depends upon extended and durable interactions, with no known end point. As soon as one party thinks it can do better by cutting and running, it will often pay it to do so, and the cooperative enterprise can quickly unravel. Robert Axelrod called this "the shadow of the future." In an unexpected way, the future reaches back in time to influence our present behavior. We can see it as a loose or statistical way of linking the fates of two cooperators. The ease with which we appreciate the influence of this expectation and incorporate it into our own actions reminds us that many of our dispositions are those we expect of a species that has evolved to live for long periods of time around sets of people we might expect to see over and over. Of course, that is precisely what the structure of our cultural survival vehicles ensures, and this makes them a powerful source for promoting cooperation.

A sense of shortening of the shadow of the future causes us almost immediately to withdraw, even if imperceptibly, from friends who announce their intention to move away, or from work colleagues who might change jobs. It is also why lame-duck political figures are so weak. An acquaintance of mine, the head of a large research organization, once told me that for only about eighteen months of his four-year term as chief executive was he able to be effective. The first year, he said, was spent learning the ropes and gaining people's confidence. The next eighteen months were reasonably fruitful. But with around eighteen months left, people started to withdraw, and knowing he would soon be replaced they became less fearful of his reproaches. It is a dilemma faced by prime ministers and presidents around the world—at least those that are elected—and it all comes back to "the shadow of the future."

Axelrod points out that marriage exploits the shadow of the future in the wedding vows "'til death do us part." In the absence of a belief in the afterlife, this is about as long a shadow of the future as any relationship can be expected to produce. Whether this is an argument for making divorce difficult to obtain, we do know that durable exchanges can even help enemies to get along. The most striking illustration of this was the live-and-let-live system that spontaneously arose during the trench warfare of World War I. Enemy combat units, facing each other from their trenches and engaging in daily bouts of deadly warfare across no-man's-land, evolved sophisticated measures to *avoid* killing each other. Artillery would be fired at the same time every day, and always a bit short. Snipers would aim high. Famously, some of these enemies even shared Christmas gifts and played soccer one Christmas Day. Commanders had to use threats of courts-martial to break up these spontaneous reciprocal relationships.

Even with a long shadow of the future, the prospect of a defection looms large in any cooperative relationship. Someone might "forget" to return your favor, or simply make a mistake and fail to return your kindness. What should you do? If you do nothing, they might get the idea they can cheat you every now and then. Experiments with volunteers and studies using computers to simulate cooperation have shown that a simple strategy of repaying kindness with kindness and betrayal with revenge is surprisingly effective. If your partner betrays you, punish them. Axelrod called it "tit for tat." It is not very costly to you, and defectors quickly learn that they will not be tolerated.

On the other hand, Mahatma Gandhi famously pointed out that this simple "eye for an eye" strategy "makes the whole world blind" as formerly happy cycles of cooperation can disintegrate into endless cycles of punishment and revenge. Indeed, anthropologists' accounts of tribal conflict cite *tit-for-tat* cycles of revenge and counterrevenge in response to homicides and thievery as the most common cause of skirmishes and warfare between groups. In *War Before Civilization*, Lawrence Keeley recounts the history of violence between two groups in New Guinea that he discreetly labels A and B:

> Village A owed village B a pig as reward for B's help in a previous war in which the latter had killed one of A's enemies. Meanwhile, a man from village A heard some (untrue) gossip that a man from village B had seduced his young wife; so, with the aid of a relative, he assaulted the alleged seducer. Village B then "overreacted" to this beating by making two separate raids on village A, wounding a man and a woman. . . . These two raids by village B led to a general battle in which several warriors on both sides were wounded, but no one was killed . . . later . . . a warrior from village B, to avenge a wound suffered by one of his kinsmen during the battle, ambushed and wounded a village A resident. The following day battle was resumed and a B villager was killed. After this death, the war became general: all the warriors of both villages, plus various allies, began a series of battles and ambushes that continued intermittently for the next two years.

One way to end *tit-for-tat* cycles of revenge and counterrevenge might be to acquire the dispositions that encourage you to exterminate your enemy in a great rush of violence. It might be just such dispositions that fuelled the brutality we saw earlier between the Tutsi and Hutus. On the other hand, if cooperation has been valuable in our past, then we might expect it to have given us strategies of forgiveness as a way of avoiding these cycles. And indeed a strategy of ignoring the first act of betrayal, then waiting before resuming cooperation, can be shown to work better than *tit for tat*. Assume the betrayal was a mistaken judgment, a moment of weakness, or maybe just a slipup. This allows groups of generous and forgiving cooperators to overcome the occasional bout of moral weakness or mere mistake from someone within their ranks.

An even better strategy is more wily and self-serving. It is sometimes confusingly called *win-stay, lose-shift*, even though it is subtly different from the straightforward version of that strategy we saw in Chapter 3. Colloquially, we might think of it as a mild form of sulking, but with an added twist. In this setting, a person responds with cooperation so long as the other person is cooperative—if you are winning, you stay. Confronted by a betrayal, you don't respond with punishment; rather you simply withhold cooperation—this is the sulking, and it corresponds to "if you lose, shift tactics." On the next exchange, though, you switch back to cooperating, and continue to cooperate so long as the other person cooperates.

This form of *win-stay, lose-shift* is the policy we follow when we have a brief argument and then make up. It gives people a second chance, and if they take that chance, cooperation is maintained. If they don't and continue to betray you, *win-stay, lose-shift* again switches back to withholding cooperation. By merely withholding cooperation rather than overtly punishing someone who has betrayed you, the strategy avoids having a series of exchanges dissolve into cycles of betrayal and revenge. At the same time it makes it clear to cheaters and others who might "free-ride" on your goodwill that their behavior won't work, and it offers incentives in the forms of glimpses of what cooperation can look like.

But what is the twist? This strategy also has a self-serving trick up its sleeve. Every now and then *it* tries defecting. Why would it do this? Remember, natural selection is not about goodness and light; it is about strategies that promote replicators. For all the *win-stay, lose-shift* strategist knows, it might be playing against a Good Samaritan who always behaves cooperatively, or simply a gullible person who always does the nice thing. The *win-stay, lose-shift* strategy cunningly exploits them by defecting. A Good Samaritan will nevertheless continue to cooperate, so *win-stay, lose-shift*, being on a winning streak, stays and exploits them again. The success of this strategy against other forgiving but less wily strategies tells us to expect that natural selection might have built into us emotions for taking advantage of the weak or gullible—an emotion that we sadly cannot easily deny is part of our species.

an expectation for fairness

OF ALL the emotions associated with getting acts of reciprocity to work, our expectation for fairness is perhaps the most intriguing and explosive. If forgiveness and generosity are like investments in keeping a cooperative relationship going, our sense of fairness is more like a police force. It is the emotion behind our belief that it is wrong for others to take advantage of us, and it might take the form in our own minds of our conscience, telling us that it is wrong to take advantage of others. It can be schizophrenic in its effects, capable of producing violence on the one hand, and startling altruism on the other. Its violent side disposes us to punish people whose actions reveal them as selfish, and for this reason it is sometimes called *moralistic aggression*. We like to think it is something only others do, but honking horns at people who cut into traffic, or heckling people who jump lines are commonplace instances of moralistic aggression deriving from a sense that someone's actions are not fair.

Once, travelling in Vienna, I was waiting for a tram, and even as it was arriving and the doors were still opening an older woman on board, wrapped in a head scarf, wagged her finger disapprovingly and hissed at me indignantly. Evidently I wasn't leaping forward quickly enough to help a younger woman with a baby in a pushchair who was preparing to clamber down the stairs to the sidewalk. In 2009, a man in the city of Guangzhou in China threatened to commit suicide by jumping from a bridge. His presence on the bridge caused traffic jams, and eventually he was approached by a passer-by who shoved him over the edge, telling a newspaper later on that he was fed up with the desperate man's "selfish activity."

On the other hand, the altruistic side of our sense of fairness can produce surprising acts of human kindness that, if we observed them in any other animal, we would think we were watching an animated Disney film. In the late 1960s, the social scientist Henry Hornstein left wallets in public places throughout New York City. The wallets contained money and identification so that if someone found a wallet, he or she could contact the owner. To everyone's surprise, around half of the wallets—along with the money—were returned. True, wallets with more money in them were less likely to be returned, but the majority of people went out of their way to return a wallet to someone unknown to them, at a personal cost of time and effort and with no promise of any recompense.

Doing the right thing is something we take for granted, even in this anonymous situation, but in no other animal on Earth would the thought that returning the wallet was the "right" thing to do even come to mind. Why do we behave this way? The economist Kaushik Basu points out that this expectation for fairness runs deep in our minds. Consider, he says, that when you ride in a taxi and you get to your destination, if you are like most people you reflexively pay the taxi driver rather than running off. It is an action you probably give very little thought to—just what we do in such situations. Even so, Basu reminds us that this is a revealing action because most of the time no one else observes us, and the thought of running off must have occurred to everyone who has ever ridden in a taxi, even though almost none of us do it. Equally, when we pay taxi drivers, they do not turn around and demand further payment. Both of you might be behaving this way out of fear of getting the police involved or of violent reprisal by the other, but still we somehow feel such behavior would be wrong or unfair.

Are we programmed somehow to do the right thing—as when we return lost wallets or pay taxi drivers—even at a cost to ourselves, just because it is the "right" thing to do, and to expect the same from others? Social scientists and economists who get people to participate in an economic exchange called "the ultimatum game" think so. In this game, volunteers are given a sum of money, say, $100. They are told they have to give some of it to an anonymous other person, but the amount they offer is up to them. The other person can either accept the offer, in which case both people keep their portions of the money, or reject the offer, in which case neither person gets anything. Volunteers are told they will not ever see each other and that the experiment involves just this one exchange.

Now, recipients should accept any offer, as something free is better than nothing. Knowing this, the person with the money should offer the smallest amount. But neither party behaves this way. The game has been played with university students, and in cultures around the world, including hunter-gatherer societies. Time after time, recipients reject as "unfair" offers below 20–40 percent of the sum given to the first person, and both parties walk away empty-handed. Those making offers seem to expect this, and the typical offer is often around 40 percent of the total. In a related exchange known as "the Dictator game," the people are told they can offer whatever amount they wish and that the recipient does not have any choice in the matter. Offers are lower, but people still give away some of their money.

What is going on? Would you reject a low offer? If so, why? Would you give more than the minimum? If so, why? Remember you are not going to see this person ever again. Some researchers interpret our behavior in these games as evidence that humans are hard-wired for altruism—that we both offer it and expect it from others. They say our actions are governed by a principle of *strong reciprocity*, a deep moral sense to behave in ways that benefit others, even when this means suffering a personal cost. According to these researchers, our strong reciprocity is evidence that human social behavior evolved by the process of *group selection*. This is the idea we saw that natural selection can choose among competing groups of people. The most successful groups in our past were those in which individuals put aside their own interests to pull together, even when that meant sacrificing our own well-being. This altruistic behavior is supposedly what we are seeing in the ultimatum game: donors give more than they need to, and recipients expect this. When donors give a small amount, recipients punish them, even though this means that the recipient gives up some money. Strong reciprocity is, according to some social scientists and economists, why we pay taxi drivers and why we return wallets, but also why we might even go to war for our country.

Group selection can work, but its effects are weak and it will always be opposed by selection promoting individual interest over the good of the group: when everyone else is pulling together, it might pay *you* to hold back. Could it be, then, that people reject low offers in the ultimatum games, not out of any sense of duty to the group but simply because a low offer is not a *fair* way for two people to divide up money that neither one really has much claim to? My hunch is that if you are like most people reading this account of the ultimatum game, you are feeling that it would indeed be unfair for someone to make you a low offer. This is especially true in the ultimatum game experiments, because the donor and recipient know their roles are arbitrary and could just as easily have been reversed. They also know—because they have been told—that the money has to be divided. In such circumstances, it is indeed only fair to divide the money equally, or at least nearly so, if we wish to acknowledge that someone is entitled to more just by the luck of the draw.

Fairness sounds good, but still, why do we think things must be fair? And why do we turn down the offer of free money? Why not just take whatever is on offer, and walk out of the experiment better off for it? What use is it turning down what you consider to be a miserly offer to "punish" someone (and thus yourself as well) you are never going to see again? Well, when something is unfair, most of us feel an emotion of indignation or anger, and it makes us want to lash out and punish the other person. We do so by rejecting their offer. But why do we do this? If we think about it, this is a spiteful act on our part. Yes, we punish that other person, maybe we feel better for it, and perhaps the punishment makes it less likely the person will behave that way in the future. But our spite also benefits anyone else who might do business with him or her. Given that it has cost us to behave this way, this is an act of altruism on our part. We don't expect this kind of altruism to evolve because your actions help others at a cost to you.

Another possibility avoids all these problems and doesn't require any notions of strong reciprocity or putting our self-interest aside. It is that rejecting the offer signals your disapproval and sends out a message that you are not someone to be trifled with. This is just what we expect of an emotion—an expectation for fairness—that has evolved to watch out for our interests. But wait, to whom are you sending this message in the ultimatum game? You have been told the experiment is anonymous, that you will never meet the donor. Maybe, but is that how people really feel in these experiments? You can be told the exchange is anonymous and that you will never encounter the person again, but that doesn't mean you can simply switch off the normal emotions that natural selection has created in us for ensuring we are not taken advantage of in reciprocal exchanges. The experimenters who conduct these studies are, in effect, asking their volunteers to leave behind at the door all of their evolved psychology for long-term relations. Robert Trivers in his "Reciprocal Altruism Thirty Years Later" put it more witheringly, saying, "you can be aware you are in a movie theatre watching a perfectly harmless horror film and still be scared to death. . . . I know of no species, humans included, that leaves any part of its biology at the laboratory door; not prior experiences, nor natural proclivities, nor ongoing physiology, nor arms and legs or whatever."

In the real world, few interactions are of the sort concocted in the ultimatum games. Our psychology is the psychology of repeated interactions, and in that context, turning down a low offer sends a message to the person you are dealing with—and to any others who might witness or hear of the event—not to try to take advantage of you in the future. Turning the offer down might cost you something now, but it pays its way as an investment in future interactions (this is why punishment is effective in *tit for tat*: it reins in cheats). What might look spiteful is actually a way of improving your longer-term prospects. Emotions that guide this sort of behavior are important for a species like us that lives in small social groups in which people live a long time, and can therefore be expected to see each other repeatedly. The experimental situation of the ultimatum game, perhaps unwittingly, elicits the sense of having an audience precisely because volunteers are told the exchange is anonymous. It seems not to occur to the experimenters who run the ultimatum and other related games that the mere fact of telling someone their actions are anonymous is a refutation of that statement! *Someone* is watching.

To return to Basu's example, why do we pay taxi drivers? One obvious reason that separates these people from ultimatum gamers is that the drivers have *earned* their payment. Riders know this and know this will make drivers more tenacious about getting their payment. Our disposition to act fairly is also most acutely switched on in face-to-face exchanges because we have come to expect reciprocity in our dealings. But that disposition is not one of behaving altruistically because this makes groups strong or because it is the right thing to do. Rather, it is an emotion that ensures we are not taken advantage of, and in any exchange we know that the other person is having the same thoughts as us. Indeed, no one should assume that taxi drivers do their jobs for *you,* or that you have *their* interests in mind. Not too many years ago the taxi rank at the airport of one of Southern California's major cities was a marketplace, not a highly regulated economy with fixed fares. Someone wishing a ride could walk among the drivers bargaining for the best offer. Under these circumstances, the law of supply and demand is at its most efficient best. When there were lots of drivers but few passengers, fares came down. But if you were to show up when only one taxi was in the rank, you could be charged more or less whatever amount the driver could get away with.

When the artificial trappings of the ultimatum games are removed, our supposed strong reciprocity fades. The economist John List got people at a baseball trading card convention to approach card dealers and ask them for the best card they could buy for $20. In another situation, they had $65 to spend. Dealers consistently took advantage of the buyers, selling cards that were well below those values. Revealingly, though, it was especially the dealers from out of town—and thus unlikely to encounter the buyer again—who cheated buyers the most. List also produced a variant of the Dictator game in which instead of telling the donors they could offer whatever they wanted, he also gave them the option of *taking* some money from the other player. List's hunch was that this removed some of the "demand" to give money (although List must acknowledge that it might also have created an expectation to take it). His hunch proved correct. Donations by Dictators fell to less than half of what they were in the conventional setting and 20 percent of the Dictators took money.

trust and the diffusion of cooperation

OUR ATTACHMENT to fairness and justice has its origins in our self-interest, and we have seen that we can respond violently when we think justice is imperiled by selfish behavior. Trivers reminds us that victims feel the sense of injustice far more strongly than do bystanders and they feel it far more than do the perpetrators. People normally raise the issue of "fair play" when they are losing. "Envy," says Trivers, "is a trivial emotion compared to our sense of injustice. To give one possible example, you do not tie explosives to yourself to kill others because you are envious of what they have, but you may do so if these others and their behavior represent an injustice being visited upon you and yours."

Still, the remarkable feature of human cooperation is that, as in the case of a suicide bomber, it is not restricted to reciprocal relations between pairs of people. Many, perhaps most, of our day-to-day interactions are reciprocal—such as when we buy a loaf of bread—but our cooperation routinely balloons in complexity well beyond exchanges between pairs of people. If most altruism and cooperation in animals can be arrayed into two levels—that driven by helping relatives and that which prospers from direct reciprocity or exchange between two parties—there is a third level to human cooperation that is diffuse, symbolic, and artfully indirect. On a day-to-day basis, we act in ways that cannot possibly be directly reciprocated, such as when we routinely and unself-consciously hold doors for people, form lines obediently and admonish those who don't, help the weak, elderly, or the disabled, return items of value, aid people in distress, pay taxes, and give to charities.

If our helping, sharing, and altruism stopped at these acts, we might be happy to let it go there as a charming peculiarity of our nature, something born of our ability to transcend our biological existence, to be empathic and to understand others' needs. But while the cost of returning a wallet or holding a door may be small, helping someone in distress might not be. As we have seen, our style of help moves beyond the *eusociality* of the social insects to the *ultrasociality* seen only in our species: the most vivid and outré form of our altruism comes to us in battlefield accounts of soldiers who fall on a grenade, charge a machine-gun nest, help others to safety under fire, or fly an airplane Kamikaze-style into an enemy ship. Few of these will have left offspring to regale with stories of their heroism, or if they have, those offspring will suddenly find themselves without one of their parents.

The next chapter asks how this kind of strange selfless behavior can arise in a Darwinian world in which it is the survivors who float to the top. We have seen here that there are some who think we are guided by a psychology of doing what is best for the group, even at cost to ourselves. But an unusual idea from genetics discussed in the next chapter shows how costly acts ranging from so-called honor killings to a disposition to suicide can, remarkably, be understood as self-interested behavior.

CHAPTER 6

Green Beards and the Reputation Marketplace

That human society is a marketplace in which reputations are bought and sold

FOR MOST PEOPLE the sight of seeing their nation's flag raised, the sound of their national anthem being played, watching their nation compete against others in international events, or the loss of one of their soldiers in battle causes a familiar emotion. We often call it "nationalism," a diffuse and warm pride in one's country or people, and a tendency to feel an affinity toward them that we do not always or so easily extend to others from different nations or societies. It has been the great and sometimes terrible achievement of human societies to create the conditions that make people share this sense. It can get us unwittingly to practice a kind of cultural nepotism that disposes us in the right circumstances to treat other members of our nation or group as a special and limited kind of relative, willing to be more helpful and trusting than we would normally be toward others. It is the emotion of encountering a stranger on holiday in a foreign land and finding them to be from your country. But it is also the emotion that gets the people of one nation to cheer while those of another suffer from a deadly act of terrorism.

These are all descriptions of our ultra-social nature, a nature that sees us acting altruistically toward others, especially other members of our societies, without expecting that help to be directly returned. That altruism ranges from simple acts such as holding doors and giving up seats on trains, to volunteering your time and contributing to charities, but also to risking your life in war for people you might not know and are not even related to. None of these acts is directly reciprocated, or not necessarily so, and the risks of exploitation by others who do not share your altruism far exceed those of simple reciprocal exchanges between two people. Our altruistic dispositions are so strong that they even extend to helping other species. What other animal would ever put in the time and effort to save tigers, or adopt an abandoned dog, or call out the fire brigade to rescue a cat up a tree?

Where does this sense of cultural altruism come from; why do we feel it so strongly and so *naturally*? How do I know whom to extend these feelings to and in what circumstances, and why am I more likely to trust someone from my own group even if I have never met them? How could it ever be in your interest to risk your life in war? We saw the outlines of an answer to these questions in Chapter 2, and here we will see just how easy it can be to get this cultural altruism to evolve.

green beards, venture capitalists, and good samaritans

IN THIS chapter we will adopt an approach that uses thought experiments and hypothetical scenarios to understand our evolved psychology. Evolutionary biologists often use such an approach in an attempt to simplify what can seem like overwhelmingly complicated situations, such as our public behavior. The risk is that the simplifications can seem to remove any realism from the examples. But the rewards of this thought experiment approach are that it often returns insights that we hadn't expected, or ones that fit so well with how we actually behave that we think the simplifications have captured something fundamental about our underlying dispositions and motivations.

We want to think about how a disposition to behave altruistically toward one another could evolve among an imaginary group of people initially lacking that disposition. This imaginary group of people could represent our "state of nature" before we learned how to cooperate with people outside of our immediate families. To see how a disposition toward cooperation might evolve in such a group, consider that a gene, an idea, or just an emotion arose in one or even a few of them that caused them to help people whom they thought were "helpers" like themselves. It could be as simple as just some good feeling that you get from helping these people. It is not a disposition to help any one individual in particular, or even to expect him or her to help you in return. It is a disposition to help people whom you think are helpers like you. You don't need to know why you have this disposition, or where it comes from; it could simply be something you feel.

Let's assume the value of the help the altruists provide exceeds the costs to them of providing it. Maybe this is something as simple as allowing someone to shelter in your house during a lightning storm. It costs you almost nothing, but it could save that person's life. If people with this disposition can identify each other and then behave as we have assumed, two things will follow. One is that people carrying the disposition will be more likely to survive and prosper from the mutual help they provide, and this will make it more likely that the gene or idea survives and can get passed on to someone else. If it is a gene, it will spread in the usual way as people who carry it will produce more offspring. If it is an idea, it could spread by people observing others and then copying this successful strategy—it would then spread by social learning. The second thing we can expect is that as this tendency to help becomes widespread in this group, cooperation will come to take on the diffuse character that we recognize in some of our own actions. We could even think of the emotion as something akin to modern feelings of nationalism, but at this stage we might think of it as a kind of tribalism, a partiality toward others in your group, or others like you. Being among these people but not being a helper would be like being at a party without anyone noticing you.

It is an idea of such simplicity that we must wonder if it could lead to anything of importance in real social life. In fact, William Hamilton anticipated our imaginary scenario in 1964. Hamilton imagined a gene that had three simultaneous effects: it causes its bearer to have some sort of recognizable external marker; the gene grants the ability to recognize others with the marker; and it gets its carriers to target assistance toward others who have it. Richard Dawkins later named these hypothetical genes *greenbeard genes* as a vivid way of calling to mind the mechanism by which such genes might recognize copies of themselves in other bodies. The idea is that those with the gene produce a conspicuous marker that allows those carrying the gene easily to spot each other and then direct their assistance toward them, and only toward them. We needn't take the green beard literally; it is simply any kind of conspicuous marker; maybe ginger hair or blue eyes, misshapen ears, or the wearing of a particular kind of hat.

It is tempting to caricature the greenbeard idea as little more than an amusing anecdote. And it is fair to say that greenbeard genes have long been regarded as fanciful playthings dreamt up by theoreticians. The charge against them is that they require an implausible combination of three effects from a single gene. Why should a gene that produces some conspicuous marker also be linked to an ability to recognize others with it, and then to behave altruistically toward them? There is no reason. If the gene did produce more than one effect, it is just as likely to be for a taste for lemonade or a preference for cloudy skies as for helping others. If it is easy for us to imagine a gene having all three properties, this might be because we have minds that have already managed to understand the power of cooperation. But we must act as if this system could arise in other animals that do not have our sophistication and therefore the single gene would have to cause all three effects at once.

In fact, some remarkable evidence lends plausibility to the idea of greenbeard genes, and hints at parallels to human behavior. There is a species of fire ant (*Solenopsis invicta*) in which workers carry a gene that causes them to kill queens in their nests, but only queens who don't carry a copy of this gene. Most ant species have just a single queen, but in this species there can sometimes be more than one. The ants recognize queens that lack this particular gene because those queens produce a chemical that appears on their outer surface. Queens that do have the gene don't produce this chemical. Workers who carry the gene recognize the chemical, and then attack and kill the queens that produce it: her green beard is her death warrant. These same workers avoid attacking queens that don't produce the chemical.

This is just the reverse of the usual greenbeard story, but the mechanism is the same. Workers carrying a particular genetic trait are able to recognize queens that don't, and execute them. Their behavior is an act of altruism toward their brothers and sisters, who will make up a sizable proportion of the workers in the nest. It is an act of altruism because by killing the queens who display the chemical, the killers ensure that she will not produce offspring that would compete with their siblings.

Whatever one thinks of this example, it is difficult to avoid the suggestion of parallels to human xenophobia and bigotry—the ants direct their hostility toward someone else solely on the basis of some identifiable external marker or characteristic. Another instance of a greenbeard gene gets those with an external feature to help each other. Among the single-celled amoebae or slime molds, individuals normally live a solitary existence. But at times of food shortages they form into towers of many thousands of individual cells. Amoebae near the top of the tower form part of the fruiting body or spore cells that will reproduce and form the next generation, but the others in the tower will die. Only a lucky few get to reproduce so this means that, for most of the amoebae, building the tower is an altruistic act. But in one amoeba species the altruism is more focused. Some individuals carry a gene that makes a protein that is expressed on their outer surface. This protein causes them to stick to other amoebae that express the same protein, but not to amoebae that lack it. Experiments show that by sticking to each other, these amoebae exclude other amoebae that aren't sticky, and this means the sticky ones are more likely to get into the fruiting body at the top. The gene for this sticky protein simultaneously fulfills the roles of producing a marker, recognizing those with it, and then assisting them. It creates a prejudice to favor those who are like you.

Still, our lives are not as simple and rule-bound as those of ants and slime molds. Our social lives are complicated: we miss things or mistakenly help selfish people, we blunder into a situation not knowing what to do, and people try to deceive us—impostors, con men, liars, and other tricksters are always lurking around looking to take advantage of someone's good nature. Can this greenbeard altruism still evolve? Evolutionary biologists often study questions such as this about the success or not of various strategies, using mathematics to represent the interactions among groups of imaginary players who suffer or enjoy imaginary outcomes. These mathematical models describe different strategies that players can adopt, and then ask how these strategies fare in competition with one another. The strategies can even be altered to see how this affects their success. I have studied just such a model for this case and it turns out we can make mistakes, or simply fail sometimes to provide help, and still the greenbeard altruism gene will prosper so long as two things are true. One is that the altruists need to help each other at least some of the time; and the second is that altruists (the greenbeards) help other altruists more often than they help selfish or non-cooperative people.

This makes sense. If altruists help each other even just some of the time, but avoid helping non-altruists or non-cooperators, collectively the altruists will be better off than the non-altruistic or selfish players who never help each other. Avoiding selfish or uncooperative people is important because helping a non-altruist means helping a competitor to the fledgling mutual aid society. This raises the possibility that natural selection has favored in us a heightened sensitivity to detecting what we might think of as social cheats or free riders, people who might take advantage of others' goodwill without intending to return it—they are the enemies of the mutual aid societies. Remarkably, the evolutionary psychologists Leah Cosmides and John Tooby have proposed just this capability in studies of human cooperation. Here is one of their examples. You walk into a bar and want to find out who is following the rule of having to be over the age of eighteen to buy an alcoholic drink. There are people of all ages in the bar; some have drinks and some don't. Which of these people should you check to work out who is following the rule? Should you target young people or people with drinks? Most of us instinctively realize that it is people with drinks—and especially young ones—who can potentially test the rule. You could find out the ages of all the people without drinks, but this alone would tell you nothing about whether any of them individually is likely or not to follow the rule. And this is Tooby and Cosmides's point: without even thinking about it, our brains instinctively know how to detect the cheats among us.

The rules that guide our cooperative behaviors toward others lead to an important "principle of information," which if followed not only makes altruism possible but profitable to the altruists. It says that if we know enough about someone, we can make a decision about whether to cooperate; but if we don't, it is better not to cooperate, because you might just be helping a selfish person. We can think of the principle this way. Let's allow i (for information) to stand for a number that can range between 0 and 1. An i of 1 says you are certain the other person is an altruist and an i of 0 says you have no confidence at all. Then the principle of information says our actions should be guided by this rule: that i multiplied by the *benefit* you provide to someone else must exceed the *cost* to you. This can be written as $i \times b > c$.

When we are certain the other person is a cooperator (i is 1), we should help if the help we provide more than pays for the costs to us of giving it. This rule makes sense: only when the benefits the altruists dole out to each other make up for the costs of providing them do they prosper as a group. The action might be something like the example I gave earlier of providing shelter to someone during a storm, or you might give someone some information or hold a door for a weak or disabled person. These actions cost you little but can be of big help to the other person.

The principle of information reveals more about our evolved psychology when we consider what it says about how we should behave when we don't know much about another person. By "should" here, we mean behave in a way that will serve our interests. Not knowing the other person well means that i is less than 1. Following the rule, it says we should never cooperate when $i = 0$. From our definition of i, when $i = 0$, you have no information about the person, and this means you are just as likely to help a non-cooperator as a fellow cooperator. For example, parents often tell their children, "Don't trust strangers." There is no particular reason to believe that a stranger is someone who will take advantage of you (a stranger to you is, after all, familiar to someone else). But not knowing what strangers are like, this has probably been a useful rule to follow in our past, and it is the rule the principle of information predicts.

The situation that is most relevant to our everyday lives is when i lies somewhere between 0 and 1. When i is a small number, it tells us that we are not certain whether the person we are helping is a cooperator. A small i also makes it hard for $i \times b > c$, and so the rule tells us to avoid cooperating with people we don't know very much about. The reason is that in effect the benefit you provide has to be discounted by the likelihood that it might go to someone who does not share your altruistic disposition. A more optimistic view is that the rule tells us our best strategy is maybe to cooperate but only on small matters. That way if we are wrong about the person, they don't gain much at our expense. This gives cooperation a way of gaining a foothold and then larger acts of cooperation can follow if the information you gain leads you to believe someone is a cooperator.

The principle of information can be seen as a psychological disposition that puts great emphasis on identifying traits in others that we think tell us something about the likelihood that they are a fellow cooperator. It might be why we are so sensitive to such things as how people dress or speak, and what their manners are like. These are not necessarily good indicators of what someone is really like, but natural selection will have favored any tendencies with a high enough accuracy rate that they have worked in the past. Our simple model tells us we don't even have to be accurate; we just have to be more likely to help someone who is an altruist than someone who is not. Natural selection seems to be telling us that it might often be useful to "judge a book by its cover," at least if that is all you have to go by. Of course, in any given circumstance, other factors can override these rules. But our all too easily felt prejudicial emotions might just be natural selection's way of making sure we follow an evolutionary rule that worked for our ancestors. If in doubt, the rule says, it is better to avoid the risk. Prejudicial emotions have no place in the modern world, but wariness of strangers or of people unlike ourselves might have deep origins in the evolution of cooperation, not so different from what we saw with the fire ants.

Incidentally, evolutionary biologists will recognize the principle of information and its simple rule as the same one that governs our altruism toward relatives. One of the most famous rules of evolutionary biology is called Hamilton's rule, named for the same William Hamilton who first thought up greenbeard genes. The rule says we should help relatives when $r \times b > c$. Here b and c are the same as for greenbeards but r is now your genetic relatedness to someone else, rather than the "information" you have about them. By "relatedness" we mean roughly the percentage of genes you have in common by virtue of sharing parents, grandparents, or other ancestors. Hamilton's rule tells us why siblings ($r = 1/2$) are more likely to help each other than they are to help their cousins ($r = 1/8$). Hamilton's rule also tells us why strangers (assume $r = 0$) are unlikely to help each other, and why clones (such as the cells in your body, $r = 1$) are only too happy to assist each other.

In fact, the connection between the greenbeard principle of information and Hamilton's rule is closer than we might expect. The quantity i in the principle of information can be thought of as your *relatedness* to someone else on the altruism gene (or idea) itself. The greenbeard is just a model of one gene or idea helping *its* relatives and the green beard is how it recognizes them! You might be unrelated to someone else on all your other genes, but you just might share this one gene (or idea). If we think about it, the familiar emotions of nationalism might be how our impulse to act altruistically toward certain others whom we think share our dispositions manifests itself. If this sounds familiar to some of the discussion in Chapter 2 about our "special and limited" form of cultural nepotism, that is so. The green beard is a way of identifying someone who is related to you at your altruism locus.

We should not be lulled by all this discussion of cooperation into thinking that the greenbeard style of altruism simply helps others. Our cooperation evolves because altruists effectively surround themselves with other altruists and thereby get back as much or more than they put in. In fact, the self-interested nature of greenbeard altruism is revealed most clearly when we realize that once cooperation spreads and becomes the norm in society, a new kind of cooperator can arise that is less discriminating than the greenbeards. This cooperator doesn't look for others like itself, it just indiscriminately helps everyone. We can think of them as *Good Samaritans*—they are kind without first judging what someone else is like. If cooperation is widespread anyway, the Good Samaritans will fare no worse than the greenbeard altruists. Good Samaritans might even do slightly better by virtue of not wasting time and effort trying to work out who is a cooperator and who is not.

However, the Good Samaritans create a problem for the altruists who follow the greenbeard rule. The presence of Good Samaritans makes the society once again vulnerable to selfish or non-cooperating people who can flourish by taking advantage of Good Samaritans. These could even be other green beards who might have a tendency to cheat when they can get away with it. Selfish people will grow in numbers, but at the expense of the Good Samaritans, whose numbers will now dwindle. The greenbeard altruists on the other hand are less vulnerable, just so long as they continue to recognize each other even just a little bit better than they recognize the reemerged selfish cheaters. If they can recognize the cheats, they will once again drive them out, or at least down to low numbers. At this point Good Samaritans can reappear, and this cyclical process will go on forever, with societies always containing a majority of greenbeard cooperators, but some numbers also of Good Samaritans and selfish individuals.

This tells us two things. One is that greenbeard altruists—and by implications ourselves—are not so much "good guys" in white hats who help everyone as "shrewd guys" who direct their aid strategically. When it comes to cooperation, we are far more like venture capitalists than Good Samaritans. We are willing to invest in other altruists like ourselves because we derive returns from having them around. Our investment is in the society of cooperators, not in anyone in particular. Cooperation itself becomes something like a "common good,"—a resource, like common grazing land, that everyone benefits from but that must continually be replenished to avoid being used up. The second thing we learn is that our cultural setting will always strongly favor those who are vigilant about and good at detecting selfish cheats because they continually deplete the common good of the cooperative society. Good Samaritans may inadvertently imperil our societies by helping people who may not always be deserving of their aid. It is no secret that the Mother Teresas of the world are often treated with a hint of ambivalence, and this could be one reason why. Another we will come to later on.

reputation and the green beard

SO FAR, we have ignored a nagging problem. If we are like venture capitalists when it comes to cooperation, how do we know whom to invest in? The idea of a green beard provides a useful image, but what if people could produce it without actually being a cooperator. These people would attract all the benefits of being a cooperator without ever having to pay the costs of being altruistic toward others. This is one of the failings of the greenbeard idea, and not just in the setting of human cooperation. If anyone can grow a green beard, then there is no way to tell altruists from selfish cheaters and altruism of the greenbeard variety will never evolve.

This criticism raises the problem of designing a signal—the green beard—that others can trust as a reliable and honest indicator that you are an altruist, and not a selfish social cheat. How *do* we know whom to cooperate with, apart from looking for clues of shared values? This is the same problem we encountered in Chapter 4 in trying to understand one possible role of extreme religious beliefs and practices. Our interest there was how observers can use these extreme acts to gauge someone's commitment to a group. Amotz Zahavi's handicap principle and Thorsten Veblen's idea of conspicuous consumption showed us that it is precisely because some of the things that animals and people do are wasteful and even reckless that we know whatever they are doing says something believable about them.

For instance, a peacock's giant tail tells us that he is healthy because it is so costly and wasteful to produce. Peahens know that only the fittest males can afford to make the biggest tails: less fit males can try but will probably die from the burdens of their giant tails. Similarly, we speculated that one possible function of religious belief was to signal your commitment to the group—religious beliefs and practices such as self-flagellation, memorizing scripture, or fasting can be costly, and for that reason can become believable signals. We also instinctively recognize Veblen's idea in the conspicuous consumption of wealthy people. A huge diamond engagement ring or an expensive car have little functional value over other rings or cars, and that is precisely why they are good signals that someone is wealthy—how else could they have so much money to waste? They are useful signals because they are expensive ways of being useless.

Zahavi and Veblen even went a step further and often emphasized that the best signals were those that were directly relevant to what an animal was trying to tell you. One of Zahavi's favorite examples was the "stotting" behavior of some gazelle species when being chased by lions. As the lion pursues the gazelle, the gazelle punctuates its escape with stots, a series of little pronks or prongs that involve jumping straight up into the air, legs stiff, and lifting all four feet off the ground. Why would they do such a ridiculous thing? Wouldn't it be far better for the gazelle to put all its energy and time into running away? About the only worse things for the gazelle would be to stop running altogether, or turn around and run *at* the lion. Another of Zahavi's examples is the curious behavior of the skylark, which when being pursued for its life by the predatory merlin seemingly looks over its shoulder and begins to sing.

All of these actions use up the very resource an animal needs to escape or protect itself, and that makes them informative about the animal's abilities. The gazelle is telling the lion that it is such a fast runner it can afford to waste valuable time by jumping vertically into the air, and yet still get away. The skylark is telling the merlin that it can waste its precious breath. Both actions are directly relevant to what the animal is trying to signal to the pursuer—fast running, or strong lungs for flying. If they are lying about their abilities, these actions are the ones that will get them killed. And indeed, not all gazelles stot, nor do all skylarks sing at the approach of a merlin, but as Zahavi would expect, those that do are more likely to get away. To understand the gravity of stotting and singing while being chased, consider that next time you are chased by a mugger, you punctuate your escape with little jumps up into the air, and even occasionally burst into song. Veblen was fond of saying that gentlemen carried walking sticks to show that they could get by without the use of one of their hands. And of course if you are wealthy, what better way to show off how much money you have than to use it up on wasteful *grands projets* such as a folly in your garden, or a private art collection? Indeed, it is sometimes said that the perfect garden folly is the one that drives you bankrupt. Why? Because to go bankrupt shows you can make the money back some other day.

The handicap and conspicuous consumption theories tell us something remarkably simple and yet profound about how altruists can go about identifying other altruists. If you want to know who is an altruist, look at who is behaving most altruistically! Altruistic acts are by definition costly because they aid someone else at your expense. They can be reckless, such as jumping into a river to save someone's life or pulling someone from a burning house. And they can be wasteful of your time and money. If earning the badge of altruist takes enough effort, then anyone who puts in the effort to do so is, by definition, an altruist.

This gives us insights into all the peculiar little acts of altruism that we routinely perform in society, such as holding doors, standing aside for people, giving up seats on buses or trains, helping the elderly, contributing to charity boxes, or even risking our lives to save animals. It is not just that we are "nice." Once it is up to others to grant you the badge of being an altruist, altruists have no choice but to try to stand out from the crowd, and it falls on you to do whatever you can to convince your societal audience to grant you the label. The value of cooperation means that something of an altruism arms race arises, forcing would-be altruists to acquire "long tails" of altruism to compete with other altruists trying to do the same. To compete at the very highest levels in the altruism competition, we have to demonstrate something akin to altruistic conspicuous consumption—we have to become altruism *show-offs*—by doing all the usual things, but also by volunteering our time, joining local community projects, tithing to a church, helping others in distress, or making large contributions to philanthropic organizations.

That we all possess these dispositions to some extent is a measure of how important cooperation has been in our past. Group action—whether it be in hunting or foraging, attacking another tribe, or simply moving to new lands—has been our species' hallmark and secret advantage over others. In turn, if cooperation has been valuable to us as a species, it is also worth protecting, and we can expect groups or just about anyone judging you to require lots of evidence before accepting you (the principle of information again). A group might subject you to long or costly periods of initiation, during which someone is held in a state of suspended worthiness as others evaluate their tendencies, consistency, and integrity. Newcomers to villages are often still remembered as "the newcomer" even many years later. We impose these trials on newcomers instinctively, even though there is no reason to believe that strangers are any less reliable per se than people we know. What these trials do is allow us to develop a good estimate of the quantity i in our model; put in more everyday terms, these trials are a measure of just how long our altruistic tails have to be to be believable.

In Sebastian Junger's example of the combat platoon in the Korangal Valley of Afghanistan, the men's survival depended upon each of them being committed to giving their lives for the others. But how could someone new to the group be trusted? The men had devised a pragmatic measure to find out. Junger describes the "blood in, blood out" rituals, in which men arriving to the platoon would be subjected to severe beatings. Junger witnessed one such beating, of a lieutenant newly assigned to lead the platoon. Without any indication of what was about to happen, the men surrounded the new lieutenant, knocked him to the ground, and beat him severely. Attacking an officer violates one of the military's deepest prohibitions. The beatings therefore signaled the seriousness with which the men took the commitment. It was a way of showing what they were willing to risk—a court-martial for their actions—and what they could do to someone who wavered in their commitment.

The lieutenant was not disposed to press charges because he knew his life depended as much on getting his men's support as theirs did on his. Enduring this painful and costly beating was a way of purchasing the men's trust. The degrading things that men in combat platoons routinely say to each other can also be seen as painful tests of commitment. If being told by another man that he would like to have sex with your mother is enough to lose your support, then it might be that bullets zipping over your head will also cause your support to waver. Of course, once men are accepted into a combat group, it is their behavior in battle that is the true measure of their commitment, because that behavior cannot be faked. Courage and valor in battle are about the longest tails of altruism one can produce. Ultimately, altruism itself is the best measure of whether someone is altruistic, and men in battle groups know this.

Thinking of our actions, and the reputations they build and maintain as green beards that advertise our altruistic nature, might help explain a curious phenomenon most of us routinely encounter in our everyday lives. It is that when people witness helpful and cooperative behavior, they are more likely to produce it themselves. If you see someone hold a door, or someone holds a door for you, if you are like most people you will look over your shoulder to see if someone is following, and if so, you will hold the door for them. The same is true of seeing people donate to the Salvation Army, give blood, or stop to help someone change a punctured tire. Witnessing any of these events makes you more likely to be helpful yourself. It is why buskers or street performers put money in their hats before they even start to play: it shows that others before you have been kind enough to contribute.

Social psychologists have long struggled to find an explanation for this effect. Couldn't it just be that it makes us feel good to do these things and seeing someone else's helpfulness or generosity reminds us of that? Maybe, but this again just raises the question of why we have minds that like to be nice or generous. A different suggestion is that when we observe others do something, it reduces ambiguity about what is appropriate behavior, or it reassures us that the behavior is safe or acceptable. This might be true in some circumstances, such as providing minor medical help. But the greenbeard idea gives a more general and natural explanation. It says we help more when others do so because their acts of altruism raise the bar for us; their actions take some of the shine off of our reputations. We must respond in kind to purchase a bit more reputation, as a way of staying competitive in the "who's an altruist?" game, and this means showing your helpful side. This could be one reason why public charitable auctions work so well, and it could be the second reason why the Mother Teresas of the world are sometimes treated with ambivalence: they make the rest of us look unhelpful by comparison.

the reputation marketplace

OUR SOCIAL systems of cooperation and helping are revealed as sophisticated marketplaces, capable of generating both individual returns and goods that benefit others. They work like a monetary system, with our personal reputations acting as the currency we use to buy trust and cooperation. Reputations are valuable, so we have to earn or pay for them. We do so by engaging in altruistic acts, costly to us but beneficial to others. Once purchased, a good reputation can then be used to buy cooperation from others, even people we have never met, just as we can use money to buy goods from people we have never met—economists call this *transferability*. In this sense, we can see how the abstract and symbolic idea of a reputation promoted trust and cooperation in our societies just as money can now. Both make it easier to purchase goods or services from someone you might not know. The transferability of our reputations occurs routinely and without us even being aware of it. For example, normally when describing someone to a third party, among the first things you will say are about that person's reputation, such as, "He's a good fellow," or, "She's very friendly," "You can count on her," or, "He's awkward and difficult."

Like money, reputations get used up. They will decay as people's memories fade; you might do something to harm your reputation; you might encounter people who have never heard of you, or someone might opportunistically attack your reputation for their gain. Malicious gossip has been used for tens of thousands of years, but has more recently ascended to uncharted heights (depths?) as people now use the Internet in strategic ways to attack others' reputations. This is particularly problematic in the hotel and restaurant industry, where a proprietor will hire people to write unflattering accounts of a rival's establishment, and the reports can be read by anyone in the world. Apart from this being a distasteful thing to do, it tells us just how valuable and transferable a currency our reputations are. Indeed, our reputations represent the first "monetary union" or single currency, working in every country. We could even see them as a form of credit, because a good reputation might allow you to negotiate an exchange on the promise you will produce your part of the bargain later (in this context it is noteworthy that "credit" derives from the Latin *credo* or "I believe"). On the other hand, our reputations' vulnerability to being attacked means we need continually to top them up by regularly engaging in altruistic acts. Such a need might lie behind much of our simple but overtly polite behaviors.

In the modern world, we can see reputation building as a fundamental part of our economic behaviors. You might buy something from me and then later return it, slightly used, meaning I cannot resell it. Why should I take it back? My act of altruism toward you might more than pay its way because you will now tell people what a fair shopkeeper I am: the transferable reputation I have earned from my altruism purchases others' trust in me and attracts their custom. So fundamental are our reputations to our social systems that comparing them to a monetary system is not merely apt, it should probably be made the other way around. Money, like reputation, is an abstract system of trust, in which possession of something that has very little intrinsic value comes to stand for something of potentially much larger value. But of course monetary systems only arose much later than our modern human social systems; they make use of the abstract and symbolic thinking that our cooperative systems might have put in place by sometime around 160,000–200,000 years ago, and they piggyback directly onto systems of reputation. And, while you can hold money in your hand, reputation can only be held in your brain.

We should recognize the transferability of reputation as one of the chief ways that the altruism we are trying to understand here differs from acts of reciprocal cooperation. That system is one of a barter economy, in which one person provides a good or service to another in exchange for a good or service from them deemed to be of equivalent value. You might give me the tanned hides from the game you have killed. I will give you back some clothing I have made from them, keeping some for myself as payment for my services. You and I both get clothing, but neither of us gains beyond our mutual exchange. It is a system that almost certainly favored the development of the psychological mechanisms that would later allow us to make the leap to buying and selling our reputations. For example, if you could acquire a reputation as someone good at hunting, many people will come to you offering to make clothing and you can choose the best deal. And if I can get a reputation for making good clothing, many people will come to me with their hides and I can negotiate a good rate of exchange for the clothes I make. An economy is born.

The returns from moving beyond mere reciprocation should also tell us of the fundamental role that language plays in our system of cooperation (and in our economies). Language unlocks the benefits of cooperation by allowing precise exchanges to be negotiated, and by allowing reputations to encourage good deals or help to block bad ones from going through. It increases the exposure of your actions—acting as an *enhancer*—making acts of philanthropy, friendliness, helpfulness, courage, and bravery play to a far wider audience than those who merely witness your actions. Indeed, there is reason to believe, as we will see in Chapter 8, that this is why language evolved: it is principally a social technology for managing and exploiting the benefits of reputation and the cooperation it enables. The lack of language is the reason why no other animal can practice this powerful form of cooperation.

To see why, imagine the following scene early in our evolution when language skills were in the process of evolving: You are good at making arrowheads but hopeless at making the finished arrows. You need at least twenty arrows to go hunting, so you are looking to exchange arrowheads that you make for the wooden shafts with flight feathers attached. Two other people you know of are good at making the wooden shafts but not good at making arrowheads. One of them has very poor language abilities. You approach that one and lay down a pile of arrowheads in front of him. You are hoping he will get the idea that you want to trade him arrowheads for finished arrows, splitting them fifty/fifty between the two of you. But he thinks the arrowheads are a gift, smiles, takes them, and walks off. You pursue him, gesticulating, a scuffle ensues, and he stabs you with one of your own arrowheads.

Replay that scene now and imagine yourself approaching the second person. His language skills are good. You lay down your arrowheads saying, "I'd like to trade arrowheads for finished arrows." He looks at you and says, "Give me your arrowheads. I'll fit the wooden shafts, and give you half the completed arrows back in a week." This is a good start, but you face a dilemma. Will the arrowmaker actually deliver on his promise or just walk off with your goods? Not knowing, you consult acquaintances, who assure you the arrowmaker is trustworthy, and the deal goes ahead. It is just possible this deal could have been completed without language; but with it, the deal could be negotiated. The reputations gained from previous exchanges with other people have eventually brought each of these people a mutually beneficial deal.

The analogy to a monetary system can also offer an alternative to a controversial view about our psychology. At a distance, our generalized and diffuse form of altruism or cooperation can give the appearance that we are motivated by a deep commitment to help others, even at our own expense, including in some extreme instances giving our lives for our nation or society in war. Some anthropologists and economists think this is an indicator that we have evolved to do things for the good of our groups. But the green-beard view is that we behave altruistically, but we don't do it for the group. It is precisely the most costly acts—the ones least likely to benefit us individually—that most efficiently purchase reputations, even if they look as if we do them for the group. Again, the analogy to money might help: when we earn money by working (analogous to earning a reputation), no one says we do it for the good of the group. Even so, the system of earning wages and then exchanging money for goods and services could be said to benefit society because it helps to build an economy; but that is not our motivation. Our motivation is personal gain. Similarly, there is nothing in the notion of buying and selling your reputation that says we do it for the good of the group.

morality, shame, honor killings, and self-sacrifice

A MEASURE of the value of cooperation is contained in what we are willing to give up to keep the system working. Norms of behavior effectively ask us to give up the right to act as we please, and some might require effort to master. Maybe it is customs in matters of dress and appearance or even some rules about how to behave or what you can and cannot eat. Following such norms becomes a signal of your commitment to the group. This is especially true when those norms are arbitrary and have little obvious value, like holding your knife and fork a certain way, or removing your hat upon meeting someone. The more arbitrary they are, the more your adherence to them reveals your commitment because of the time you have had to spend to learn them. Their arbitrariness also makes them useful for identifying outsiders. I was once at a formal dinner at an institution that followed a particular rule about how to arrange your knife and fork across the dinner plate to signal that you had finished that course of the meal. A guest to the institution that evening didn't know the rule and no one had told him, perhaps because he did not speak English—the language of the rest of the diners. The waiters, obedient to the rule, would not remove his plate and bring everyone their next course until he had done so. He and the rest of us sat there for some time until eventually someone discreetly showed him what to do.

A conformity to arbitrary norms, like the commitment to an unknowable religious proposition, demonstrates that you are the kind of person who can be counted on to make a commitment, and the more so the more mysterious or arbitrary is the norm that you follow. But it might be in our adherence to some so-called morals that we see the true value of our societies, at least to the cooperators. Morals are accorded a special status that exceeds that of mere norms. They are seen as offenses against reason, truth, and good conscience, and usually include such things as adultery, stealing, and murder. But there is nothing special about morals—they don't exist "out there" in some Platonic heaven—except that they are often things on which there is widespread and vigorous agreement, and then only perhaps because we would not like the behaviors they prohibit to happen to us. Most people would agree that stealing and killing are wrong, and even morally wrong, but it is never very clear what we mean by *morally* wrong: humans beings kill each other regularly. We do so in brutal and violent ways, often in large numbers, sometimes sadistically, and we even heap praise on people who kill when it is done against someone we deem to be an enemy.

So, morals must confront the charge that they are rules that society merely makes up to serve its interests. But what or whose interests? Stealing and killing, far more than most acts, imperil the altruism that flows from mutual trust and good reputation. Societies need to banish them from people's minds so that trust can develop and then flow. Elevating these actions to the level of morals is a way a society overtly advertises that the loss of reputation and freedoms from being ostracized, punished, jailed, or even executed will be great. The ease with which most of us can be persuaded to give up the potential benefits that might come from violating them—from stealing or killing—is a measure of just how valuable our societies can be to us. It behooves our parents to transmit this valuable information to us. The proof that morals are of local value and not some deeper metaphysical entity is that as soon as we move outside that cultural wall of our own particular survival vehicles, the "morality" often comes crashing down.

Shaming people is just a way of taking their reputations away, and once again shows us how powerfully motivating the desire to cling to a good reputation can be. During World War I, some older women in English villages used a threatened loss of reputation to shame young men to go to the deadly trenches of that war. What these reluctant young men feared nearly as much as going to war was the white feather of cowardice these women dispensed when they encountered young men in their villages. This was a form of moralistic aggression, a rebuke for a perceived shirking of a young man's cooperative duties. And it shows that our societies are perched on a shaky pillar of conflicting interests. What was best for these women—being protected by vulnerable young men—was less attractive to the men asked to provide the protection. Or, as the World War I poet Siegfried Sassoon put it in his poem "Suicide in the Trenches":

You smug-faced crowds with kindling eye
Who cheer when soldier lads march by,
Sneak home and pray you'll never know
The hell where youth and laughter go.

A disturbing measure of the value of a good reputation is revealed in its ability to motivate some of our most extreme and repugnant behaviors as a species. Human parents rarely kill their own children. When they do, it is often the result of some sort of provocation, or maybe temporary insanity sometimes brought on by divorce, separation, or depression. But one form of killing our own children arises with such predictability as to call for a different explanation. So-called honor killings are common in several societies around the world. Young women are especially vulnerable if it is deemed that they have done something to dishonor their family. Here is a profoundly uncomfortable illustration of both the power of (perceived) reputation in human society, the value of our societies to us, and our tendency to treat reputation as if it is heritable and runs in families: we are willing to kill our offspring to pay off the reputation debt of their perceived transgressions and therefore keep our own reputations intact. Killing their own offspring is the most costly and direct thing parents can do to harm their own reproductive success, and therefore the most informative act they can undertake to regain their loss of reputation. A reputation is worth a human life. Here is the marketplace of reputation again, but now printed onto a bank note of grotesquely high denomination.

Self-sacrifice is the most costly and therefore most believable signal of one's commitment to a cause—be it family, nation, tribe, religion, or some abstract political idea. Self-sacrifice can therefore send large benefits toward one's relatives. This means that once a system of reputation has evolved and established itself in our psychology, the thought that thousands or even millions of people are watching or listening must be enough to drive some people to extreme acts. This could be the same psychology that when it slips just a little bit shades into suicidal terrorism. Or into self-sacrifce in war: no higher honor could be bestowed on a Japanese Kamikaze pilot in World War II than to give his life, and equally no greater shame could be had than by failing to do so. In the mind of the terrorist or pilot, either action would get broadcast to an entire nation, and then the transferability of reputation would ensure that its effects would flow back to his family. Or, imagine you have been selected to be sacrificed to some god, a great honor in your society. Thousands have gathered to watch this great event that can bring better luck to your people. Could you refuse, knowing what the repercussions would be for you and your family? Societies seem to have appreciated this psychology instinctively and have often used it to their advantage with great and awful effects. Similarly, why do we honor our war dead so highly, especially as it is too late for them to benefit? They deserve our highest respect for their sacrifice, but somehow in doing so we acknowledge that some form of payback—reputation enhancement—is needed to keep families willing to send their sons off to battle. No ant, bee, or wasp would make such a request.

common and public goods: competing to cooperate

ECONOMISTS DEFINE a "common good" as something like common land for grazing. Everyone is allowed to use it, but use by one person reduces the amount left for others. Public goods are the same but don't get used up. Air to breathe is a public good, or very nearly so, as are radio and television signals, or streetlights at night. But so are police forces and fire brigades, and things such as lighthouses and foghorns. All of these can be used and in most instances have only negligible effect on what is available to others. Getting people to contribute to public or common goods poses the problem of how to get people to behave selflessly or altruistically. If I choose to use our common grazing land less than you, my benevolence is to your advantage. When I pay my taxes but you don't, I am helping to create a common or public good that you might benefit from as much as or more than me. You are cheating or free-riding on our social contract.

Economics also identifies something called a "second-order" public goods problem. Sometimes people will punish others they suspect of being free riders, as in the case of the Englishwomen who dispensed white feathers to young men. But we also mete out this moralistic aggression ourselves every day when we honk our horns at people or shout at those who might jump a line. What interests economists is that punishing free riders, like contribution to a public good, is an act of altruism. When I punish a free rider, everyone else enjoys any benefits that might flow from my actions. Maybe the person becomes less likely to free-ride, or maybe others witnessing my actions will decide not to step out of line themselves. Either way, everyone benefits.

So, why should anyone ever contribute to commons and public goods, and why do we punish free riders, especially as punishing someone can be a risky business—they might retaliate. In fact, cheating on public goods is rife, and many of us cannot be bothered to punish free riders. The phrase "the tragedy of the commons" refers to the all too frequent tendency for people to overuse their share of common or public goods. When we exceed the speed limit in our cars, allow our flock of sheep to overgraze the common land, or don't recycle as much of our rubbish as we might, or we cheat on our taxes, eat more than our share of food, lie about how much food we found from a day's foraging, or jump lines, we are part of the problem of controlling and maintaining commons and public goods. The problem of global warming is looming as our most daunting exploitation of a vital commons, and one that most of us cheat on extensively by using far more resources than are sustainable.

These examples are just another way of saying that people seek to use the benefits of the cooperative society for their own gain, rather than putting themselves second to the good of the group. But we must also acknowledge that cheating and free-riding are not nearly as widespread as they might be. We do maintain common land, and most people generally obey speed limits and pay taxes, and some of us (as we saw in Chapter 5) even return wallets. Much of this of course can be attributed to laws and police forces; but as we said, these are public goods themselves and so someone has to contribute to them—someone has to contribute enough taxes to pay for the police forces. So, we return to the problem that even if we allow that there is an incentive to cheat, and many of us do or have done at some time in our lives, we are to some extent self-policing. Put this way we can see that there is a measure of altruism to be explained.

But public goods and second-order public goods do not raise any new questions about why people can behave altruistically. We can expect the same motivations that get people to cooperate in general will spill over into these public goods. So long as the public goods pay their way by producing something individuals on their own could not achieve, people in general should be content to maintain them. Contributing to public goods might also in some circumstances be a way to buy reputation points—another way of lengthening your altruistic tail—because these are acts that cannot be interpreted by others as ones done for your own gain. If contributing to a public good enhances your reputation, that particular public good need not even benefit you directly; it is the reputation points you earn that are your reward. If reputation really is the currency we use to buy others' trust, then we might even be expected to compete to cooperate on public goods as a way of demonstrating our commitment to the group. The problem of explaining at least some public goods and second-order public goods problems is then turned on its head. What seems like a cost or burden becomes a virtue, something we wish to parade in front of others.

Among human hunter-gatherers certain kinds of food are almost universally shared out between all members of the tribe, independently of who has acquired them. This is especially true of meat acquired as a result of men going out on a hunt. There is often little relationship between who actually kills the animal and how much food they get from it, or even how much their immediate family gets. Instead, meat is often carried back to the camp or village, where it is shared out among members of the tribe. The exact extent to which this sharing occurs varies from tribe to tribe and even depending upon the particular kind of food to be shared; but it is a reasonably safe generalization to say that hunter-gatherers everywhere they have been studied practice some form of food sharing.

The anthropologist Kristen Hawkes points out that when food is shared this way, it becomes something like a common or public good because people can benefit from it without having to contribute to the shared pot. So, what maintains the public good? We intuitively grasp that what makes the practice of food sharing fascinating and even somewhat incredible is that it seems to fly in the face of what we expect from Darwinian natural selection. Why, for example, do hunters or foragers not just keep what they find? And why should someone bother to engage in the risky and arduous practice of hunting when they know the spoils will have to be shared out among many people? Some anthropologists even go so far as to call the practice of allowing the meat to be shared out "tolerated scrounging" or "tolerated theft." These questions pose the fundamental problem of living in a cooperative group. There will always be a conflict of interest between doing what is best for you and your kin and doing what is best for your group. The more you do for yourself, the worse your group might perform; but the more you do for your group, the worse off you and your family might be.

Careful anthropological fieldwork supports the idea that food resources, especially meat, really are shared, so we can dispense with the idea that somehow hunters are stashing food on the side and returning to it later for themselves and their families. This isn't to say the practice doesn't occur, or that people don't try, just that it is hard to get away with because everyone is watching. Among the Hadza hunter-gatherers of Tanzania, honey is a much-valued resource, and when it is found it is shared out like meat. The anthropologist Frank Marlowe, who spent years living among the Hadza, realized that because honey is easy to hide and does not need to be cooked, those who find it will often try to keep it for themselves or give it preferentially to their family members. He says, "I have seen a man sneak honey into his house, where he shared it with his wife and child plus two young single women he discreetly signaled. Still, it can be difficult to hide. Once a man slipped into camp and put his honey under my Land Rover to wait for an opportune time to fetch it and give to his children. But when he finally retrieved it, others saw it, and he had to share it with everyone present."

This leaves people in a quandary. If you have to share your resources, and you really can't get away with cheating, or at least not very often, there is very little incentive to get up off the ground and do much. In the modern world, we might recognize this problem in the controversy surrounding public benefits. Do they maintain people at a reasonable level until they can find a job, or do they act to remove people's incentives for working? Hawkes, who has also lived among the Hadza, as well as among the Aché of Paraguay, suggests that, at least among the hunter-gatherers,

> the incentive for providing widely shared goods is favorable attention from other group members. If those who provide public goods are listened to and watched more closely than others and favored as neighbors and associates, they have a larger, readier pool of potential allies and mates. When this is so, foragers face a trade-off between increasing their families' food consumption and increasing the attention they will get from other members of their group. Public goods will be provided when, for some individuals, the fitness value of the latter outweighs that of the

former. *This solution to the public goods problem relies on identifying a fitness benefit that depends directly on the consumption value of the public good, but is distinct from it.* [italics added]

The message is that if you can get more out of being an altruist than you can out of looking after your own well-being, you will choose the former. You are willing to contribute to the public good because the benefits you derive from doing so are distinct from and thereby in addition to any "good" you might get from the public good: you gain reputation points and attention in amounts related to the meat's value and how hard or risky it was for you to get it.

Here, then, is another example of acquiring a long tail of altruism, and it is one that fulfills the Zahavian ideal of sending a message that is directly relevant to something you are trying to say about yourself. A hunter-gatherer giving away meat is like a gazelle stotting, a skylark whistling while being pursued by a merlin, or you giving away your hard-earned cash: it directly reduces your chances for survival and therefore says you have energy and skills to spare. Having said this, it also helps that meat is perishable. If you kill a large animal, it is likely to go bad before you or your family can consume it all anyway. In the circumstances, the best thing you can do is saunter back to the village and play the role of the community-minded do-gooder. The loss to you in meat you could have consumed is small, but the rewards can be great.

The wider picture of this public-spirited behavior is that our psychology has been shaped by the fact that our cultures have acted as units for our survival in a way that strongly linked our fates, and these are the conditions that favor public-spirited dispositions. In our natural state of living in hunter-gatherer cultural survival vehicles, everybody would have realized that everyone's success at nearly everything they did would in some way depend upon the group pulling together. Sure, the hunter goes out and kills the meat himself (nearly all hunting of large prey is done by men), but think of the larger sphere of cooperative activity that makes his actions possible. Someone else is foraging while he is out hunting. Someone is guarding the village from raids, someone is looking after the children, and there will be other men with him assisting him on the hunt. Finally, someone had to make the various spears and nets and snares that are used in the hunt, not to mention all the other paraphernalia that make daily life possible: huts, vessels for carrying water, cured skins for clothing, baskets, and tools.

Crucially this is not to say that we behave in a "group-selected" way or that we behave "for the good of our groups." Rather, the nature of the food-sharing problem shows us how the conflict between what is best for you and what is best for your group can be resolved. I like to think of this as *enhancement selection*. You can actually do things that provide direct benefits to the group and at a cost to you, because those costs are more than erased by the enhancement of your reputation. As soon as this becomes possible, you might be encouraged to become public-minded about just about anything you were particularly good at because you can sell it for reputation points in your community. Maybe it is making good arrows or better hand axes, being good at fishing or knowing where to find the best berries. What looks like altruism on your part is actually just another stall in the reputation marketplace.

parochialism, xenophobia, and the principle of information

SADLY, GIVEN our demographic success as a species, conflict has been inevitable throughout our history. We have seen the terrible violence that human groups are capable of directing at one another, and speculated that parochialism might be an emotion that arises to make it more likely you will vanquish a competing group in battle. And we have seen how false beliefs about another group can indeed make it easier to conduct violence against them.

We can also see parochial views toward other societies arising from the way we have envisioned that cooperation works within human societies. Our cooperation depends upon acquiring high-quality and up-to-date information about others' reputations. Lacking that information, as we saw from the principle of information, it has paid us throughout our evolutionary history to withhold cooperation simply as a way of avoiding providing help to people who might not share our dispositions. We acquire information about others' reputations from observation, gossip, and word of mouth, and so the information we have on people from other groups will normally be well below that required to know whether we should cooperate. But it gets even worse than this. You and the members of your cooperative group know that the people in another group have formed their own allegiances: these people by being committed to their group have positively demonstrated their lack of commitment to yours.

Our success as a species has lain in using our cooperative group to generate shared knowledge and technology. This shared resource clothes and feeds us, it protects us, it puts a roof over our heads, and it allows us to project ourselves into new territories. Would you give away things of such value without a fight or even just inadvertently to the wrong people? The costs are potentially so great that we might expect that natural selection has acted to make us wary of cooperating with strangers, and especially ones from other groups. Self-interested discrimination toward others based on group membership is not something we should condone, but it may have been a tactic that served us throughout our evolution.

We don't know if these scenarios are true, but there is little doubt that our all too common feelings of parochialism or xenophobia are directed at others because they are members of a particular group, not at them per se. Indeed, we could not direct our xenophobia at someone we don't even know. Violence or aggression against members of our own group normally violates the trust that is the foundation of our social nature. But this same aggression, violence, and even killing members of other groups can both demonstrate a commitment to one's own group and deliver valuable resources. In any other animal species we would not be surprised that these factors positively favor some forms of violence and murder. It bothers us because as human beings it conflicts so directly with our self-image as good and cooperative people.

CHAPTER 7

Hostile Forces

That we owe our big brains less to inventiveness than to conflicts of interest among social minds engaged in an arms race to be the best at manipulating others

IF OUR BRAINS were our necks, we would resemble a giraffe. In fact, so out of proportion are our brains to the size of our bodies, we would resemble a giraffe with an elongated neck. This is a book about how culture has sculpted our minds and behavior, not about our appearance and other physical traits. But it might just be true that it is in the nature of our brains that culture has had its most significant effects. We can say this because in many respects human beings are rather ordinary animals. We are not particularly fast runners, not particularly good at climbing trees, not all that strong, and not very tough—our odds would be poor in a fight against an angry baboon, which is just a fraction of our size. But we are intelligent. In fact, it could be said we are infinitely more intelligent than any other animal we know of, at least in this narrow sense. We are the only animal that continually constructs new lifestyles and ways of life on a scaffolding of progressive cultural evolution. Based on what we know about the other animals, we could wait forever and not only would they fail to make the progress we have made in the past 80,000 or so years, they would not even have changed their behavior substantially from what they do now. Were we to go away, returning in a million years, the chimpanzees would still be behaving as they almost certainly have for the previous millions of years, fishing for termites from the ground with the same kinds of twigs, blades of grass, or pieces of bark. The same is true of other animals. But not of us, and it is all down to our brains.

Among all the millions of different species of animal, the mammals have unusually large brains for the size of their bodies, bigger than, say, those of comparably sized fish, or reptiles, or birds. Among the mammals, the primates—monkeys and apes—have larger brains for their size than other mammals. Among the primates, the Great Apes—orang-utans, gorillas, chimpanzees, and humans—have the largest brains, but even among the Great Apes, humans are exceptional. Our brains, at around 3.25 pounds (1,400–1,500 grams), are three to four times the size of a chimpanzee's or a gorilla's brain. They are the largest brain for a given body size of any animal known, roughly seven to eight times larger than expected of a typical 130–180-pound mammal. A giraffe's neck, long as it is, is not seven or eight times longer than that of a comparably sized grazing animal such as an eland.

Our species last shared a common ancestor with chimpanzees around 6 to 7 million years ago. That common ancestor probably had a brain that weighed around 300 to 400 grams, or less than 1 pound. The evolutionary route our lineage took for the first 3 million years after separating from this common ancestor is not well known. But already by around 2.5 million years ago, brains had enlarged to about 600 grams in the species called *Homo habilis* or handy man. The species we call *Homo erectus*, the first truly upright ape, arose around 1.8 to 2 million years ago in Africa, and had a brain size of around 800–1,000 grams, or about two thirds the size of our own, and already roughly twice the size of a chimpanzee's brain.

Homo erectus' brains only slowly enlarged over the next million years, reaching around 1,200 grams by around 500,000–700,000 years ago in the species called *Homo heidelbergensis*. Not long after, two lineages split off from this species, one leading eventually to modern humans, the other to the Neanderthals and their sister species the Denisovans. By 200,000 years ago, human brains weighed around 1,250 grams, and they reached their modern size of 1,400–1,500 grams by around 150,000 to 100,000 years ago. This means that in just the last 1 million years or so, the human lineage enlarged its brain by an amount equivalent to the entire volume of a chimpanzee's brain.

Brains are costly organs to own and maintain. They account for only about 2 percent of our weight, but they require perhaps 20 percent of our need for energy when we are at rest. When you come home from the office having used your brain all day, you can blame your tiredness on having to feed this hungry gas-guzzler. This means that the increasingly brainy ape we were becoming must have been paying for its intellectual profligacy by improved survival and success at reproduction. We might have received some help in supporting our large brains from an unlikely source, starting perhaps 250,000 to 300,000 years ago. Around that time, premodern humans stumbled upon the idea of using fire to cook. As Richard Wrangham describes in his book *Catching Fire*, cooking food enhanced the amount of energy our ancestors could extract from it. Even if fire alone does not tell us why we were on a trajectory toward having large brains, the ability to cook might have made them easier to feed and maintain.

Some anthropologists suggest that inventiveness is the hallmark of modern humans and the obvious force that would have paid for our large and hungry brains. Neanderthals had brains nearly the same size as ours, but their brains were different, and one of the chief differences is given away by the Neanderthals' appearance. They have a tremendously pronounced brow, but very little by way of a forehead. As E. H. Gombrich said of Neanderthals in his *A Little History of the World*, written for young audiences and first published in 1936, "Now, if all our thinking goes on behind our foreheads and these people didn't have any foreheads, then perhaps they didn't think as much as we do. Or at any rate, thinking may have been hard for them."

Gombrich's delightful explanation might be right. It is indeed true that humans are perhaps the only species with foreheads, and we now know that those foreheads conceal a part of our brain known as the *cortex*. All mammals have a cortex but ours is unusually well developed. It is a region of our brain in which the connections among neurons are especially complicated and dense. Brain scientists link the cortex to imaginative, creative, symbolic, and other "high-level" thinking, and indeed we have seen that *Homo sapiens* was showing the first glimmerings of symbolism by 160,000 years ago. By sometime around 100,000 years ago, armed with our new brains, we became vastly more creative and inventive than our predecessors, and by 50,000 years ago we were building complex tools with more than one part, musical instruments, and bone tools with specialized functions. We had developed an aesthetic sense, as seen in portable art and jewelry, cave art and carvings, ceremonial burial and body adornment. Put it this way: the Stone Age didn't end because people ran out of stones; it was replaced by a new age of inventiveness.

Surprisingly, though, another feature of our intelligence might tell us that our inventiveness is not necessarily evidence of a great increase in raw intellectual capacity. As we have also seen, we appear to be the only species able to copy and imitate others' ideas, an ability we called social learning. It makes cultures evolve rapidly by gathering together the talents of many individual brains, and making the contents of those brains available to everyone. But the irony is that social learning might also reduce the advantages to any one of us of being clever and imaginative, because your inventiveness can bring the same benefits to others as it does to you. This means there are opportunity costs to having a large brain if that large brain exists to be inventive. Why should I build and then support a large and hungry brain just to be inventive if I can get by copying your best ideas? I could use the energy needed to power my large brain for something else. And, if useful innovations are difficult to produce, the time we spend on creative but fruitless acts might be better spent gathering food, hunting, defending our territories, or repairing our shelters, and especially so if someone else can put in the effort of innovation for you.

The thought that we might not be very inventive is one that most of us will dismiss outright. But when we think about it, we really are not all that good at being inventive. How do we generate new ideas? No one really has very much understanding of this question, but what we do know is that innovation is not just a matter of thinking hard until you get *the* answer. Imagine yourself 50,000 years ago trying to design a better spear. You want it to fly further than the ones you have, but also penetrate deeper into the animal you wish to hunt. What shape and weight should you give it? Most of us probably would not know the answers to these questions, so we would experiment by trying out lots of different options, and then choose the best. There would almost certainly be many false starts and wrong turns. Even today, our best engineers armed with centuries of accumulated knowledge can still be surprised by what they build. The Millennium Bridge is a stylish pedestrian suspension bridge that crosses the Thames in London near St. Paul's Cathedral. It was opened on June 10, 2000, and was closed later the same day having already been nicknamed "the wobbly bridge." Despite the best efforts of the engineers who designed it, the bridge swayed and twisted vertiginously as people walked across it.

But these and other wrong turns won't really have mattered too much in our past if a species has social learning. Like natural selection acting on genes, social learning will examine everyone's attempts at some problem and act as a remarkably efficient sorting machine or "algorithm." Someone making spears will have, even if just by chance, come up with a good design, and all of you trying to make better spears can copy it. This process even speeds itself up all on its own, as the pace of change of modern life attests. As more of you copy the good design, there will be more people using it, and so this design will become even more likely to be copied and improved upon. Over long periods of time, cultural evolution relying on social learning will be able to build objects of great complexity by a series of small incremental changes. It can do so even if at any given moment the proposals for how to improve the existing form are no better than random. There never have to be any great leaps of imagination or understanding, because just by chance one of these proposals will lead to an improvement, and social learning will find it. Engineers eventually solved the Millennium Bridge's swaying—although not at the first attempt—and now everyone building bridges has access to this information.

If you still cling to the view that humans must be highly inventive to have produced their great works of culture and technology, you would be in good company, even if mistaken company. The theologian and philosopher William Paley gave one of the more memorable arguments for the existence of a creator in his *Natural Theology*, written in 1802. Paley imagines himself out in the English countryside:

> In crossing a heath, suppose I pitched my foot against a stone, and were asked how the stone came to be there; I might possibly answer, that, for anything I knew to the contrary, it had lain there forever: nor would it perhaps be very easy to show the absurdity of this answer. But suppose I had found a watch upon the ground, and it should be inquired how the watch happened to be in that place; I should hardly think of the answer I had before given, that for anything I knew, the watch might have always been there. Yet why should not this answer serve for the watch as well as for the stone? Why is it not as admissible in the second case, as in the first? For this reason, and for no other, viz. that, when we come to inspect the watch, we perceive (what we could not discover in the stone) that its several parts are framed and put together for a purpose. . . . This mechanism being observed (it requires indeed an examination of the instrument, and perhaps some previous knowledge of the subject, to perceive and understand it; but being once, as we have said, observed and understood), the inference, we think, is inevitable, that the watch must have had a maker: that there must have existed, at some time, and at some place or other, an artificer or artificers who formed it for the purpose which we find it actually to answer; who comprehended its construction, and designed its use. . . .
>
> Every indication of contrivance, every manifestation of design, which existed in the watch, exists in the works of nature; with the difference, on the side of nature, of being greater or more, and that in a degree which exceeds all computation. I mean that the contrivances of nature surpass the contrivances of

> art, in the complexity, subtility, and curiosity of the mechanism; and still more, if possible, do they go beyond them in number and variety.

Paley's assertion is that complex objects imply a maker because we can't imagine them assembling themselves. His exposition is worth reading in full (the quotation above is just an excerpt of a longer passage), and not just for his elaborate description of the watch. Paley takes his reader on a gentle walk through the mind of someone confronted by watches, but also by marvelously complicated works of nature.

By implication, Paley thought it implausible to believe that a complicated thing like a biological species could ever have arisen on its own. But his argument is often used to illustrate a misunderstanding of natural selection in the biological world. The last half century of evolutionary biology has told us that it is precisely the great sorting power of natural selection that means it can produce objects of vast complexity; it can do so without any designer; and it can do so despite the fact that the variety it has to act on at any given moment has been generated by a purely random process. Genes don't know how to mutate to produce some desired outcome. In fact, the overwhelming majority of mutations are damaging to the organism that produces them. But all that is needed is that one of these mutations from among many leads to an improvement, and that the lucky individual who has it survives and passes it on to its offspring. Over time, individuals that inherit this fortuitous mutation will also tend to leave more offspring and eventually everyone will come to have the trait. This is natural selection, and if this process is repeated over thousands of generations, many small modifications can accumulate, one on top of another, to produce complex traits.

Natural selection and Paley's argument were the inspirations for Richard Dawkins' metaphor and book *The Blind Watchmaker*, an account of how natural selection "blindly" produces complex organs and other traits from random mutations. Our discussion of social learning should make us realize that Paley's argument is also wrong when applied to our cultural world, or at least it is not nearly as right as we would like to think. This is because the great sorting power of social learning is also, in principle, a blind watchmaker, or nearly so, and in this case it really is building watches, not complicated organisms. Like genes mutating, our capacity to be "inventive" is just a way of generating varieties that social learning can sort through. Just so long as we have a way to generate a variety of outcomes for cultural evolution to act on, and an ability to recognize a good outcome when we see it, social learning can blindly do the rest, even if the mechanism that generates the variety—our so-called inventiveness—is random. Even our ability to recognize good outcomes need not be very good; in fact, it also hardly needs to be better than random itself for good ideas to spread. Social learning and the cultural evolution it drives is responsible for our cars and toasters, our comfortable sofas and beds, pencils and paper, trains, alarm clocks, breakfast cereal, anything made of metal, computers and space shuttles—in short, just about everything in our daily lives—and yet few of us "comprehend their construction."

The Oxford chemist Peter Atkins in *The Creation* elegantly summed up evolution by natural selection, saying, "A great deal of the universe does not need any explanation. . . . Once molecules have learnt to compete, and to create other molecules in their own image, elephants, and things resembling elephants, will in due course be found roaming the countryside." To paraphrase Atkins, once ideas have to compete in our minds, things like toasters, computers, and space shuttles will in due course just appear—they are inevitable and do not need any explanation. Still, in practice, we would expect to evolve to be better than random both at generating innovations and at recognizing good outcomes, because we must compete with others, and especially with other groups. But the message of social learning is that we can be far less inventive than we give ourselves credit for and yet still expect toasters and computers to appear if we wait long enough. Even Einstein purportedly said, "I have no special talents. I am only passionately curious." We can grant Einstein this conceit, and still realize that because of social learning, being sneaky or at least shrewd at using what others have invented might have tempered our inventiveness.

I was once in the Australian Outback and met, through an interpreter, an Aboriginal man called Sammy. Sammy was showing me how his tribe made a hafted knife by heating a plant over fire to extract a sap or pitch that they used to glue sharp pieces of stone to a wooden handle. Once glued, it was further secured by wrapping it with a long fiber from a local plant. The interpreter said that Sammy was happy to answer questions. I had noticed that the way Sammy notched the handle to hold the blade didn't seem likely to produce a strong bond. I asked him why he didn't cut a slit in the wooden handle to insert the blade down into before gluing it. He replied that his tribe had always done it his way. I don't tell this story to suggest that I had a better alternative, or that I had somehow spotted a flaw in a piece of cultural technology that his tribe might have used for millennia. There was probably a very good reason why they produced the tool as they did. My point is that Sammy didn't seem to know what that reason was (or he simply couldn't be bothered to tell this ignorant interloper one of his tribal secrets).

There is no reason to think that Sammy is unusual. Social learning provides a niche or role for innovators in society, but it might be a small one, because innovation isn't easy and the rest of us can get by just fine copying them. That niche will stay the same size even as population sizes increase because social learning means a little bit of creativity goes a long way. It is a sad commentary, but most of us today might be little more than glorified karaoke singers in most aspects of our lives. Indeed, it is possible today to be almost entirely lacking in any sort of ingenuity and yet still get on in society just fine. Most of us spend nearly all of our lives doing things others have taught us and using things others have made. The people who made the things we use almost certainly had little understanding of what they were doing when they made them, having inherited most of the information from their ancestors or predecessors.

So complicated are machines like NASA's space shuttle or the software operating systems of computers that no one knows how they work. Entire teams of people are needed to keep them going, and even then they rely on protocols, checklists, and sets of instructions carefully worked out by others before them. In 2002, NASA even went scavenging on eBay to find a large quantity of old medical equipment containing the Intel Corporation's 8086 computer chip. This is the chip that IBM used in its personal computer in 1981. It is millions of times slower and less efficient than modern computer chips, so why did NASA want it so badly? NASA's computer software for testing the space shuttle's critical booster rockets had originally been written for the 8086 chip, and the shuttle fleet would have to be grounded unless this software could be run. Once such critical software has been written, cleared of bugs and other errors, and rigorously tested to see if it works, engineers are almost superstitiously fearful of altering it or trying it out on a different chip for fear it might fail. These pieces of software are so complicated that no one can simply rewrite or alter them, and no one can be sure they will work with 100 percent reliability on a new machine. In fact, NASA put out plans in 2002 for a $20 million project finally to upgrade this system.

It has probably been like this for tens of thousands of years. In *The Evolution of Technology*, George Basalla rejects the heroic innovator view of technological changes in favor of a Darwinian view of gradual change. He documents how well-known technologies, including transistors, steam engines, hammers, and chopping tools, all trace their history back through many small successive changes, or combine elements of other technologies. And it is true: how many truly innovative thoughts have you had, thoughts that might have made a difference in the history of cultural evolution? Or if that sets the bar too high, how many thoughts have you had that others would wish to copy, like a better way to shape a hand ax, or how to weave, or make a better spear or soufflé? How many people do you know who have?

Our individual capacity for inventiveness, even in the sciences, might be far less developed than we like to think. In the entire history of science and natural philosophy, the list of people whose ideas have profoundly shaped our lives is short. The remarkable ability of social learning to sift ideas has meant that a few great innovators can go a long way, so that most of us simply aren't very good at it. The historian of science David Edgerton has written: "It is not sufficiently recognized that creation, scientific or otherwise, is a tragic business. Most inventions meet nothing but indifference, even from experts. Patents are little more than a melancholy archive of failure. Most ideas of every sort are rejected, as would be clear if there was a repository for abandoned drafts, rejected manuscripts, unperformed plays and unfilmed treatments."

The Romans are not known as great innovators, but they were skillful copiers of art and culture, and often to their benefit. The second-century BC Greek historian Polybius in his *World History* recounts how the outcome of the First Punic Wars between Carthage (roughly present-day Tunisia) and Rome might have been determined by a fortuitous accident that put Carthaginian technology into Roman hands. Polybius describes a Carthaginian ship in hot pursuit of Roman ships:

> On this occasion the Carthaginians put to sea to attack them as they were crossing the straits, and one of their decked ships advanced too far in its eagerness to overtake them and running aground fell into the hands of the Romans. This ship they now used as a model, and built their whole fleet on its pattern; so that it is evident that if this had not occurred they would have been entirely prevented from carrying out their design by lack of practical knowledge.

With their fleet of new ships modeled on those of the Carthaginians, the Romans went on to defeat them and occupy Sicily.

Imitation is hard-wired into our brains and available to us from infancy. Try sticking your tongue out at a baby and you might be surprised that it returns the gesture. But consider what even this simple act of imitation requires. The baby's eyes have to transduce the light rays that bounce off your tongue into electrical signals that get sent via its optic nerve to its brain. Then somewhere in the brain and by means that no one yet understands, that information about your tongue causes a new set of electrical signals to get sent down nerves to the baby's mouth and tongue, where they cause precise muscular movements that rely on coordinating the actions of many muscles. The process is not "linear"—the baby's brain has made up a set of instructions that cannot be derived from a simple alteration of the input from its eyes. And yet, the baby does all this on its first go and without practice. In fact, there are some suggestions that humans are hyper-imitators. We seem to imitate so precisely that we sometimes imitate actions that are not strictly necessary to the task to be accomplished. The machinery to produce such imitation must be complicated and expensive to own and maintain, and this tells us that imitation and copying have played an important role in our species' survival and prosperity.

Before we leave this topic it might be useful to point out that any shortcomings we have at being inventive or innovative are likely to be magnified in our modern world. This is because it is not necessary for the numbers of innovators in society to keep pace with increases in the size of the population—many people can happily get by copying just one good innovator. This effect is enhanced by language and writing, both of which transmit ideas and innovations well beyond those who came up with them. And this raises a serious question about the kind of dispositions and temperaments that our modern world will encourage. As our societies become ever more connected and "globalized" it will become increasingly easy for most of us not to innovate at all—to become intellectually lazy and docile, at least in matters of inventiveness. The irony is that this might be happening at a time when more innovation is needed than ever before just to maintain the levels of prosperity many of us already enjoy, and to raise it for those who have, up to now, been less fortunate.

big brains and the social arms race

IF INDIVIDUAL inventiveness per se hasn't played the role in shaping our intelligence that we might have thought, a capability that is difficult for others to copy by social learning might have. That capability might turn out to be our social intelligence. To see why, we need to introduce the idea of a "moving" or "unpredictable target." We use our brains, as do all animals, to confront our environment, to exploit animals and plants, and to compete with each other. Finding food, fighting off disease, managing the vagaries of weather, and avoiding being eaten by other species are all things Darwin would have called "hostile forces of nature." But many of these hostile forces are predictable or slowly changing, and so natural selection can respond by giving organisms ways to defend ourselves against them. A plant's spiny surface or the toxins it produces in its leaves to discourage grazing animals can normally be neutralized by evolving some sort of defense. Maybe an animal acquires a particular gut enzyme that breaks down the toxin or develops a tough lining of the mouth. The plant will then normally respond by changing its spines or toxin to thwart the grazing animal.

In other instances, animals must use their brains to outwit other animals, but the process can still be quite predictable. No matter, for example, how many crocodiles are in a river, if a gazelle gets thirsty enough, it will go to the river for a drink. Here outwitting becomes a matter of adopting a strategy that takes advantage of the other animal's predictable behavior. For the crocodile, it is to wait motionless for long periods of time. For the gazelle, it is to pick areas of the river where crocodiles are more easily spotted, such as shallows. These strategies and counterstrategies can normally be achieved by following rules, and this is probably what governs the behavior of most animals—rules, schema, or "algorithms" that the brain can follow in a contingent way. The brain is programmed to do one thing in one circumstance and a different thing in another. It is the sort of behavior that we can program into robots. And it is the sort of behavior that when the circumstances that the algorithm evolved for change ever so slightly, the algorithms can cause animals to do things that seem stupid to us. A good example is that of dogs straining at a leash to get to something just out of their reach. Natural selection has equipped dogs with an algorithm to charge at things they want. It never considered they might be on a leash.

Carnivore species have to outwit their herbivore prey, and in general they have larger brains for the size of their bodies than their prey do. Being a carnivore means getting up each morning having to think hard about how to find the next meal, and in particular having to outwit the animals you seek to kill and eat. By comparison, a grazing animal such as an antelope, upon awakening, normally finds its next meal in front of its nose. That, and the relatively larger numbers of grazing animals in a typical herd compared to their predators, means that each individual grazing animal need not think as hard about avoiding predators as the predators have to think about catching them. Indeed, this is one of the chief reasons that grazing animals form into herds, so much so that they have even been described as "selfish herds." As we might expect from this asymmetry, fossil skulls of carnivores and their herbivore prey show that throughout history the carnivores have had larger brains relative to the size of their bodies. But the fossils also show what looks like an arms race—when one move is met by a countermove—between the brains of these two groups. Over a period of around 60 million years, herbivore brain size has increased, and these increases have been met by increases in carnivore brain size, with carnivores always maintaining their relative edge. The arms race has been competitive enough that an average herbivore today has a larger brain for its body size than an average carnivore did 40 to 60 million years ago.

But even in this arms race neither herbivore nor carnivore brain sizes come close to those of primates, much less our own. This might be because the arms race between herbivores and carnivores is still somewhat predictable, and it only occupies a fraction of each animal's life (herbivores spend most of their time grazing and larger carnivores may sometimes go days without hunting). As herbivores get brainier they might become more cautious and vigilant about carnivores, or more stealthy, but when it comes to escaping from one that is chasing you, about all you can do is run faster, and maybe bob and weave with greater flair. The same holds for the carnivores, and the predictability in both animals' behavior means that they can be controlled by relatively simple genetic programs that employ rules of thumb or algorithms to get their jobs done.

On the other hand, rule-based behavior such as in these examples works less well when the thing to be anticipated is itself flexible and imaginative in its behavior. When this happens, natural selection is presented with a moving and unpredictable target, and this often means building in redundancy and complexity because now the possibilities multiply endlessly. For example, primates have evolved more complex social behavior than most other animals. They live in societies with fluid dominance hierarchies, complex and shifting networks of relationships and coalitions, power brokering and conflict resolution, and many repeated interactions over long periods of time. Now, instead of a brain following simple rules, it becomes necessary for it to look ahead, to consider alternatives, and weight them according to their probabilities of success at the particular time and place. The sheer number of possible contingencies and outcomes means that the old rules have to give way to broad strategies that must be constantly modified in the light of recent information. Success in these complex societies goes hand-in-hand with social success, and this means being good at using and manipulating those around you: if you have more friends than I do, when it comes to a fight, you will have more people on your side; if you have more friends, someone might be more attracted to you as a mate, or less likely to challenge you; if you are going hungry, I might be more likely to help; or if you are unwell, I might be more likely to remove ticks from your fur.

The primatologist Alison Jolly writing in 1966 and the psychologist Nicholas Humphrey writing in the 1970s suggested that in these kinds of complex social societies, any trait that allows you to outwit your neighbors in this social competition will grant benefits and soon spread to your offspring and theirs. To remain competitive, others must match your social sophistication with their own; and when this happens, the species is set off on a social arms race in which increases in social intelligence in some must be matched by increases in others. This is almost certainly a genetic rather than cultural evolutionary process. The reason is that social intelligence, unlike watching someone make a better hand ax, is not easy for someone else simply to imitate or teach to others, because the situations in which it is used never appear twice in exactly the same form. If you can't simply copy social intelligence, then your only alternative is to become a social innovator, working out for yourself how to manipulate society, and this requires a big socially intelligent brain.

This is just the opposite of what we concluded for inventiveness, where most of us can get by just fine copying others. Humphrey and Jolly were aware that the nature of human societies also meant a new hostile force emerged in our lives—other humans. Humans increasingly had to deal with other fellow humans whose ingenuity in social matters and desire to manipulate society grew with their increasingly large brains. We got locked into a back-and-forth struggle with each other's minds, and these minds presented a constantly shifting and unpredictable target. It was an arms race that can be likened to the arms races between our immune systems and the viral and other infectious diseases they attempt to thwart. Diseases such as colds, influenza, malaria, or HIV all change in unpredictable ways governed only by the number of different configurations of their genes, and this is an effectively infinite number. Even today, no one can predict how these diseases will change from year to year. If we could, we could prevent epidemics. Our immune systems dedicate trillions of cells of every variety they can muster, each one trying out some new configuration on its cell surface in an attempt to recognize and destroy the invaders. The only other single organ of such complexity and investment might be our brains.

But our brains must do even more: rather than merely playing a defensive game as our immune systems do, our brains attempt to stay one step ahead of their rivals, and this means getting inside their minds to try to anticipate what they might do next. It becomes necessary to develop what have been called "theories of mind"—a sense of knowing what you think another animal knows, and being aware that it is having similar thoughts about you. Our brains can effortlessly think about situations like "I know that she wants to buy that work of art." Psychologists call this first-order social intelligence. With a little more effort we can think, "I know that she wants to buy that work of art, and that she knows I am thinking I want to buy it." This second-order intelligence can easily extend to a third level in the form of "I know that she wants to buy the art, and that she knows I am thinking I want to buy it, and that she is aware that I am aware of her desire to buy it." Some psychologists say that some of us can routinely deal with even higher-level orders of social intelligence! But what is most revealing about stating them in a sentence is that it makes the mental calculations sound torturous and lengthy but this is something we do almost without thinking.

This pressure to be able to think about and imagine what is going on in the mind of your competitor may have been the force that gave rise in humans to what we now recognize as our conscious minds. With consciousness, an animal can "bring to mind" the things it ought to be thinking about, consider alternatives, and devise plans. In *The Selfish Gene*, Richard Dawkins suggested that consciousness might have been the final stage of animals becoming ever better at simulating in their own minds what must be going on in someone else's mind, and then sifting among the alternatives that the simulation throws out. Ultimately to make the simulation complete it must include a model of itself interacting in the world it is attempting to simulate, and *poof*, this is self-awareness. (Dawkins cautioned not to take this idea too seriously because he thought it might lead to an infinite regress. The difficulty is that if consciousness arises from some virtual observer who reads and interprets the simulation, how do we account for that observer? The answer might be that a simulation of oneself needs another to "see" it, and so on.)

The "persistence hunting" style of the San Bushmen of the Kalahari Desert might be a case of humans using mental simulation to great advantage over a less intelligent adversary. San hunters use a combination of running and tracking to pursue their prey across the South African veld, and this can require them to run many miles for extended periods of time. Endurance running of this sort is only seen in humans, and is thought to have been one of the earliest forms of human hunting, having evolved perhaps as early as 2 million years ago. If true, it could go some way toward explaining why humans seem to be so good at running long distances in marathon foot races. Among the Sans' favorite large prey are the kudu and eland, both of which are large grazing animals that can stand over five feet at the shoulder. Like most large grazing animals, kudu and eland have evolved to escape from predators that can put on bursts of speed, but not run long distances. This makes them vulnerable to the San who, remarkably, seem capable of keeping up their dogged pursuit in the hot sun in hunts that can last sometimes eight hours, eventually running their prey to exhaustion.

Now, the link to mental simulation is this. Like any hunter, the San creep up on their prey in an attempt to surprise them. Normally they are spotted and the animals move off by trotting away to a safe distance. The San continue to pursue them this way, forcing the animals repeatedly to move, tiring them out. Sometimes the animals move far enough away that the San lose sight of where they have gone. When this happens, the San are forced to follow the animals' tracks. On some occasions they even lose the tracks and it is at these times they seem to rely on a mental simulation to work out where the kudu have gone. San hunters report that they try to think like a kudu (or eland) would think. They will get down on all fours and try to put themselves into the mind of a kudu, even reenacting how it might behave in an attempt to work out which direction it has gone. If this works, and their simulation leads them to the animal, it might move off again. The hunt then becomes a test of who will collapse first, the man or the animal. But the man always has the one-sided advantage of his mind simulation. Just imagine if the eland or kudu could think like him.

Of course, other humans do have the ability to simulate what is going on in their adversaries' minds. When competition among brains is fierce, as is suggested by the idea of an arms race, there are many losers, and so it would not have been sufficient merely to keep up: one had to stay among the leaders to survive. If, as we have imagined, the competition was mainly centered on this newly emerging consciousness and sophistication in social and psychological traits, the sieving or filtering out of survivors from non-survivors each generation would have sorted people most strongly by their brains, rather than sheer brawn. Among the evolving human lineage, a social and psychological arms race provides a plausible mechanism, then, not just for the enlargement of our brains but also for the rapid pace of that enlargement over time. Increasingly, the survivors would have been those able to deploy sophisticated psychological strategies emboldened by strongly perceived emotions, and able to engage in strategic and fluid coalitions. These would have to be met by yet more sophisticated strategies in others.

Once a species starts evolving along this trajectory, it might be difficult or impossible for other similar species to keep up. This might then also provide the answer to a question that has long puzzled anthropologists: why we alone emerged as the sole survivor of our lineage, and why no other species acquired intelligence like ours. The other *Homo* species that had been spawned along the way, including *H. habilis, H. erectus, H. ergaster, H. heidelbergensis*, and *H. neanderthalensis*, would each have had to compete with the next large-brained and shrewd species that was about to emerge; perhaps that next one just got one step ahead and the others never recovered. It has even been suggested that our dominance of our particular niche in life is what kept chimpanzees from moving on from theirs. As for the rest of the animal kingdom, few have entered the social corridor the primates did which set our social competition into overdrive.

the dna that makes us human

IF THIS story about our emerging brains is broadly correct, then we might expect to find genes that cause our brains to enlarge but also change the nature or structure of our brains, not just add more brain material. A computer with more processors is not necessarily cleverer than one with fewer processors, just faster. The publication of the entire sequence of the human genome in 2001 made available a list of our genes. This alone was of limited value for understanding our evolution. But as similar lists became available for other species, it became possible to ask questions about which of our genes differed most from them. The answers display the exquisite precision with which natural selection can sculpt our genes, even though it only gets to see their actions via our behavior.

Darwin's great idea of "descent with modification" teaches us that these comparisons can be used to answer questions about what has happened along an evolving lineage of species over vast stretches of time. Darwin realized that species evolve and give rise to one or more daughter species that then go on to do the same. This means that any pair of species we care to examine will have, at some time in their past, shared a common ancestor from which they both descend. If we wind the clock of evolution back around 300 million years, a species existed that, although no one would have known it at the time, would become the common ancestor to both present-day mammals and birds. A comparison of the same gene in a contemporary bird and a contemporary mammal then records the evolutionary changes that have occurred in one of these lineages, plus those in the other, or 600 million years altogether.

If we wind that clock back just 6 million years or so, we find the species that was the common ancestor to ourselves and chimpanzees, and this is why comparisons between these two species are so informative about the question of what makes us genetically human. But the task of finding differences between humans and chimpanzees is made daunting by two factors. One is that both species have about 3 billion of the chemicals called bases or nucleotides that make up our DNA genetic code. These 3 billion bases are strung out along the structures we call chromosomes. Worse, we are about 90–95 percent identical to chimpanzees in the sequence of these bases along the chromosomes, and more like 98–99 percent in the sequences that we call genes—sequences that carry the codes or instructions for making proteins. What this means is that in the 12 or so million years that separate the two living species (6 million years each since our common ancestor), only about 1 percent of our protein coding DNA has changed.

Still, powerful computers make it possible to check these billions of bases, and when they were checked for humans, forty-nine regions of our genome emerged in which the pace of evolution had dramatically accelerated in our lineage compared to chimpanzees and to other animals. The regions were called *Human Accelerated Regions*, or *HAR*s, and they should be the Holy Grail genes that make us human, the parts of our genome that really distinguish us from chimpanzees. This is because accelerated change is an indicator that natural selection has been acting particularly strongly on a gene. It means that each time some new helpful variant has arisen, it has quickly spread through the population until everyone has it, and then this process has been repeated as new and different variants arise.

The *HAR*s were ranked from *HAR1* to *HAR49*, with *HAR1* being the most rapidly evolving of all the segments. Comparisons between chickens and mice showed that *HAR1* hardly evolved at all in the

300 million years from the common ancestor of birds and mammals, changing in just two of its 118 bases. But then there was an abrupt acceleration. In just the 6 million years from the common ancestor of chimpanzees and humans to modern humans, *HAR1* managed eighteen changes. This translates to a 450-fold increase in this bit of DNA's rate of evolution. But the best part of the story is that this most rapidly evolving segment of our DNA is active in human brain cells. *HAR1*'s high rate of evolution tells us that it must have granted substantial benefits. Those lucky enough to carry copies of it that had one or more of the beneficial mutations must have enjoyed clear advantages over their less lucky and somewhat dim friends.

What makes *HAR1* even more extraordinary is that at just 118 bases long, it is about one tenth the length of a typical gene. Genes, as we said, are segments of DNA that contain the code to make proteins, which in turn are the building blocks of bodies. Our hair and fingernails, muscles and eyes, hearts and kidneys, and our skin are all made of proteins. *HAR1* does not make a protein; in fact, it is not even a gene. Instead, *HAR1* turns out to be a segment of our DNA that influences or regulates how other genes are expressed, and this might be why it can make such a difference to our brains. In *HAR1*'s particular case, it influences the ways that neurons in our brain develop and project into new areas. Rather than simply making our brains bigger, it changes the structure, density, and complexity of our brains and the connections their neurons make with each other. The changes that *HAR1* brings are just such as we might have hoped of a species that supposedly finds it easier to think than the Neanderthals did. Merely adding processors to a computer is not the same as getting these processors to talk to each other and share their information more effectively.

Many of the genes that influence the size of our brain are located in our genomes near small regulatory segments like *HAR1*. So, the real wonder of our evolutionary changes since the chimpanzee is how few changes have led to such profound differences and over such a short period of time. Natural selection hit upon just those changes that could make a big difference to our behavior. Many more of these segments and the genes they affect will come to be implicated in our brain growth and development, but one stands out for its possible link to language in humans. *FOXP2* is a segment of DNA that, like *HAR1*, regulates the expression of other genes. All mammals have it, and it is expressed throughout the body, including in brains. Unlike *HAR1*, it has not changed so dramatically since we split from our ape ancestors, but it has recently acquired what appear to be two critical changes that affect the control of facial muscles that are involved in producing speech. Even mice fitted with a copy of the human form of *FOXP2* are said to squeak differently from those with the normal mouse form!

Very recent evidence shows that the Neanderthals had this same variant of *FOXP2*, leading many people to conclude that they also had language. But this is premature. They might have had language, but simply having this variant is not proof. *FOXP2* affects our brains by altering the expression of at least one hundred genes: it is thought to cause about fifty of them to be expressed more and another fifty to be expressed less. So, for *FOXP2* to work similarly in the Neanderthal brain, we would have to find the same one hundred other genes, and presume that they worked the same way in both brains. But this seems unlikely. My car has an engine and so does a Ferrari. But my car is no Ferrari. We know our brains differed from the Neanderthals' in having a more fully developed and highly interconnected cortex, the uppermost layer of our brains. But we also have reason to suspect that the social arms race we have suggested is responsible for our unusual brain might not have been so pronounced in the Neanderthals. The archaeological record points to a species with far fewer artifacts, hinting at little social learning of the sort that is so prevalent in our species. And, as we saw in Chapter 2, it is the presence of social learning that established the need for systems of exchange and cooperation; in short, that established the need for a social brain.

domestication by brain genes

ONE OF the most surprising effects of our big socially charged brains was to preside over their own diminution. Having steadily enlarged for roughly 2 million years, they have shrunk by around 10 percent in the last 30,000. We also became less robust or more *gracile*—thin-boned—during this time, so it might just be that our brains were adjusting to a reduced stature. But one of the most reliable differences between domesticated animals and their wild ancestors is that the domesticated ones have smaller brains: as a rule, domesticated animals are just a bit dim, or less "street smart." Could our brains have domesticated us as well?

Domestication is like taking up residence in a protective bubble, and right across the history of evolution it is linked to things becoming simpler. Single-celled organisms that have taken up residence inside the cells of other organisms normally have many fewer genes than their wild ancestors. They jettison genes they no longer need, genes that served functions in their wild state but that are now provided by their host. The structures called *mitochondria* that exist inside each of our cells and that produce energy are thought to be ancient bacteria that took up residence inside cells like ours over 1.5 billion years ago. They probably had around 3,000 genes when they moved in; now they have 16.

The same sort of protective bubble is erected when an animal is domesticated. Now your shepherd looks after you, sees off predators, finds or steers you to food and water, and keeps you warm. The animal's response seems to be to jettison features it no longer needs, and this includes shedding some of its costly and expensive brain. And why not? It no longer needs so much of it now that the shepherd is doing the thinking. Domesticated rats, mice, mink, cats, dogs, pigs, goats, sheep, llamas, and horses all have smaller brains than their wild ancestors. Wolves outperform dogs on searching tasks. Domestication also tames animals directly as their human handlers preferentially breed the less aggressive ones. Dogs become less packlike; sheep and cows become calm and relaxed around humans, and more dependent on them. They also become less tuned in, or less switched on, because their human masters are doing all the hard work.

The word "bovine" technically just means cowlike but is used as an adjective to mean stolid, slow-moving, and dim-witted. Maybe our big brains have made us more bovine by cosseting us in technologies built from social learning. Remember it was this social learning that we thought removed much of the need for us to be inventive in the first place. Who among us is good at tracking game, lighting a fire without matches, or finding edible plants in the local wood? Technology and mastery of the environment are great levelers of people, and this acts as a further boost to domestication. The anthropologist Robert Lee reported that among the San Bushmen of the Kalahari, when disputes reach a point where there might be open conflict, someone will often declare, "we are none of us big, and others small; we are all men and we can fight; I am going to get my arrows."

Our brains might also have domesticated our outward appearance, making us one of the more peculiar-looking of the mammals. Mammals are really only distinguished from the other animals by two key traits. Mammals have fur—no other kind of animal does—and we evolved the ability to lactate or nurse our young, again something no other animal does. But nearly all of our other features we share with other animals. Now, fur, like feathers in birds, is a valuable invention for its combination of being breathable, its ability to shed water and snow, and for being an excellent insulator, and an insulator that works even when wet. These are just the qualities you might expect of a material natural selection devised to cover warm-blooded animals that sweat, get rained on, and often live in cold climates.

But humans are naked. We have jettisoned this valuable fabric. An even greater irony is that having done so, we now hunt mammals and kill them so we can wear *their* fur. Even in our modern world, fur spun into the form of wool is still the preferred fabric as a base layer for skiing and other cold weather pursuits, and wolverine ruff is still the preferred insulator on the hats of polar explorers. So, why would humans have got rid of their fur, especially as being naked makes us more susceptible to cuts and bruises and to exposure to the sun? Surreptitiously look around you now if you are reading this book somewhere in public. Those bare patches of smooth skin you can see, and that appear self-evidently normal and even attractive in a human, seem ludicrous in any other mammal. Imagine your dog sheared of all its fur or a polar bear naked as a human. The ridiculous-looking image you have in your mind is what you look like to other animals.

One suggestion is that we lost our fur when we moved out onto the savannah, early in our evolution, so that we could cool our bodies more effectively. Our *Homo erectus* ancestors were upright hunters and foragers, who would have spent much of the day baking in the hot sun, and so losing their fur and becoming naked might have been beneficial. For instance, no one knows how far back in time the persistence running style of modern-day San Bushmen existed, but if *Homo erectus* used it, then shedding their fur might have been an important way to protect themselves from overheating, although as it is males that do this hunting, this explanation fails to tell us why both males and females became naked. Another difficulty with this idea is that becoming naked should also then have benefited those other animals that lived in the noonday sun, and especially those that we chased, but they haven't lost their fur. There are also some calculations that animals without fur would lose more heat at night—when they need to retain it—than they could give off during the day, returning a net loss for being naked. Finally, there is the problem that when we left the savannah and travelled around the world, our hair did not grow back as we inhabited colder climes. Natural selection would have had plenty of opportunities to pick people out who were slightly more hairy than others—they are easily spotted at the nearest swimming beach or public pool.

A different idea to explain our nakedness gets our brains involved, but the link won't be immediately apparent. It is a little-known fact that the single largest cause of death in animals, and probably all organisms, is parasites—viruses and bacteria, but also things like biting flies, or lice, ticks, worms, and other infectious organisms that transmit disease, suck our blood, and live in our guts where they eat our food. Because these parasites reproduce so quickly they can always stay one step ahead of our immune systems. Malaria, transmitted by a mosquito bite, still kills far more people around the world each year than wars. Recent estimates put the figure at around 1 million per year in Africa alone. The human immunodeficiency virus or HIV kills even more people than malaria, currently around 3 million per year. Few of us realize just how much energy we give over to our immune systems to fight these and other parasites, but farmers know that cows fed antibiotics get bigger. What? They get bigger because they do not have to use up precious energy fighting off infectious diseases, and freed of that burden they turn the excess energy into growth.

Fur is a convenient and safe home for the parasites known as *ecto-parasites*, creatures such as flies and ticks that plague us either by living on the outside of our bodies or by biting us and transmitting diseases such as Lyme or malaria. A measure of the burden of these flies is that many animals spend large portions of their days—over 25 percent in some cases—trying to remove ectoparasites. Monkeys huddle in groups to groom each other, not as a way to look better but to remove parasites. So bountiful is the harvest that one possible motivation for grooming someone is to be able to eat the parasites you get. But the more probable reason is to "curry favor"—the word "curry" here appropriately being an old expression for brushing a horse—by removing someone else's ticks. Grooming is not always an option, and horses, cows, bears, and just about all large mammals rub themselves against trees at least partly to remove parasites. Some animals—horses among them—even have specialized muscles for twitching their skin to make flies jump off, or they can switch their tails to swat at them.

Now, the connection between our brains and our nakedness is that our lack of hair might be a form of domestication by intelligence. Walter Bodmer and I have suggested that humans might have lost their fur as a way to reduce the burden of these annoying and disease-carrying ectoparasites. Our great intelligence—or our abilities at social learning—would have made us uniquely suited to replace the functions of hair or fur with our technologies. We can build fires, create shelters, and make clothes as ways of avoiding the loss of heat when we needed to retain it, or as a way of blocking the sun when its baking rays become too hot. Unlike fur, clothes can be changed and washed or even discarded if they get infested. Having these technological replacements for fur available at our fingertips might have set us on a trajectory of becoming less hairy as natural selection favored people who carried around fewer infections.

It might not be a coincidence that where humans do still get ectoparasite infections, it is on those parts of our bodies covered with hair. We get head lice, and some unfortunate or adventuresome ones among us even get pubic infestations, but no one gets arm lice, or colonies of them on their legs, face, or backs. Another hairless mammal is the apparent exception to the mammalian norm of having fur, but it might just prove the ectoparasite rule. Naked mole rats, as their name implies, have lost their fur. They live in large and densely populated underground burrows and rarely venture out. The burrows are warm and humid places where parasites can easily multiply and move from host to host. Quite possibly, then, naked mole rats have been able to lose their fur as a way to control parasites, but without the threat of freezing to death.

Lice might even allow us to put a date to when we became naked. Body lice plague all animals with fur, but humans, uniquely it seems, are blessed with two distinct varieties. One is specifically adapted to living in thick hair—this is the so-called head lice species—but the other has evolved adaptations for living in clothing. The molecular biologist Mark Stoneking ingeniously recognized that the two species might have separated when humans adopted the habit of wearing clothes. If this habit arose because, owing to a lack of fur, we were getting cold at night or perhaps shivering in strong winds, then clothes put a date on our nakedness. Because clothes do not normally remain preserved at archaeological sites for more than a few thousand years, there is no good evidence as to when this happened. But by careful comparison of slight differences that had accumulated in the genes of these two otherwise closely related species, Stoneking was able to infer that they separated about 107,000 years ago. It is remarkable to think that our burgeoning nakedness around that time created not just modesty but an industry that accounts for billions in sales worldwide today. And it is all down to our ingenuity at ridding ourselves of parasites.

(Just to show how difficult research on these topics can be, Alan Rogers with S. Wooding undertook a different approach to trying to estimate when we became naked. Rogers studied the melanocortin 1 receptor, which is a gene that influences skin color. People of African descent all share a particular variant of this gene that produces their darker skin and confers a strong resistance to sunburn. Non-Africans have many different varieties of this gene, but none of them produces dark skin. The version of this gene in chimpanzees, which are covered with fur, differs from any human form but also fails to provide resistance to the sun. This led Rogers to wonder if the sun-resistant form of the gene might have arisen when we became hairless and thus exposed to the sun. His analysis suggested the sun-resistant form might have appeared around 500,000 years ago. Rogers has gently teased me that if we accept his analysis and the clothes-lice story is correct, we might have stood around naked and cold for about 400,000 years! The comment might not be as far-fetched as it sounds. When Darwin visited Tierra del Fuego at the southern tip of South America, he found the native Fuegians essentially naked in this cold and harsh climate. They seemed not to have a tradition of making clothes; to stay warm, they smeared their bodies with seal fat and slept curled up together in groups. They also made fires. Indeed, in 1520 Magellan called this region the "land of smoke" after the hundreds of beach fires he observed from offshore in his ship. Only later was the name changed to Tierra del Fuego or "land of fire.")

If having less hair did grant the naturally selected advantage of reducing our burden of parasites, it would probably quickly have become part of our tastes in a prospective mate. That big hairy guy might just be carrying ticks. Sexual selection is the process by which natural selection favors traits that make it more likely you will attract a mate. Sexual selection normally acts on one sex more than the other, and often it is males. The reason is that males typically have greater reproductive potential because a male can easily produce lots of children with many different females, while females are limited by how many babies they can gestate in their reproductive lifetime. This difference in reproductive potential means that males will normally be forced to compete with each other to attract females because there are more males than are needed. Females, in their turn, can then afford to become choosy about who they mate with. And this is why the large and gaudy ornaments, songs, trills, odors, and sexual displays of most animals are found in males—they are all ways of persuading females to mate with them rather than some other fellow.

Indeed, we might expect that females will have evolved expensive tastes, and all because of this difference in reproductive potential.

But when it comes to hairlessness, both women and men are expected to prefer less hair in a partner for the simple reason that both want a healthy mate, and neither wants to catch the other's parasites. These considerations could explain why many people find hirsuteness unattractive, and why products for removing hair from our bodies are such a big industry, sending huge advertising revenues to television stations all too happy to broadcast commercials about the latest five-bladed razor or hair removal cream for women. Even some of our preferences in fashion might reveal ancient tendencies to avoid people who could be carrying parasites. One of the most enduring features of women's fashion is the backless dress. We do not normally think of backs as secondary sexual characteristics like breasts or hips, so why all the interest in them? It might not be an accident that our backs expose the single largest patch of bare skin on our bodies. A backless dress, without our even being aware of it, acts as a billboard broadcasting one's healthy—and hairless—skin.

Our nakedness exposes another genetic trait that we have lost, or nearly so, and again domestication by our brains might be the reason. Among the most striking artistic or symbolic objects that our ancestors produced were the Venus statues that have been found from as far back as 24,000–30,000 years ago. They depict women with exceptionally large thighs and bottoms, and some also have large breasts and other hypertrophied sexual characteristics. The statues are often interpreted as exaggerated or symbolic forms, representing ideals to be upheld or sought out. Throughout most of our history as hunter-gatherers, starvation or near starvation was a daily fact of life, at least until the invention of agriculture. A woman who could store enough fat to attain a shape like those depicted in the Venus statues would have been a walking advertisement for her ability to acquire food and to provide for her children. This might seem trivial to us today when food stores for many of us are often no more than a few minutes away, but not to hunter-gatherers.

There is good reason to believe that the Venus statues might not have been exaggerations, or not just exaggerations. A now rare morphological trait known as *steatopygia* produces nearly exactly the hip shape depicted in Venus statues. Women throughout our history might have been more at risk of starvation than men because they would normally be providing food for themselves but also for any child they might have been gestating, nursing, or rearing. Indeed, a hunter-gatherer female would have been in one or more of these circumstances for nearly all of her adult life. Steatopygia is an example of natural selection not just providing these women with insurance against starvation but an exquisitely fine-tuned one. Fat stored on the hips requires less energy to carry around because this is where our center of gravity lies. Steatopygia might very well have been the normal shape of some African and Andaman Island women until as recently as 10,000 years ago. But ever since our brains came up with the idea of agriculture, it has not been as advantageous to store fat and the trait has nearly disappeared. Even so, we still see ancient remnants of it in our tendency to store excess fat on our hips when we gain weight.

Our domestication continued when the plants we domesticated, also beginning around 10,000 years ago, turned around and changed us. The *alcohol dehydrogenase gene* or *Adh* helps animals, including humans, to metabolize alcohol. This protects our livers but also our brains. Common fruit flies carry this gene because they are regularly exposed to naturally fermenting fruit. Genetic studies of Han Chinese and Tibetan populations show that around 10,000 years ago natural selection began to act strongly on these people, favoring a variant of the *Adh* gene that improved their ability to degrade alcohol. This corresponds to a time when rice crops were being domesticated and rice production was spreading across what is now southern China. It is not known whether these people acquired their *Adh* genes from regular consumption of rice wines they produced or simply to protect them against routine exposure to alcohol from naturally fermenting rice. But among contemporary Han Chinese, those carrying the selected variant are less likely to suffer from alcoholism.

The trend for culture to select our genes by domesticating us means that modern humans are far more closely related on many of their genes than the passage of time might suggest. The reason is that "natural selection" is really just a euphemism for selective death. Strong selection means those who lack certain combinations of genes are more likely to die before reproducing, while those lucky few who have them become the progenitors of the rest of us. Modern humans entered Europe sometime around 40,000 years ago, but owing to selection, that does not mean that if you are of European origin your genes are separated by that amount of time from those of other Europeans. On many of your genes you will share common ancestry with *all* other Europeans as recently as a few thousand years ago, or even more recently than that. If you are reading this book on a train or airplane, the stranger next to you might be far more closely related than you think, at least on some of your genes.

One of the best-known examples of this is the ability to digest milk as an adult. The 5,000 or so animal species that make up the mammals are the only animals that produce milk. All mammal infants can digest it because they have a gene that makes an enzyme called lactase, and this enzyme breaks down the lactose sugars found in milk. After weaning, mammals no longer have access to milk, and so the gene that makes the enzyme gets switched off. This would have been true of humans throughout our evolutionary past, but an ability to digest lactose milk sugars as adults is now common in people of European and African descent. What they share is a cultural history of having ancestors who, sometime around 10,000 years ago, began domesticating animals. In what would prove to be a double act of domestication, these animals went on to domesticate their owners. Cows, sheep, and even camels could provide a ready supply of meat, but also of milk. The meat was edible, but the first groups of people to domesticate these animals would have found the milk largely indigestible, at least to the adults.

But then someone in Europe and someone else in Africa each got lucky. It was in fact a 1 in 3 billion chance for each of them. Our genome is made up of about 3.3 billion of the chemicals called bases or nucleotides. It turns out that a mutation or change to a single one of them confers the ability to digest milk as an adult. The solution was simple: ensure that the genetic switch that turned off the infant's ability to digest milk sugar got disabled, meaning that the ability to digest milk persisted throughout life. Natural selection found just the right switch in both of these people, and they are in slightly different places on the same gene. The two variants of this gene are among some of the most rapidly evolving that have ever been studied. They might have arisen only around 6,000 years ago, but so great is the advantage of being able to digest milk as an adult that all of us who can do so are recent descendants of these two lucky people who were around at that time.

Chances are that if you see someone near you drinking a latte and you are both either of European or African ancestry, the two of you will share a very recent common ancestor—someone who probably lived in the last few thousand years and was lucky enough to have had this gene. There could not be a clearer demonstration of the power of human culture to shape and select our genes. Nothing in our evolutionary history or in the entire history of the mammals would have seen this coming. Indeed, new evidence confirms that Neanderthals were not tolerant to lactose as adults, so the ability arose only in our lineage. But like domestication in general, this one has also made us more juvenile or even infantile, as we now somewhat lazily rely on animals for energy that in the past we would have had to use our brains to hunt and forage for.

still evolving

IT IS sometimes said that the question of whether we are still evolving can be rephrased as, Are we all having the same number of offspring? Looking around the world, some groups are having more children than others. But that alone is not sufficient to say human populations are evolving. Natural selection maximizes the number of *grandchildren* you leave, not the number of children. I might produce ten children and you might produce three. But if I cannot provide for my ten, they might not produce as many of their own children as your three. Time will tell who among us is leaving the most grandchildren, but for now the question is not easy to answer.

Still, there is reason to believe that humans are still evolving, and the reasons come back to our brains: the real message of our evolution is written into our responses to the cultural changes our brains have unleashed. *HAR1* will almost certainly prove to be just one of many genes affecting the structure and organization of our brains. Two that have already been identified are *ASPM* and microcephalin. A variant form of microcephalin that arose 37,000 years ago is currently sweeping through human populations. The timing of its probable origin corresponds to the full flowering of culture in fully modern humans. A variant of the *ASPM* gene, also sweeping through human populations, arose just 5,700 years ago, coinciding with the spread of agriculture and animal domestication, the development of cities, and early writing. Its remarkably young age implies that the human brain is still evolving and evolving rapidly. If you are of European, Middle Eastern, or Far Eastern descent, including Iberians, Basques, Russians, North Africans, and South Asians, chances are you have this variant in your brain. But if so you should not conclude that you have higher intelligence—the variant form might simply confer some sort of metabolic or energetic difference. No one yet knows.

Beyond brain genes, genes that affect the size of our teeth and jaws and genes related to our appearance and attractiveness to others are still evolving. Even where we don't have direct genetic evidence, there is good reason to expect that as we change the nature of our society, many new kinds of abilities will be favored, and this will drive new trends in our evolution. The ability to concentrate for long periods, as is required of many workers who use computers, or as is increasingly asked of school-age children, might be one such trend. Language and communication abilities continue to grow in importance as more and more of our business is about trades and exchanges. The number of children someone produces is statistically linked to their wealth, and so if there are any genes that might be related to temperaments or abilities that affect our earning power, those genes will be evolving. Current trends among women to have children later in life will favor women with a genetic predisposition to delay their menopause.

Crime has been falling steadily for at least 1,000 years, and we have every reason to believe that it has been falling for the last 10,000 years, or ever since we began to live in larger communities. Remarkable statistics kept for the last millennium in the United Kingdom show that crime rates have declined throughout that period, and have shown especially sharp drops at periods in history when militias or other gun-bearing groups have been disarmed. Just as there is selection for tameness in the domestication process of wild animals, modern society has increasingly selected against excessively aggressive people within communities. Violent and antisocial people are increasingly being pushed to the margins of society, where they have fewer job prospects and fewer opportunities to reproduce. They may be put in jail, or even executed. The world can sometimes seem a violent place, but we are steadily becoming a more democratic and peaceful species, more cooperative, kinder, more empathetic, and more generous, descended from more aggressive ancestors in our not-too-distant past. We are still being tamed by culture, and if there are genetic differences among us that map onto these more peaceable natures, they will be finding favor in our ranks.

PART III

THE THEATRE OF THE MIND

Prologue

PART II SHOWED US the complex network of social relationships, alliances, reputations, and identities that make the cooperative venture of human society work. But those chapters ignored an opportunity that lurks in the background of our cooperative societies. That opportunity is to manipulate them to serve us just that little bit better than average, to somehow get a bit more than our share of the spoils. How have our minds been structured to do this, especially as other minds will be trying do the same as ours?

Before we can begin to answer these questions, we need to confront the greatest mystery of our minds, and that is consciousness itself and just who or what is in charge of it. For most of us, our day-to-day existence is taken for granted and led by an actor we know well. The actor is us, the "I" that is doing the speaking when we say, "I did this," or, "I did that," the "me" that we see when we peer inside our minds, and the "you" that is reading this text. If you are like most people, your connection to your inner self is so intimate that it defines you—you cannot imagine yourself as someone else, and you never go away (even though sometimes you might wish you did), except perhaps when you are asleep.

Remarkably, there is reason to believe that this part of our self, this "I" or "me," is something of an illusion. The great eighteenth-century philosopher David Hume, thinking about personal identity, wrote that "for my part, when I enter most intimately into what I call myself, I always stumble on some particular perception or other, of heat or cold, light or shade, love or hatred, pain or pleasure. I never can catch myself at any time without a perception, and never can observe any thing but the perception."

What Hume had stumbled on was the realization that he could not find a "watcher"—no homunculus residing in his brain summoning the next perception or idea; no self that was independent of his thoughts. He concluded that we

> are nothing but a bundle or collection of different perceptions, which succeed each other with an inconceivable rapidity, and are in a perpetual flux and movement . . . the mind is a kind of theatre, where several perceptions successively make their appearance; pass, re-pass, glide away, and mingle in an infinite variety of postures and situations. . . . The comparison of the theatre must not mislead us. They are the successive perceptions only, that constitute the mind; nor have we the most distant notion of the place where these scenes are represented, or of the materials of which it is composed.

Hume's ideas can be difficult to accept. Are we really just a collection of different perceptions that come gliding along into our minds, and if we are, who or what decides what gets presented? Our perception of consciousness or what it means to have a mind is that there is precisely a sort of inner screen on which our life plays out, and that "we" are there not merely to watch it but to control what gets played on it. But we now know that Hume's instincts were right: there is no central place where our brains collect up all our thoughts and present them for "you" to watch. If there were an inner "you," then that you would also have to have an inner you, and so on, ad infinitum. For no less a giant of European philosophy than Immanuel Kant, Hume's ideas on the mind and metaphysics were enough, in Kant's own words, to break "my dogmatic slumber."

There is a proposal from modern brain-imaging studies that

would have interested Hume, and forces us to take seriously the idea of our minds as a place "where several perceptions successively make their appearance; pass, re-pass, glide away, and mingle in an infinite variety of postures and situations." It is that even at rest our minds are not merely blank, springing into action only when something requires their attention. Rather, our minds spontaneously engage in what has been called "stimulus-independent thought." They naturally wander, flitting from one thought to another with fluidity and ease, and without being "asked," as if their default setting is to be busy and occupied. Who or what is causing this, and for what end? No one really knows, but regions in the brain's cortex that cause this activity have been identified. Stimulus-independent thought might be why our minds daydream, rather than merely going into idle when we are not using them, like the impression we might get of a cat sitting on a windowsill with its eyes slowly falling shut. Stimulus-independent thought might be why we are easily bored, do crossword puzzles and brain-teasers, play chess, read books and visit art galleries, watch films, play or listen to music, or even just draw circles in the sand or toss stones into the air if only to see how they fall. Stimulus-independent thought might also be why we dream and could be a reason why some people suffer from insomnia.

It is as if our minds have appetites for thinking, and this might be just what we expect of a mind that rides inside a body that has a vastly greater capability for changing its world than does the body of any other mind. On the other hand, some might speculate that our stimulus-independent thought is merely a byproduct of thoughts vying for our attention. This is not the radical idea it might sound. Ideas have to compete for space in our minds, and those best at somehow making their way into our consciousness will have a much higher chance of being transmitted to some other mind. It could even be that these "selfish" ideas created consciousness itself as a way of getting us to tell others about them. What appears as a region of our brain that acts independently of our thoughts—the part of our brains responsible for stimulus-independent thought—might just be a launch platform they created.

But it is doubtful this is the whole story of the contents of our minds. Stimulus-independent thought might provide the raw input to Hume's theatre, but the idea of natural selection, that the fittest survive, may help to frame its contents. Writing in 1995, Richard Dawkins called our attention to the fact that "not a single one of our ancestors was felled by an enemy, or by a virus, or by a misjudged footstep on a cliff edge, before bringing at least one child into the world." Dawkins's simple but compelling point was that we inherit our genes from a long line of survivors—and not from those who lived alongside our ancestors but failed to survive; there is real selection in natural selection. It is why the current generation of survivors have genes that are good at making bodies: over many generations, natural selection has concentrated into the survivors' bodies genes that make bodies good at surviving. It is why our hands and feet, eyes and ears, are so good at being hands and feet, eyes and ears.

Natural selection, weeding out the survivors from the non-survivors, is also why brains are so good at being the central processing and relay centers of our bodies. But while other animals have brains and they think, it is our conscious awareness, that self-reflective sense of self, that we think sets us apart. A dog sitting at the base of a tree, peering up into its branches, might be thinking that there is a squirrel up the tree, but we don't believe it is having the thought, "I think there is a squirrel up there." Dog owners will disagree with this remark, but try this out with your dog. Drive a stake into the ground and then another maybe 20 feet away. Put a long leash on your dog and tie one end of it to the first stake. Then loop the dog by its leash around the second stake so that it is facing back toward the first. Now put a plate of food just out of reach. People who study animal behavior call this "the detour task." Experiments show that your dog will strain and strain against its leash "thinking" that if it just pulls harder it will get the food.

Most dogs do eventually work out what to do, but a dog that was "having thoughts" about its predicament would solve the problem immediately, as we would expect of a human set this task. So to Dawkins's list we can tentatively add that not a single one of our ancestors' *perceptions*—in Hume's broad sense—let them down. None of our ancestors was felled by misperceiving "heat or cold, light or shade, love or hatred, pain or pleasure." But we can also add to this that none of our ancestors failed to survive from misjudging an opponent's rage, misunderstanding the collective mind of their group, by failing to communicate their own intents, needs, or desires, or by misunderstanding these in others, not at least before bringing a child into the world. Our social and psychological framework, even the broad outlines of our aesthetic and religious preferences, may have been molded this way. It is easy to dismiss this contention as unimportant in our modern world, but that is not the world in which our minds evolved. That world was one, to paraphrase John Irving in *The World According to Garp*, with an "undertoad," and in which "the dog really bites."

Still, I say "tentatively" in the preceding paragraph because some may wish to exempt our conscious minds from any sort of control or shaping, allowing us "free will," or the ability to think about what we want to think about, when we want to, and why. But this may be to ask the wrong question or expect too much. What we want to ask of our minds is what someone with a conscious mind who wanted to be a survivor would use that mind for, and how would they use it. Returning to Hume's notion of a theatre, then, we can ask just what will appear in it. The answers are the topics of chapters 8 and 9, but we can anticipate them in a general way. We should see that our minds are especially finely tuned to manipulate social systems for their own ends because that is how they will reap the most benefits. But remember, if our mind is a theatre, you might have less control over its contents than you think.

CHAPTER 8

Human Language—The Voice of Our Genes

That our language evolved because we are the only species with something to talk about

something to talk about

YOU POSSESS the most powerful, dangerous, and subversive trait that natural selection has ever devised. It is a piece of neural-audio technology for rewiring other people's minds. You have a way to implant thoughts and ideas from your mind directly into someone else's mind, and they can attempt to do the same to you, without either of you having to perform surgery. Instead, natural selection has equipped you with an apparatus for producing action at a distance. It is not a roar or a bark, but something far more sophisticated. It is your language, of course; its apparatus is your brain and vocal cords and its medium is sound waves that travel through the air. When you speak you are using a form of telemetry, not so different from the remote control of your television; it is just that your language relies on pulses of air pressure rather than pulses of infrared light. Just as we use the infrared device to alter some electronic setting within a television so that it tunes to a different channel that suits our mood, we use our language to alter the settings inside someone else's brain in a way that will serve our interests. Language is your genes talking, getting things they want.

Language's subversive power has been recognized through the ages in censorship, in words you cannot use and in phrases you are not allowed to say. According to the Bible story of the Tower of Babel, early humans developed the conceit that by working together they could build a tower that would take them to heaven. God, angered at this attempt to usurp His power, destroyed the tower, and to ensure that it would not be rebuilt he forever confused the people by giving them different languages. Even today, saying the wrong thing at the wrong time or place can get you accosted, thrown into jail, or even killed, and all because of a puff of air emanating from your mouth. Our training in the arts of using this powerful instrument starts early on. Imagine the amazement in the mind of a child when it first discovers that just by making the right sound or sequence of sounds, it can get objects, as if by magic, to move across rooms, and into its hands, or maybe even into its mouth.

We instinctively recognize that our language is unique among all forms of biological communication, but what do we mean by that? All animals communicate, but humans are the only animals with language. Human language is distinct in having the property of being *compositional*: we alone communicate in sentences composed of discrete words that take the roles of subjects, objects, and verbs. This makes our language a *digital* form of communication as compared to the continuously varying signals that typify the grunts, whistles, barks, chest-thumping, bleating, odors, colors, chemical signals, chirruping, or roars of the rest of life. Those familiar sights, sounds, and smells can only be more or less intense, or more or less persistent. They might signal an animal's status, or intentions, or indicate its physical prowess; they might tell a predator it has been spotted, or send a message to nearby relatives of an imminent danger. But lacking subjects, verbs, and objects, these acts of communication do not combine and recombine to produce an endless variety of different messages.

Thus, your pet dog can tell you it is angry, and even how angry it is, but it cannot recount its life story. By comparison, we can use our language to look into the future, share the thoughts of others, and benefit from the wisdom of the past. We can make plans, cut deals, and reach agreement. We can woo prospective mates and warn off our enemies. We can describe who did what to whom, when they did it, and for what reason. We can describe how to do things, and what things to avoid. We can express irony, surprise, and glee, worry, pessimism, love, and hate or desire. We can be witty or grave, we can be precise or deliberately vague.

No one knows when the capacity to communicate with language evolved, but we can narrow the range of possibilities. Our closest genetic relatives the modern chimpanzees cannot speak, but we can, and so this tells us that language evolved sometime in the 6 to 7 million years that separate us from our common ancestor with them. *H. habilis* and *H. erectus* skulls reveal the impressions of two slightly protruding regions of the brain—known as Broca's and Wernicke's areas after their discoverers—that neuroscientists have identified as being involved in speech, at least in humans. This had led some researchers to suggest that a capacity for speech, even if perhaps a rudimentary one, might have existed 2 million or more years ago. But Broca's and Wernicke's areas are also enlarged in some apes, so their presence is not by itself a trustworthy indicator that a species had language. There is also no evidence to suggest that these early *Homo* species had anything even remotely close to the complex societies, tools, or other artifacts that we recognize as fully human.

A question that excites great differences of opinion is whether the Neanderthals spoke. This excitement has been fanned by the recent discovery that the Neanderthals had the same variant of the segment of DNA known as *FOXP2* that we do, and that has been implicated, among many other effects, in influencing the fine motor control of facial muscles that is required for the production of speech. But, just as was true of Broca's and Wernicke's areas in *H. habilis* and *H. erectus*, having the same variant of *FOXP2* as modern humans do doesn't tell us that the Neanderthals had language—they might have but, as we have seen in earlier chapters, we cannot conclude that they did on the basis of this short segment of DNA.

We cannot of course be sure that the Neanderthals lacked language, but this line of genetic reasoning combined with the Neanderthals' relative lack of cultural sophistication compared to our own—sophistication that we associate with social learning—points to the conclusion that language evolved with the arrival of our own species. Our capacity for language was almost certainly present in our common ancestor because today all humans speak and speak equally well. There are no languages that are superior to others and no human groups that speak primitive as opposed to advanced languages. This would make language no more than around 160,000 to 200,000 years old, although some anthropologists think language arose even later than this, pointing to the sharp increase beginning around 70,000 to 100,000 years ago in evidence of our symbolic thinking and the complexity of our societies. It is just possible that all modern humans alive today trace their ancestry back to common ancestors who lived around that time, so language evolving at this later date, though unlikely, is possible.

Whenever language emerged, it probably built on what had been a series of developments in our ancestors of what we might think of as *proto-languages*, backed up by the physical apparatus to make languagelike sounds. Our language capacity might have grown out of ancient primate abilities to vocalize and gesture, and probably required a capacity to string ideas together in chains, and to anticipate what others might be thinking. Tecumseh Fitch in *The Evolution of Language* describes how the early proto-languages might have contained just a few isolated words with very little in the way of grammar. Thus, pointing and making a few sounds might have slowly evolved into that sound coming to be associated with whatever was being pointed at, and the first noun was born. Making a different sound while perhaps stabbing a spear at that same thing—maybe a wildebeest—could have given rise to the first verb. Put these two together and the first primitive sentence was born: "Stab wildebeest!" It wouldn't take long for pronouns to arise as a way of getting someone to do your stabbing for you, and "You stab wildebeest!" might have been heard echoing across the savannah.

Still, why did language wait to make its appearance in our species, and not earlier or in any other species? A hint might be found in the quite remarkable lack of social complexity of the *Homo* species that preceded us. The lesson of these species is that in trying to explain the appearance of language we should look for the need for it, not just to pieces of anatomy or to genes. Evolution doesn't just produce complex adaptations like language, hand them to a species, and then sit back and see what they do with them. There must be something that is pulling the trait along so that it pays its way by granting some advantage to its bearers.

To most people, the advantage of having language is obvious—it allows us to communicate. This is true, but we need a theory or explanation that tells us why we have language and no other species does. Indeed, it is not even clear that "communicating" on its own is sufficient to get language to evolve in us or any other species. This is because it is only by granting its bearers some individual advantage that natural selection can get a trait to evolve. But when it comes to language, this leads to two evolutionary predicaments. One is that much of what I have to say might benefit you, and potentially at a cost to me. Thus, if I tell you what plants can be used to make poison to put on your arrowheads, you might not only kill more prey than me, you and your family might even manage to kill what few there are before I do. Natural selection never promotes naive altruism, so surely this will favor people who keep their mouths shut, and language will die a silent death. On the other hand, maybe I can instead use my language to mislead or trick you. Now my actions are no longer altruistic and indeed might help me at *your* expense. But this poses the second predicament. If you know that my acts of communication are designed to benefit me, surely this favors people who don't listen. And even if you do listen, why should you believe me? Talk is, as they say, cheap.

But wait, what if I tell you I will teach you how to get poisons from plants if you teach me how to make a better arrow? Or, what if one day out on the tundra you have acted with great courage in bringing down a mammoth, and back in the village, I make sure word of your exploits travels fast, hoping you will give me more than my share of the meat? Or maybe you and I hatch a plot along with seven of our friends to raid the neighboring village. These are all social actions that involve the exchange of ideas, striking deals, and coordinating activities, and they depend on trust and reputations. They are also things that only we humans do; in fact, cooperation of this sort has been our species' secret weapon, returning enormous benefits and making our cultural groups formidable and fearsome competitors. We have acquired an entire psychology geared to making our style of cooperation work and to protect us from the risks inherent in even the simplest acts of reciprocal exchanges with others who might try to take advantage of our good will.

And this now gives us an insight into why language evolved because we have quietly ignored the fact that we couldn't undertake any of these acts of cooperation without it. Language evolved to solve the crisis that began when our species acquired social learning—probably some time around 160,000 to 200,000 years ago—and immediately had to confront the problem of "visual theft." Remember that was the crisis that arose when humans became able to copy each other's best ideas. Language solves this crisis by being the conduit that carries the information our species needs to reach agreements and share ideas, and, as we saw earlier, it makes the "marketplace of reputation" possible. It disarms our conflicts and turns them toward our advantage. This requires something more than the bleats, chirrups, roars, chest thumping, odors, and bright colors of the rest of the animal kingdom. Our social complexity depends on language: without it we might still be living like the Neanderthals were when we first entered Europe sometime around 40,000 years ago.

Both evolutionary predicaments are solved. You and I can use language to arrange and negotiate the terms of an exchange of information, goods, or services that brings benefits to both of us. These are reasons for me to speak and for you to listen. But words *are* cheap, so how can you trust mine? You can't, and you will know that I might try to use language to acquire more than my share of the benefits of our exchanges. Still, language provides you with a trait that you can use to "police" my actions. If I fail to keep up my end of the bargain, you can break off our partnership and quickly, widely, and *spitefully* spread the word that I am not to be trusted. This spite is cheap for you and might bring benefits your way by convincing other cooperators to work with you rather than with me. On the other hand, you must exercise your spite carefully lest you develop a reputation as someone who spreads false rumors. Actions might speak louder than words but words travel faster and further. Language "watches" and regulates our social behavior, and in a way that dictators armed with arrays of CCTV cameras could only dream of.

Contrary to what most of us probably take for granted, then, it is not the main function of language merely to communicate, as when, say, two computers send information back and forth. Rather, language evolved as a self-interested piece of social technology for enhancing the returns we get from cooperation inside the survival vehicles of our cultures. And this answers the question of why we are the only species to have it: we are the only species with language because the unique nature of our social systems means we are the only species with something to talk about. Thus, when we watch animated films with talking squirrels, deer, bears, rabbits, and other animals, it is not just their language that is out of place, it is their behavior: these animals are acting like humans. They are making plans, coordinating their actions, cooperating, and trading with each other, the very things other animals don't do. In Kenneth Grahame's *Wind in the Willows*, when Mole, Badger, and Ratty go to Toad Hall to try to convince Toad to change his errant ways, they are not acting like badgers, water rats, and moles. Real badgers, water rats, and moles don't visit country homes, row in boats, tend gardens, live in houses, or do many of the other things in that charming book—they don't even have names!

Thinking of language as a trait for promoting cooperation might give us insight into what the great, but sometimes exasperatingly gnomic, philosopher Ludwig Wittgenstein meant when he said, "If a lion could speak we could not understand him." Maybe it is this: if a lion could speak it would not be a lion, but something more like us—we wouldn't understand it as a *lion*. We wouldn't understand it as a lion because lions behaving as lions don't really have much to discuss beyond what they already achieve with their forms of communication. If they did have more to discuss, they wouldn't be lions, but something more like Mole, Badger, and Ratty. In fact, most of the communication in animal social systems is about signaling location, or who is dominant to whom, or disputes about food, territory, or mates. These are issues that can be settled by signals of grunts, chirps, whistles, odors, chest-thumping and head-butting, bites and grimaces, and that is why animals have these systems of communication rather than language. Why talk when a roar will suffice?

An unforgettable passage from Alexander Kinglake's *Eothen* captures the sense in which language is not merely for communicating. Meaning "from the East," *Eothen* is a first-person account of this young Englishman's Grand Tour of the Near East in 1834. Kinglake decided to travel by camel across the Sinai to Cairo, accompanied by a servant. After being out in the desert for several days, he describes seeing

> a mere moving speck on the horizon. . . . Soon it appeared that three laden camels were approaching, and that two of them carried riders; in a little while we saw that one of the riders wore European dress, and at last the riders were pronounced to be an English gentleman and his servant. . . . As we approached each other, it became with me a question whether we should speak. I thought it likely that the stranger would accost me and in the event of his doing so I was quite ready to be as sociable and chatty as I could be, according to my nature; *but still I could not think of anything particular that I had to say to him.* Of course, among civilized people, the not having anything to say is no excuse at all for not speaking, but I was shy and indolent, and felt no great wish to stop and chat like a morning visitor in the midst of those broad solitudes. The traveler perhaps felt as I did, for, except that we lifted our hands to our caps and waved our arms in courtesy, we passed each other quite as distantly as if we had passed in Bond Street. [italics added]

As it turns out, the two protagonists eventually did speak, but only because their camels slowed up and turned around! Even then, the other man's greeting was the meager "I dare say you wish to know how the Plague is going on at Cairo?"

It is little short of extraordinary that these two ships could almost pass in the night without speaking to each other, and strains credulity when we appreciate the setting. Kinglake's only partly ironic use of the word "accost" to describe being spoken to tells us something of the diffident and private character of Englishmen of his day (not entirely extinct even now). But there is something more to this example than reserve. Kinglake is right: there is an element of being accosted when one is spoken to out of the blue, and this is precisely because language can zero in so quickly on matters that we might wish to keep private, matters that were they made public could damage our reputation, give away our plans or simply provide information to someone that could benefit them at our expense. Not knowing the other man on the camel, Kinglake might have concluded that he really didn't have anything to say to him. Confronted with the same realization, the other man did what we all do when we have nothing to talk about: his remark was the equivalent of talking about the weather.

language, dna, and regulation

IN THIS section I want to examine a remarkable similarity between the nature of human language and an unusual feature of our genetic systems. That similarity is that both evolved specifically to promote replicators' interests within their vehicles—genes in their bodies on the one hand, and people in their societies on the other. Thus, we will see that the nature of our language is precisely what we would expect of a system that evolved to allow us to vary our expression—or the way we are seen—inside our cultural survival vehicles, and in a way that benefits us, just as genes can vary their expression inside our bodies in ways that benefit them. Understanding this similarity will explain why human language had to be different from all other forms of animal communication and why it had to adopt a form that we also find in our genes.

To begin this story, we need to appreciate a puzzling feature of our genetic inheritance. Our genomes represent the sum total of the genetic information on our twenty-three pairs of chromosomes. In our species, these chromosomes contain somewhere around 21,000 genes. What is perhaps surprising is that this is scarcely more than the 19,000 genes in *Caenorhabditis elegans,* a small worm about 1/32 of an inch long that lives in soil; the 15,000 genes of a fruit fly; and only four times more than the number of genes in a yeast, which is just a simple microscopic single-celled organism that causes your fruit to ferment. In spite of having similar numbers of genes to the two animals, our bodies are almost unimaginably more complicated, comprising trillions of cells making an uncountable number of connections, and building hundreds of different kinds of tissues, from eyes and muscles, to hearts (a special kind of muscle), to kidneys. Our brains alone account for many hundreds of billions of these cells, and counting the connections they make to each other is something like trying to count all the stars in the universe. By comparison, *C. elegans'* body is composed of fewer than 1,000 cells. And yeast of course doesn't even have cells (just the single cell), much less arms or legs, digestive tracts, blood vessels, or brains.

We learn two unexpected lessons from this: one is that we are woefully underspecified. Our genes alone cannot possibly carry enough information to specify all the connections that make up our bodies. The other is that an organism's complexity seems not to be related in any obvious way to how many genes it has. How, then, do we achieve our vastly greater biological complexity than a simple worm or fruit fly? The answer appears to reside in how we *use* our genes, and not in how many we have. For instance, all mammals have about the same number of genes, and yet they differ remarkably in how they have used them to produce their different sizes, shapes, and capabilities. How we use our genes is why we can share over 98 percent of our genes with chimpanzees but differ so utterly from them. We even have over 80 percent of our genes in common with mice, and 75 percent in common with the much-loved monotreme mammal, the platypus. Moving outside of the mammals, you share 60 percent of your genes with a fruit fly, and even 50 percent with bananas!

A mysterious feature of our genomes that we might think of as their "dark matter"—that as-yet-unidentified substance that is thought to account for a majority of the matter in the universe—is emerging as a principal reason why we can share so many genes with these other species and yet be so different. The common view of our genomes as being packed with genes that provide the instructions to make our bodies turns out to be only a small part of the story. Our genomes contain vast stretches of DNA that are not organized into genes, and are not used for making the protein building blocks of our bodies. This is the dark matter, better known as "junk DNA," and it comprises a startling 99 percent of our genome, and the genomes of most other "higher" organisms. It is extraordinary but true that only about 1 percent of our genomes is made up of the things we normally think of when we talk about genes. Junk DNA's existence had been appreciated since the early part of the twentieth century, but the discovery in the late 1970s that it was present in such vast quantities was seen as "mildly shocking" even by the editor of the prestigious scientific journal *Nature.*

Much of the junk DNA exists in the form of small genetic parasites called transposons that can infect our genomes in much the same sense that a virus infects our bodies. They go by names such as *LINE-1* (*long interspersed nuclear element), SINE* (*short interspersed nuclear element*), *P-elements*, and *Mariner.* They derive the name *transposon* from their capability to make a copy of themselves that gets inserted at a different place in the genome. They have been present in plants and animals for millions of years, being inherited from generation to generation and even from species to species as new species arise from old. What these genetic parasites all have in common is, like any good replicator, they are good at getting themselves copied. This is the way they reproduce, and they will do this simply because it is what they have evolved to do in competition with other transposons. They are doing this inside your body as you read this, and it is an inevitable fact of natural selection that, once transposons exist, your genome will fill up with the ones best at reproducing themselves.

It has long been a puzzle why we put up with junk DNA rather than evolve ways to remove it. One answer is that most of the time the parasites don't affect us, they merely accumulate inside our genomes. Even then, we still have to carry this extra DNA around and replicate it along with our genes whenever one of our cells divides. Worse, on rare occasions, when one of our transposons makes a copy of itself, that new copy can get inserted inside one of our genes. This can disrupt the gene's normal function and sometimes cause harmful effects. Some unusual cases of hemophilia and even of bowel cancer have been blamed on *LINE-1* transposons moving around inside our genomes. In fact, we now know that we have evolved "genomic immune systems" to help control transposons, much as we have bloodborne immune systems that help us fight off diseases and infections. Still, the junk DNA accumulates in our genomes because our genomic immune systems don't catch them all and because they typically don't remove the junk DNA, just render it inert.

But modern genomic studies have begun to supply yet another answer to why we might allow at least some of the junk to build up in our genomes. It appears our genes use it to help them make our bodies. Our meager supply of genes is not capable on its own of specifying all the necessary parts to build a body, much less the precise times they are needed and all of the connections these parts make with each other. Genes make proteins, and proteins are the building blocks of our bodies. Our hair, muscles, nerves, fingernails, blood cells, and skin are all made of different kinds of proteins. Nearly every cell in your body carries a complete copy of your genome, but only some of your genes are used in any given place in your body and at any given time. This means there must be something that tells the genes that make your eyes not to switch on in the back of your head, or genes for teeth to stay silent in your toes. Something has to provide the instructions to get genes to team up to produce complex structures such as hearts and kidneys, or the chemical networks that create our metabolism. There is no little homunculus perched on a stool inside us calling out instructions from a manual. Instead, the vast quantities of junk DNA in our genomes fortuitously seem to contain a nearly unlimited variety of different messages whose language our genes have learned.

The complexity of our bodies might be built, then, on the enormous expansion of junk DNA that began hundreds of millions of years ago in our distant ancestors. Careful research shows that by complicated chemical means a gene can use stretches of the junk DNA to regulate or control when, where, and how much it is *expressed*. A gene is expressed when it makes a protein, and just as a person uses language to express a thought, genes seem to use the vast vocabulary of sequences of junk DNA to express themselves in exactly the right amounts and at exactly the right times and places in our bodies. Our genes are using the junk DNA to promote *their* interests within our bodies because genes that make better bodies are more likely to survive and be passed on to the next generation. Incredible as it might sound, junk DNA has been a source of great evolutionary innovation, and we all might just owe our existence, at least in part, to it. Differences in gene regulation are why two animals like chimpanzees and humans can be about 98.5 percent identical in the sequences of their genes and yet be so different on the outside. Junk DNA might even have enabled the ancient biological transition from single-celled organisms like yeast to complex multicellular organisms such as ourselves, elephants, lizards, and monkeys. Yeast and other single-celled organisms don't have arms and legs, brains and eyes, and it turns out they have very little junk DNA. This is not to say junk DNA exists for *our* good, or even for the gene's good. It exists because it is good at making copies of itself, and genes have merely been able to exploit its presence.

Before we can grasp the significance of gene regulation to the story of human language, we need to understand one more feature of our genetic system. That feature is that our DNA is a *digital* system of inheritance. A digital signal is one, like Morse code, in which information is transmitted in distinct packets. Morse code relies on just two of these packets, a dot and a dash, and for that reason is a *binary* digital signal. Our DNA's digital signal, rather than having just two packets, uses four distinct molecules called *bases* or *nucleotides*, and these are normally abbreviated *A*, *C*, *G*, and *T*. Digital signals are needed whenever a system requires great fidelity and great variety. The first of these, fidelity, refers to the ability to be transmitted over and over without losing the signal. Digital signals have fidelity because they have a measure of being self-correcting. If when transmitting a dot in Morse code a little bit of error creeps in, the receiver can usually recognize from the context of the broader message that the signal is still a dot. This means that when that dot is transmitted again, the error will have been removed.

By comparison, if the signal to be transmitted is a continuous rather than digital one, receivers cannot know what part is error and what part is the true signal. Loudness, brightness, or length are all continuous signals, and if, for example, I send a loud signal to you, and make it just a bit too loud, you will not know that I have made an error and thus you will be likely to repeat the error if you transmit the signal to someone else. Over long periods of time and many transmissions of this signal, it can drift well away from its starting point because no receiver ever knows what part of the signal is correct and what part is error. Similarly, if I ask you to copy a drawing I have made, you might be able to produce something like it, but it will not be an exact copy. But if I made it from a set of different-colored dots—like the Georges Seurat pointillist paintings—you probably could copy it exactly, even if with some difficulty if it were a Seurat.

There is a beauty, then, in the simplicity of our genetic system. Its digital nature means we can expect it to have high fidelity, and this is just what we require of a system that has to be transmitted repeatedly over millions and millions of years. But we also require our genetic systems to have variety or the ability to produce a wide range of different messages. And here digital systems excel. A Morse code message of just two of its binary signals already yields four possible combinations, each one of which is a different message (*dot-dot, dot-dash, dash-dot*, and *dash-dash*). A message of length three can specify eight different instructions ($2 \times 2 \times 2$); there are sixteen possibilities with just four signals and thirty-two with five, and in general a message of length n can produce 2^n (two multiplied by itself n times) varieties. A message of just twenty binary signals can therefore specify over a million different outcomes (1,048,576, to be exact) and each message is easy to distinguish from the others.

A DNA signal of length n can produce 4 raised to the power n different messages. This is because every position in the gene can now take four different outcomes (*A*, *C*, *G*, or *T*). By stringing together long chains of these bases, our genes can make a variety of different proteins. Equally, though, this means the vast stores of junk DNA in our genomes can also produce an effectively unlimited variety of different instructions for regulating our genes, and this might be just what our bodies need to function effectively. In fact, a measure of the importance of the digital *regulatory sequences* that reside in our junk DNA is that their precise sequences are often highly conserved. Thus, for example, a mouse and an elephant might share some exact or nearly exact regulatory sequences, despite being separated by many millions of years of evolution.

Now we are in a position to understand what is so special about human language. Like our genes and Morse code, human language is also a *digital* communication system. No other animal's is, and if I have done my work in preparing you, you will realize this tells us that human language, unlike the continuously varying signals of the other animals, must have evolved for a task that requires both great fidelity and great variety. Our language is digital by virtue of being built from discrete entities we call *words*, and those words are made from simpler building blocks of discrete sounds—the DNA of our language—called *phonemes*. The digital information that forms a word is sent through the air in pressure waves that we form with our tongue, mouth, and breath. Your ear upon hearing one of them sends the signal to your brain, where it is decoded, just as your television decodes the signals it receives from your wireless remote control, or your body decodes the message in a gene. The system to a large extent "snaps back" into place or is self-correcting when small errors are made because if a word is mispronounced or given an unusual accent or emphasis, receivers can still normally work out what word was intended, and thus this alteration is not retained in the next transmission of that word.

Being digital, our language can easily transmit millions upon millions of different messages, not—like most animal signals—limited to just a few dimensions such as bigger or smaller, louder or softer, or more or less fierce. Imagine you utter just a three-word sentence with a subject, a verb, and an object—such as *I kicked [the] ball*. If you can choose from 20 different words at each position of that sentence, you already can make 8,000 different sentences (20 × 20 × 20). Of course, there are many more words than this, and sentences can be of any length. This means our linguistic systems can effortlessly produce many more messages than the digital color systems of your television or computer can produce different colors on their screens. In fact, our language can produce an unlimited variety of messages because a sentence can be of any length.

Edmund Rostand's engaging play *Cyrano de Bergerac*, about a seventeenth-century French nobleman and duelist with a large nose, provides an amusing illustration of the nuanced expression language can achieve—just the right expression for every circumstance. When the Viscount Valvert tells Cyrano, "Your nose is a trifle large," Cyrano, amused but indignant, responds:

> Young sir you are too simple, you might have said at least a hundred things by varying the tone . . . like this, suppose. . . . *Aggressive*: "Sir, if I had such a nose I'd amputate it!" *Friendly*: "When you sup it must annoy you, dipping in your cup; You need a drinking-bowl of special shape!" *Descriptive*: "'Tis a rock! . . . a peak! . . . a cape!—A cape, forsooth! 'Tis a peninsular!" *Curious*: "How serves that oblong capsular? For scissor-sheath? Or pot to hold your ink?" *Gracious*: "You love the little birds, I think. I see you've managed with a fond research to find their tiny claws a roomy perch!" *Truculent*: "When you smoke your pipe . . . suppose that the tobacco-smoke spouts from your nose—Do not the neighbors, as the fumes rise higher, cry terror-struck: 'The chimney is afire'?" *Considerate*: "Take care . . . your head bowed low by such a weight . . . lest head o'er heels you go!" *Tender*: "Pray get a small umbrella made, lest its bright color in the sun should fade!" *Pedantic*: "That beast Aristophanes names Hippocamelelephantoles must have possessed just such a solid lump of flesh and bone, beneath his forehead's bump!" *Cavalier*: "The last fashion, friend, that hook? To hang your hat on? 'Tis a useful crook!" *Emphatic*: "No wind, O majestic nose, can give THEE cold!—save when the mistral blows!" *Dramatic*: "When it bleeds, what a Red Sea!" *Admiring*: "Sign for a perfumery!" *Lyric*: "Is this a conch? . . . a Triton you?" *Simple*: "When is the monument on view?" *Rustic*: "That thing a nose? Marry-come-up! 'Tis a dwarf pumpkin, or a prize turnip!" *Military*: "Point against cavalry!" *Practical*: "Put it in a lottery! Assuredly 'twould be the biggest prize!" Or . . . parodying Pyramus' sighs . . . "Behold

> the nose that mars the harmony of its master's phiz! blushing its treachery!"—Such, my dear sir, is what you might have said, had you of wit or letters the least jot. . . .

"Wit or letters the least jot" is precisely what Cyrano himself would have lacked had he been any other animal apart from a human because none of them is blessed with our digital communication system. Jared Diamond describes how in America in the late 1940s, Keith and Catherine Hayes brought up a chimpanzee called Viki in their household. But instead of bringing her up as a chimpanzee, they attempted to raise her like a human infant. They played and talked to her, fed her at the table, sang to her, and generally just treated her like another member of the family. Viki excelled at playtime but struggled despite heroic efforts on the part of her owners ever to speak. Even after hours and hours of intensive training she could manage just three words, "mama," "papa," and "cup," and then it must be acknowledged that the Hayeses were better than outsiders at recognizing these "words"—something to which any proud parent of a two-year-old will also have to confess.

Many years later, Diamond imagined Viki trying to speak, using at most the two vowels and two consonant sounds in these words (charitably, Viki could produce four phonemes, an *a* and *u*, and a *c* and *p*, and even this probably overestimates a normal chimpanzee's abilities). Diamond, who was giving a lecture at Trinity College in Dublin, asked his audience to "try seeing how many different words you could speak if you could only pronounce the vowels *a* and *u*, and the consonants *c* and *p*. If you wanted to say 'Trinity College is a fine place to work,' all you could manage would be 'Capupa Cappap up a cap capcupap.' Your attempt to say 'Trinity College is a bad place to sneeze' would result in identical sounds."

What task requires the fidelity and variety that human language possesses? Our social phenotypes or cultures are immensely complex compared to those of any other animal, and we derive gains from varying our "expression" inside it much like a gene does inside a body. Thus, we deploy our digital language to manipulate this social phenotype in ways that promote our survival and reproduction, and in the same way that genes use their vast digital repository of junk DNA to vary their expression inside our physical bodies. In fact, we could say that deep down language might just be the latest form of gene regulation—the voice of our genes. We rely on the great variety and vocabulary of language to share ideas, to promote cooperation, to build alliances, and to enhance our reputations, just as genes rely on the great vocabulary and variety of junk DNA to regulate how much, when, and where in our bodies they are expressed. We use our language to be deceptive or charming, kind and forgiving, or spiteful and vindictive. We use it to manipulate or bewitch others, to collude with them, or to foster or defuse factional disputes. We use language to embroider and exaggerate our own dossiers and gently diminish or disparage those of others. We alter our speech or dialect strategically—linguists call this code switching—to signal our connection to a group or individual.

As we saw with Viki, none of this could be achieved with the simple continuously varying signals of the rest of the animal kingdom, limited as they are to "more" or "less" of some quantity. Instead we needed a digital mechanism capable of great variety and fidelity. And so, just as junk DNA might have enabled the transition to large complex multicellular organisms, the evolution of language might have enabled the transition to our complex societies. Not being able to speak in the newly emerging human societies would have been like being a bird that could not fly. Just as wings open up an entirely new sphere to be exploited, language opened up the sphere of cooperation, and genes for human language would have quickly spread.

Before leaving the topic of digital systems and regulation, I would like to mention that it was yet another kind of digital regulation that enabled our modern electronics revolution, and that has changed the way we lead our lives. John Mattick points out that up until recently airplanes, clocks, and even computers were *analogue* as opposed to digital devices, manipulated by continuously varying signals such as levers, springs, heat, or pressure. For example, airplanes were flown with a stick, and large springs were wound up to drive clocks. But once our engineers discovered digital regulation—instructions encoded in strings of binary numbers arbitrarily long, and hence precise—these machines and a host of new ones, such as digital cameras and music players, could become more complex.

The U.S. Air Force's Stealth fighter planes provide an extreme example of a machine that could not exist without digital regulation. These planes achieve their stealth in part from having a shape that makes them reflect enemy radar in such a way that the radar image loses its coherence. The trouble is that this shape makes them virtually impossible to fly, and they rely on millions of split-second adjustments to keep them airborne. Making these adjustments is beyond the capabilities of mere (analogue) human pilots—so much so that the Stealth planes are flown only with a large input from onboard computers. Little did the designers of these machines know that nature had beaten them to this digital regulation by hundreds of millions of years.

words, languages, adaptation, and social identity

IF LANGUAGE is "the voice of our genes," we should see evidence of this important role in how we use language and in how elements of language evolve. One way to do this is to ask what words we use most often in our everyday speech, and when we do this, two surprising results emerge. One is that there is a huge disparity between how often different words get used, with some being used hundreds and even thousands of times more often than others. So great is this disparity that somewhere around 25 percent of all our speech is made up from a mere twenty-five words! According to the *Oxford English Dictionary*, the English language's top twenty-five include *the, I, you, he, this, that, have, to be, for*, and *and*; but *they, we, say*, and *she* make it into the top thirty. All of these are used thousands to perhaps 30,000 times per every million utterances, whereas most of the remaining 200,000 or more other English words that great dictionary catalogues tend to be used only very infrequently, some of them only a few times in every million utterances. When, for example, was the last time you used the words *indefatigable* or *expository* or *behoove*?

The second surprise is that speakers of different languages all use more or less this same subset of words frequently in their speech and a different set infrequently. It seems that around the world we all talk about the same things and in roughly the same amounts. This holds whether the speaker is French or Bantu, Chinese or Hungarian, Basque, English, Turkish, Finnish, Greek, or using a Polynesian language. We know this from studying a common set of words that the American linguist Morris Swadesh, working in the 1950s, proposed as a "fundamental vocabulary." Swadesh's goal was to identify a list of words that should be found in all human languages. It includes words such as *what, where, when, mother, father, fish, bird, hold, count, throw, float, say, day, night, bite, eat, sky, drink*, and *louse*, along with number words, names of body parts, pronouns, colors, and common verbs, adjectives, and nouns. It deliberately excludes technological words or words that describe specific environments or features of them. *Louse* might surprise modern readers, but Swadesh was interested in the history of language use, and lice infections have been a common feature of our history—and still are in many parts of the world (and annoyingly among children of school age).

Linguists and linguist-missionaries have translated this list into thousands of languages around the world. The frequently used words in nearly every language are typically *I, you, he, she, it*, and other pronouns; the verb *to be*; number words like *two, three, four, five*; and *who, what, where, why*, and *when*. Other words in the Swadesh list such as *scratch, guts, stick, throw*, and *dirty* are typically used much less often, no matter what culture is studied. Putting this together with the results from the *Oxford English Dictionary*, we not only seem to be using language the same way around the world, but for the same reasons—principally to talk about each other, to refer to quantities of things, and to what people are doing, when they did it, to whom, and how much. These are also the words that we would expect to be used if language is for promoting and monitoring social relations, making exchanges, and advertising and assessing reputations. Because this information comes from a worldwide sample of languages, it is a reasonable step from there to guess that this will have been true throughout the history of human language use.

The genes you pass on to your children will have been replicated or copied only a small number of times between you acquiring them from your parents and then your children inheriting them from you. Even so, each time a gene is copied there is a chance that a mutation or error will creep in, and over long periods of time genes can change out of all recognition. But for a word to last for even a generation it will have been spoken—its sound replicated—thousands or even millions of times, and of course this will be especially true of the frequently used words. If words could not be stably transmitted from mind to mind, then something I told you about someone else might get corrupted by the time it makes it way around the tribe—like what happens in a game of *Whispers* or *Telephone*. And something my mother said in her youth might not be intelligible to me by the time I am old enough to appreciate it. If this happened, our cultures would erode and decay.

But this is not what happens. In fact, our languages can demonstrate quite extraordinary degrees of stability over long periods of time, far longer than is necessary for us to be able to communicate with each other throughout our lifetimes. Here, for example, is 1,000 years of the evolution of the familiar Lord's Prayer, spanning Old to Modern English:

> Fæder ure þu þe eart on heofonum; Si þin nama gehalgod to becume þin rice gewurþe ðin willa on eorðan swa swa on heofonum.
>
> OLD ENGLISH—11TH C.

> Oure fadir that art in heuenes, halewid be thi name; thi kyndoom come to; be thi wille don in erthe as in heuene.
>
> MIDDLE ENGLISH—1380

> Our father which art in heauen, hallowed be thy name. Thy kingdome come. Thy will be done, in earth, as it is in heauen.
>
> EARLY MODERN ENGLISH: *King James Bible*—1611

> Our Father, who art in heaven, Hallowed be thy Name. Thy kingdom come. Thy will be done, On earth as it is in heaven.
>
> LATE MODERN ENGLISH *Book of Common Prayer*—1928

Most of us have great difficulty now in reading the Old English, but what we see from one version of the Lord's Prayer to the next is the gradual process that Darwin called "descent with modification." We saw this earlier as a description of how species, over long periods of time, give rise to somewhat different daughter or descendant species. Darwin appreciated that a similar process was true of languages, saying in *The Descent of Man*, "The formation of different languages and of distinct species, and the proofs that both have been developed through a gradual process, are curiously parallel. . . ." Ancestral languages evolve to give rise to somewhat different daughter languages, which in turn do the same. Each of these daughter languages retains some of the features of its ancestral or mother tongue, but differences creep in as people fan out to occupy new areas.

When this process is played out over a large area, and among different sets of ancestral and descendant languages, entire family trees of related languages evolve. This was recognized as early as the late eighteenth century, nearly a hundred years before Darwin, by an English judge, Sir William Jones, working in colonial India during the reign of George III. To process court papers, Sir William found it necessary to learn Sanskrit, and in doing so he became aware of curious parallels between Sanskrit, Latin, and Greek. Jones described these to a meeting of the Asiatic Society in Calcutta on February 2, 1786, noting that the Sanskrit language bears "a stronger affinity . . . [to Greek and Latin] . . . both in the roots of verbs and in the forms of grammar, than could possibly have been produced by accident; so strong, indeed, that no philologer could examine them all three, without believing them to have sprung from some common source, which, perhaps, no longer exists." For instance, the Sanskrit word for the English "brother" is *bhratar*; the ancient Greeks used the word *phrater* (φρᾱτηρ) to mean something akin to a fraternity or brotherhood; and *frater* is the Latin word for "brother." Another example is the Sanskrit word for "three" or *tri*, which is *tria* (τρία) in Greek and Latin. The familiar-sounding *khanda* in Sanskrit is the French word *candi* and the English "candy."

What Jones had identified would later be recognized as the Indo-European language family, and it includes the languages currently spoken all over Europe, parts of Central Asia, and the Indian subcontinent. The archaeologist Colin Renfrew in his *Archaeology and Language* links the spread of these languages to the origin of agriculture sometime around 9,000–10,000 years ago in the region of the world known as the Fertile Crescent (roughly present-day Turkey and Iraq). Farming reset the world's carrying capacity to a higher level, allowing a greater number of people to survive in a given area. The growing populations meant that farmers and their ideas spread out in all directions from the Fertile Crescent. Those that went north and west formed what we recognize today as the Greek, Germanic, and Romance or Latinate languages of Western Europe; those that went north and east produced the Slavic languages; and those that went south gave rise to the languages of Iran, Afghanistan, Pakistan, and India. The Basque language of northern Spain, although in Europe, is not an Indo-European language. In fact, it might be an isolated remant of the languages that were spoken by the hunter-gatherers in Europe before farming and farmers arrived. Russell Gray and his colleagues were later able to confirm Renfrew's arguments by applying dating techniques to the Indo-European languages, confirming that this family probably did arise sometime around 9,000 years ago. And it is because of descent with modification within the separate branches of this family that today we recognize similarities in the Romance languages of Spanish, French, Italian, and Portuguese, just as we recognize similarities in the Germanic languages of German, Danish, Dutch, and English.

But how long can the words causing these similarities last, and are there some in particular that turn up over and over in different languages as ones that get retained for the longest periods of time? It turns out that it is the small subset of words that we use most frequently in our everyday speech—and especially those we have suggested are related to social relations—that can show truly startling resistance to changing, sometimes being conserved in a related form across all Indo-European languages. For example, linguists recognize that the number word for *two* of something is probably derived in all Indo-European languages from the same shared ancestral sound that has been conserved for thousands of years. Thus, in Spanish the word is *dos*, it is *twee* in Dutch, *deux* in French, *due* (doo-ay) in Italian, *dois* in Portuguese, *duo* (δύο) in Greek, *di* in Albanian, *do* in Hindi and Punjabi, and Caesar would have said *duo*. This leads to the proposal that the original or proto-Indo-European word was also *two*-like in its sound, and indeed some scholars suggest *duwo* or *duoh*.

A handful of other words, including the words for *three* and *five*, *who*, *I*, and *you*, are also highly conserved like the word for *two*. For example, the English *three* is *tre* in Swedish and Danish, *drei* in German, *tre* in Italian and *tres* in Spanish, *tria* in Greek, *teen* in Hindi and *tin* in Panjabi, and *tri* in Czech. The proto-Indo-European word for *three* might have been *trei*. These conserved words are closely followed by other pronouns such as *he* and *she*, and by the *what*, *where*, and *why* words, all of which can show a striking degree of similarity among many Indo-European languages. Remembering that the Indo-European languages all derive from a common ancestral language, this tells us these words have been retained in separately evolving branches of this family tree, each one of which represents up to 9,000 years of language change. If we add up the time in each of these branches, we see that some words have not substantially changed their forms for what is, in effect, hundreds of thousands of "language-years" (thus, for example, the sound for *three* has been evolving separately in the Germanic, Romance, Slavic, and Indian languages).

This degree of conservation tells us that the people alive in the parts of the world where the Indo-European languages arose might have been using forms of these slowly changing words that we could still recognize today. Their words would not have been identical to our modern words, but close enough that if we travelled back in time and encountered a gang of three of our proto-Indo-European ancestors, we might be able to point to ourselves and say, "I, one," then pointing to them, say, "You, three," and be understood. Now, this might not be a wise or even very imaginative thing to do, and you might think it a limited conversation anyway. But consider that if you are of Indo-European origins and you are reading this book on an international flight, you might have less in common linguistically with the person seated next to you than you would with a linguistic ancestor who lived close to 10,000 years ago.

How far back in time can we go with words? The existence of a set of highly conserved words raises the possibility that we might be able to find some evidence for what our mother tongue, or the language of the very first humans, was like. The linguist Merrit Ruhlen has proposed twenty-seven "global etymologies," or words left over from our original or proto-language, citing evidence that they are found in many language families from all over the world. Ruhlen's list includes words for *who*, *what*, *two*, *water*, and *finger* or *one*. For instance, Ruhlen points out that the sounds *tok*, *tik*, *dik*, or *tak* surface repeatedly in these language families as a word for the number *one* or *toe* or a *digit*. Ruhlen's proposals have always been highly controversial among linguists, but it should not escape our attention that he includes words we have seen are among those that are both frequently used and highly conserved. They are also those we might have expected if indeed we have used language to monitor and manipulate social relations throughout our history.

Whether or not we can ever reconstruct the human mother tongue, we should be astonished that words can be retained and conserved over thousands of years and potentially millions of speakers. For a word to be transmitted, a sound I make must travel through the air and enter your ear, where it is turned (transduced) into an electrical signal that travels to your brain. Then, at a later time when you want to use the word, your brain must somehow send messages to your mouth and lungs to get you to produce the same sound. That sound will then travel to someone else's ear, where the process will be repeated. The opportunities for corruption or loss of signal are many. We should also remember that, unlike for genes, there is seldom any necessary connection between a word's form (its sound) and its meaning—I might just as easily call a *tree* a *table*, and vice versa. Where a gene's chemical form is directly related to the protein it makes, this "form-function" connection is generally only true in language of the so-called onomatopoeic words that imitate sounds, like *bang*, *meow*, *moo*, *woof*, or *pop*.

We might have suspected that the more a word is used, the more likely it is to acquire mistakes or errors as it passes out of successive speakers' mouths, into others' minds via their ears, and then out again. Instead, it is the words we use less often that are more prone to changing. Consider that in contrast to a highly conserved word like *two*, what English speakers call a *bird* the Germans call a *vogel*, the Spanish say *pajaro*, the Italians *ucello*, the French *oiseau*; to Aristotle and modern Greeks after him, the word is *pouli*, and Caesar would have said *avis*. Words like *dirty* and *guts* are even more variable. For instance, compared to the conservation of sounds for *two* of something, there are at least forty-five different words for *dirty* among the Indo-European languages (four of these are the German *schmutzig*, Dutch *vuil*, French *sale*, and Spanish *sucio*). In fact, a given word for *dirty* gets replaced by some new sound every 800 years or so, compared to the many thousands of years a word like *two* can last.

Are these infrequently used words more prone to change because they are somehow less important and so it doesn't matter so much if someone comes up with a new word or mispronounces an existing one? This might be part of the answer, but only indirectly. When we think about it, we realize that because there is seldom any connection between a word as a mere sound and that word's meaning, different words (sounds) have to compete with each other for the privilege of carrying a particular meaning. We can pick and choose among them, and this competition for our affections is natural selection acting on words, sieving out the winners from the losers. Among English speakers, nearly everyone uses the word *kitchen* to describe the place where they prepare their food. That sound has won in competition with other sounds we might produce—such as *scullery* or *galley*—to describe that particular meaning. In other cases, we can see the competition for space in our minds going on in front of us in common words like *sofa* and *couch* or *living room*, *sitting room*, *reception room*, or *parlor*.

Which of these forms will win? There are no simple answers, but precisely because there is seldom any necessary connection between a sound and its meaning, the competition often focuses on characteristics of the sounds themselves, and we can expect the competition to be more intense the more we use a word. One of the most common ways that frequently used words adapt to our minds is to get shorter. Do you say *automobile* or *car*? *Refrigerator* or *fridge*? *Cannot* or *can't*? The Harvard linguist George Kingsley Zipf had recognized this relationship between frequent use and shorter words by the 1930s and it is now enshrined in the eponymous Zipf's law. We give people "diminutives," or shorter names, for names like Alexander (Alex) or William (Will, Bill), Arthur (Art), Richard (Dick), David (Dave), Michael (Mike), but we seldom if ever lengthen a name. We see this too in the contractions—*can't, won't, don't*, and in familiar usages like *g'morning* or *g'day*. So strong is the effect Zipf's law describes that none of the words that English-language speakers use more than a few thousand utterances in every million is longer than around five letters (words like *the*, *and*, and *is* might be used 10,000 to 30,000 times per million). If a sound "wants" to be highly used as a word, it can't afford to be long. When we say someone uses "big words" to mean he or she uses words we don't know, we are, without knowing it, recognizing that the words we don't know are typically ones we don't hear very often, and so natural selection has not pared them down to size.

We learn from this that out of all the possible sounds we could make as words, only a far smaller number has ever been used. The *Oxford English Dictionary* catalogues around 250,000 English words, of which around 50,000 are considered extinct or no longer used. This might sound like a large number, but even 250,000 words is just a minuscule subset of all possible words. Consider just for sake of discussion that English conventionally defines five vowels and twenty-one consonants. Let's consider how many six-letter words we could make from this set of twenty-six letters, restricting ourselves to words with two vowels and four consonants, as in a word like *letter* or *friend*. If words are unrestricted, then each of the four consonants could be any one of the twenty-one that we can choose from and each of the two vowels can be any one of five. This leads to $21 \times 21 \times 21 \times 21 \times 5 \times 5 = 4{,}862{,}025$ possible six-letter words with two vowels. Let's now move to seven-letter words also with two vowels: say, a word like *letters* or *friends*. There are $21 \times 4{,}862{,}025 = 102{,}102{,}525$ possible varieties of these!

This simple exercise tells us that the survivors among all the possible sounds that could have become words—that is, the words we actually use—are a highly rarefied set of winners of an overwhelming competition to occupy our minds. For example, there are no words in English that begin with "ng" and words that begin with "q" are nearly always followed by "u." Adding to this already intense competition among sounds is that our minds are remarkably good at remembering words, meaning we can compare them with ease. In fact, so good are our minds at remembering words that we easily recognize words that are used less than once per million utterances in normal speech, even if we cannot always remember their meanings. *Adumbrate*, *dilatory*, *feculent*, *parvenu*, *fractious*, and *traduce* are all used perhaps one to five times per million utterances. *Eponymous*—mentioned earlier—is another. How many do you recognize? How about *demanate*? Don't recognize it? Good, because it is not recognized as an English word. If you got all of these right, or even most of them, then your mind can routinely discriminate words it hears at only very low frequencies from so-called words like *demanate* that it should have never heard.

We are beginning to get an answer to our question of why frequently used words might be so stable. It is not necessarily their importance to communication per se. Rather, the frequently used words might be stable over long periods of time because they have become so highly adapted to our minds that it becomes difficult for a new form to arise to outcompete or dislodge them. It is no accident, then, that all of the highly used and highly conserved words that we use in talking about social relations—*who*, *what*, *where*, *why*, *when*, *you*, *me*, *I*, *he*, *she*, *we*, *it*, and the number words from *one* to *ten*—are short, mostly monosyllabic, and easy to pronounce. This isn't just a property of English or Indo-European languages more generally. Listen to the numbers one to nine for the New Guinean language of Mangareva: *tahi, rua, toru, ha, rima, ono, hitu, varu, iva*. The frequently used words also evolve to be distinct from each other and therefore less prone to being misunderstood. Try saying the words *mail* and *nail*, *fail* and *veil*, or *bale* and *pail* over the phone to someone, and there is a good chance you will be misunderstood. Now, imagine *fail* means *two* and *veil* means *ten*, and you are using these words to order shirts over the phone from a clothing retailer: chances are if you wanted two, your bill might be higher than you expected. But this sort of thing seldom happens with the number words because we don't mishear them—they have evolved to be distinct. And this is also true of the words we have said are related to social relations more generally. Humans have domesticated language to make it efficient at communicating what they want to talk about the most.

Incidentally, this discussion of how languages adapt to our needs as speakers can help us to understand a proposal that has never caught on. Esperanto is a designed or made-up language that was proposed in the late nineteenth century as a politically neutral and easy-to-learn universal system of communication. But despite attracting a loyal band of followers, it has never become popular. One reason might be that it is not really politically neutral, with many of its words linked to or drawn from Indo-European languages and especially Latin and Greek forms: "yes" in Esperanto is *jes*, "hello" is *saluton*, and "no" is *ne* (which would be especially confusing to Greeks as *ne* [ναι] means "yes" in Greek). "Good morning" is *Bonan matenon*. A second reason is that Esperanto was probably never really necessary, especially now as English is used so widely, not to mention that adopting Esperanto meant giving up one's own linguistic heritage. But the final and most interesting reason from the perspective of our discussion of how words evolve is that to succeed, Esperanto has to win in competition with other languages. But unlike these competitors, Esperanto is made up. It has not had to evolve to suit our minds and the ways we speak. There is reason, then, to suspect that it is not as "good" a language (not as easy to learn, use, or speak) as natural languages that have had to go through the sieving of cultural selection.

If our natural languages can be so stable and easy to use as a result of their evolution, why are there so many around the world? There are about 7,000 different languages currently in use, and this is almost certainly far fewer than there were in our past. Even 7,000 is more different languages for a single species of mammal than there are mammal species. By comparison, all human beings share a more or less identical set of our other great digital system of inheritance, our genes. Why the difference? The Babel myth we saw earlier provides one explanation and yields the amusing irony that our many languages exist to stop us from communicating! In Chapter 1 we even saw that there might be a kernel of truth to this. Groups of people often change their languages as an act of asserting their social identities, or as a way to be different from others living nearby.

But could there be another reason, one that is closer to the theme of the Babel story of restricting the flow of information? If our languages really do act as conduits of important social information, then it might be useful to protect this information from others, and especially so when they live nearby. Changing your language might be a way to avoid eavesdroppers who could be lurking in the bushes. The many different languages spoken in the tropics might come about because this is where people are the most tightly packed and have the greatest need to protect themselves.

Speaking a different language has the additional value that it allows people quickly to identify others who are not members of their group. There is the poignant scene in the movie *The Great Escape* when the British and American prisoners of war have fled from their German captors and two of them find their way in disguise to a German railway station. A German officer at the station is suspicious but unable to identify them. Suddenly, he shouts out in German for everyone on the platform to get down. All of the German civilians immediately comply, exposing the fleeing prisoners, who are gunned down. Or listen to this passage from the Old Testament:

> The Gileadites captured the fords of the Jordan leading to Ephraim, and whenever a survivour of Ephraim said, "Let me go over," the men of Gilead asked him, "Are you an Ephraimite?" If he replied, "No," they said, "All right, say 'Shibboleth.'" If he said, "Sibboleth," because he could not pronounce the word correctly, they seized him and killed him at the fords of the Jordan. Forty-two thousand Ephraimites were killed at that time. (Judges 12:5–6)

Languages that have gone through large numbers of what we might think of as cultural splitting events or divorces in their past—where one group of speakers, for whatever set of reasons, divides into two—tend to accumulate more changes than languages whose history records fewer social upheavals. It is as if the many-times-divorced languages have suffered a series of distinct bursts of rapid change, each one an attempt to distinguish the language from the form it is separating from. One such divorce might have influenced the form of American English. American English drops the *u* compared to British English in words such as *colour*, *behaviour*, and *honour*. These changes didn't arise haphazardly or over long periods of time. Rather, they were introduced overnight when the American educator and compiler of dictionaries Noah Webster (1758–1843) produced his *American Dictionary of the English Language*. In preparing that work he insisted that "as an independent nation, our honor requires us to have a system of our own, in language as well as government." To Webster at least, spellings of words as a way of marking out a distinct identity had been elevated in importance to the philosophical issues that lay behind the American War of Independence.

language extinction

WE STARTED this chapter by describing language as the most powerful, dangerous, and subversive trait that natural selection has ever devised. It is also, perhaps, our most intimate of traits, being the voice of the "I" or "me" that defines our conscious self. It is the language of our thinking, and it is the code in which our memories are stored. So it is not surprising that one of the greatest personal losses a people can suffer is the loss of their native language. And yet, currently, somewhere around fifteen to thirty languages go extinct every year as small traditional societies dwindle in numbers, get overwhelmed by larger neighbors, and younger generations choose to learn the languages of larger and politically dominant societies. Whatever the true numbers, the rate of loss of languages greatly exceeds the loss of biological species as a proportion of their respective totals.

Some projections say that only a handful of languages will see out this century. This raises the question of what language will win, if ever a single language should succeed all others on Earth. Currently, three languages are spoken by a far greater number of people than any of their competitors. Somewhere around 1.2 billion people speak Mandarin, followed by around 400 million each for Spanish and English; and these are closely followed by Bengali and Hindi. It is not that these languages are better than their rivals, it is that they have had the fortune of being linked to demographically prosperous cultures. On these counts Mandarin might look like the leader in the race to be the world's language, but this ignores the fact that vastly more people learn English as a second language—including many people in China—than any other. Already it is apparent that if there is a worldwide lingua franca, it is English. Once, in Tanzania, I was stopped while attempting to speak Swahili to a local person who held up his hand and said, "My English is better than your Swahili."

Still, English itself might be transformed as it is bombarded by the influences of such large numbers of non-native English speakers who, when they use it, bring along their own accents, grammar, and words. On the other hand, English's willingness to take in so-called foreign words has—for at least the millennium since the Norman conquest of the English in 1066 brought an influx of Norman French vocabulary into the English language—been the key to its adaptability. Just as we have seen how words must adapt to be competitive in the struggle to gain access to our minds, languages have to adapt as a whole to remain useful to their speakers, and those that do so will be the survivors. Self-appointed human "minders" in the form of reactionary grammarians, sticklers for spelling, or those who deliberately try to exclude some words and phrases (like the officials of the French ministry we saw in Chapter 1), will succeed in controlling the rate at which their languages naturally change, but in doing so might consign them to the backwaters of international communication. Already this might be happening to French and German. The alternative to this control is not the free-for-all that some might fear. If communication is important, languages will never change at rates that imperil the very reason for which they exist.

CHAPTER 9

Deception, Consciousness, and Truth

That our minds might have evolved more to manipulate others and ourselves than to perceive the truth

the "i and thou"

WE TAKE IT for granted that the most intelligent mind on Earth is designed to perceive the truth and act on it. But an evolutionary perspective tells us to expect that our minds have evolved to be good at promoting our survival and well-being and truth might only be part of what they use as their currency. In a brief passage from his Foreword to the first edition of Richard Dawkins's *Selfish Gene*, Robert Trivers wrote that if deceit

> is fundamental to animal communication, then there must be strong selection to spot deception and this ought, in turn, to select for a degree of self-deception, rendering some facts and motives unconscious so as not to betray—by the subtle signs of self-knowledge—the deception being practiced. Thus, the conventional view that natural selection favors nervous systems which produce ever more accurate images of the world must be a very naive view of mental evolution.

Increasingly, cognitive science teaches us that our perceptions and memories are not just fallible; they are stories our brain concocts to prop up our egos, justify our decisions, and condone our actions. They are the stories we want to see and hear and they often bear little resemblance to what "really" happened. Julian Barnes quotes Stravinsky as saying, "I wonder if memory is true, and know that it cannot be, but one lives by memory nonetheless and not by truth." If you doubt Stravinsky, gather together your family or a set of old friends and reminisce. You will be surprised, and possibly even distressed, to find out that not everyone agrees with your memories.

For the religious philosopher Martin Buber, the "I and thou" expressed our relationship to the eternal thou or God, but we can easily imagine that "thou" to be our genes. They are the truly eternal players that our minds will have had to answer to, having been engaged in a struggle for survival since long before the continents separated into what we know as our modern Earth. If the last sixty years of experimental psychology, personality, evolutionary, and neurological studies have shown us anything, it is that the minds that have proven useful in that struggle are far more bewildering than we might have expected. For one, the inner "I" that you think you know so well probably doesn't exist. It is an illusion, the construction of a mind that is in turn a construction of its genes, genes that have been selected to produce brains that further their ends. Those brains will use false beliefs, copying, lies, deception, self-deception, and just about anything else they can lay their neuronal hands on to promote our—and consequently their—survival and reproduction. The genes that create this mind sometimes don't even agree among themselves how it should "feel," and this can lead to internal conflicts. But a second reason is equally sobering: even if we pretend that some little homunculus exists in our heads, we will see it often acts without "us" even knowing, often only bothering to tell us of its decisions later.

Indeed, the whole concept of personal identity is so tenuous that John Locke devoted an entire chapter to it in his *Essay Concerning Human Understanding*. Locke came to the startling conclusion that, apart from our memories, we cannot even be sure we are the same person we were in the past. But of course as Stravinsky reminds us, those memories are fallible. The unease or alarm such a thought can cause is only alarming so long as we cling to the reality of this "I" that we think exists inside us, rather than concentrating on what really matters to genes, which is to develop strategies that promote their survival.

the co-evolution of intelligence and deception

ON FEBRUARY 17, 2004, a man called Peter Bryan was granted temporary permission to leave the ward of his "low support accommodation" hostel in London. He went to a shop where he bought a claw hammer, a Stanley knife, and a screwdriver. Later that day he encountered a Mr. Brian Cherry, whom he battered to death with the hammer, then sawed off both his arms and left leg, and scooped his brains from his skull and ate them, but not until he had first fried them in butter. Peter Bryan was arrested, covered in blood, after Brian Cherry's neighbors alerted the police. Earlier that day, before releasing Bryan, staff at the hostel had conducted an hour-long interview with him. *The Times* later reported that at that meeting staff described him as "calm and jovial," and that there were "no concerns regarding his mental state." The police doctor who examined Bryan after he had murdered and eaten Mr. Cherry said his mental condition "did not necessitate an urgent transfer to hospital." Instead, he was remanded to Belmarsh Prison in London, where he assaulted staff and behaved "unpredictably."

By April of that year he was finally diagnosed as mentally ill and sent from Belmarsh to Broadmoor mental hospital in the countryside near London. Three days after arriving there, he was placed in a medium-security ward and left alone with other patients. Within a week, he assaulted a patient, Richard Loudwell, who later died of his injuries. When asked why he had assaulted Loudwell, Bryan said: "I wanted to kill and eat him. Cannibalism is natural. . . . If I was on the street, I'd go for someone bigger for a challenge."

(As an aside, cannibalism might indeed be "natural," even if now taboo. Archaeologists working in northern Spain have unearthed twelve Neanderthal individuals, including both men and women.

Their bones show cut and scrape marks consistent with being eaten—and probably by other Neanderthals. In his sixteenth-century essay "Of Cannibals," Montaigne quotes a taunt from a prisoner to the tribespeople holding him captive and about to kill and eat him: "these muscles, this flesh and these veins are your own, poor fools that you are. You do not recognize that the substance of your ancestors' limbs is still contained in them. Savour them well; you will find in them the taste of your own flesh." Cannibalism was so prevalent among members of the Foré tribe of New Guinean highlanders right up through the early part of the twentieth century that entire villages came down with the debilitating and fatal brain disease kuru.

Kuru is caused by a protein known as a prion that is found in brain tissue, and is similar to the brain-wasting dementia in humans known as Creutzfeld-Jakob disease. The disease is spread when an altered form of the prion in the brain tissue that someone consumes makes its way into that person's brain, where it converts their prions to the altered and disease-causing form. About 40 percent of people worldwide carry a form of the prion gene that provides some protection against kurulike brain disease, suggesting it has been prevalent in our history. But so widespread and evidently long-standing was the practice of cannibalism among the Foré that a novel genetic variant of the prion protein which granted even further protection from developing kuru arose among those Foré who had been most exposed to eating human flesh.)

Peter Bryan had been sent to the London hostel in 2002 after serving eight years for the violent murder of a shop assistant. Over the next year at the hostel, he threatened staff and other residents. In spite of this, after one of Bryan's routine assessments his caseworker wrote that he was making "good progress" and "does not present any major risks." Later enquiries into his case reported that it had been difficult for even experienced health professionals to detect just how dangerous Bryan was. *The Times* for September 3, 2009, reported what the health authority responsible for him had found:

> Peter Bryan clearly had a very severe and complex mental illness. In his lengthy contact with a range of services and a range of

> professionals, he was able to function at a high social level and did not display any of the typical behaviour or symptoms one would associate with a severe mental illness. We accept that elements of the care provided to Mr. Bryan could have been better but we also note that the independent report does not say the killing of Mr. Cherry could have been predicted.

How could someone on the verge of bludgeoning a person to death, sawing off their limbs, and cooking and eating their brains be described as "calm and jovial," "able to function at a high social level," and not displaying behavior typical of severe mental illness? We should resist the temptation merely to blame incompetent social services staff or to pass this story off as a bizarre but rare misfortune. It could be either or both of those things. But Peter Bryan's case forces us to accept the unsettling possibility that otherwise pathologically mad human beings have the capacity to deceive even those trained to spot abnormalities and to do so repeatedly over a number of years. This story is not an isolated one. Reports of serial killers, con artists, fraudsters, impostors, charlatans, and mountebanks whose acts, if not as gruesome as Peter Bryan's, display the same sophisticated abilities to deceive others are never very far from the headlines.

Bryan's deceptiveness becomes less surprising when we recall that our large brains probably evolved in a social arms race with other brains in which increases in social intelligence in some were met by replies in others. Traits that allow you to outwit your neighbors in this social competition will grant benefits and quickly spread to your children, and then to theirs. This will have meant, among other things, developing the capacity to "get inside" our rivals' minds, to try to anticipate what they might do next. It became necessary to have a "theory of mind"—a sense of knowing what you think another animal knows, and being aware that it is having similar thoughts about you. In these circumstances, one of the best ways to thwart someone else's theory of (your) mind is to develop sophisticated abilities at deception. In fact, we have reason to expect that deception is a normal—in the sense of routine—feature of our societies. Our social organization, based on cooperation with people we are not related to, means that we engage in a nearly continuous stream of exchanges of favors, goods, and services. Successfully deceiving others can therefore produce a nearly continuous stream of rewards as we tip these exchanges in our favor, getting just that little bit more of our share of the benefits of cooperation from each of them.

Natural selection will have equipped us not only with the emotions to take advantage of others, but with a bag of tricks to enhance those deceptions—charm, flattery, lack of empathy, self-deprecation, and an ability to hide or fake emotions. It will also, as Trivers realized, have favored an ability to hide our emotions from ourselves, the better to deceive others. An averted gaze, a strained voice, or sweaty palms can give us away, but if we can deceive ourselves about our own motivations, we might be able to hide even these tell-tale signs. This evidently was not lost on the makers of the fourth film in the *Terminator* series, the story of an epic struggle between humans and machines. The machines deploy robotic and humanlike "Terminators" to despatch key people in the human defensive army. One of the Terminators in that film was so convincing that he was able to charm his way right into the humans' inner sanctum. His secret weapon was that even he did not know he was a Terminator.

On the other hand, if deception can return the rewards we think it does, then natural selection will also have favored keen abilities for detecting it. We saw earlier that we might even have finely tuned mechanisms that operate in social situations to help us spot who is a social cheater and who is following the rules. And indeed, knowing that deception is wired into our nature, most of us are continually on guard against it, and other kinds of cheating, in others. Is that the right taxi fare? Is the dinner bill correct? Did the shopkeeper give me the right change? Why did he avert his gaze when talking to me? Is that the item I ordered or a cheaper imitation? Is the person walking toward me, or who has just sat down next to me on the train or bus, someone I can trust? Should I open the door to someone who knocks at my door late at night asking for help? Can I trust my spouse?

Who wins this arms race between deceivers and those trying to detect them? That is the wrong question, because as we have seen, both capabilities will reside in all of us and be routinely deployed by all of us to a greater or lesser extent. On the other hand, deception is a parasitical strategy that feeds on cooperation, and this might limit its reach. If our abilities to deceive others were good enough, they could undermine cooperation and even cause our systems of reputation and exchange to collapse. If that happened, people would instinctively fall back to cooperating principally with relatives whose genetic commonalities make them less likely to deceive each other. That hasn't happened, and it is probably because cooperation is the part of our social organization that has returned riches unimaginable to any other species. Whereas deception is a "zero-sum" interaction—someone must lose when you gain—cooperation is a "win-win" strategy.

This might suggest to us that natural selection has acted more strongly on abilities to detect deception than on the art of practicing it. It might even have favored tendencies that limit our own levels of deception lest we jeopardize the very cooperation from which we personally derive so many benefits. Given this, the best way for deception to survive is probably to keep its ambitions in check, and generally not call too much attention to itself—which indeed might be one of the roles of our consciences. If this reasoning is correct, then most of our deception—although by no means all—probably acts on the margins, closer in its effects to petty theft than to violent murder. In this form it is probably all around us, almost as if it seeks some sort of outlet in our everyday lives and we can't always resist its minor thrills and temptations.

We might be good at practicing as well as detecting deception, but a bold experiment that the psychology professor David Rosenhan undertook in the early 1970s reminds us that even trained mental health professionals can be fooled. Rosenhan's report on the experiment, which appeared in *Science* in 1973, describes how eight sane people gained admission to psychiatric hospitals by showing up and complaining of "hearing voices." Other than this and lying about their true occupations, everything they told the staff at the hospitals about their lives was true. All were admitted to hospital with a diagnosis of schizophrenia, and told they had to remain there until their conditions improved.

Immediately upon admission all eight stopped complaining of

any symptoms and acted, to the best of their abilities, normally. Staff described them as "cooperative" and as exhibiting "no abnormal tendencies." But what followed was a surprise to everyone. All of the fake-insane were given medications (which they secretly did not swallow) and hospitalized on average for just under three weeks, although one of the eight was detained for over seven weeks. Upon discharge they were each given a diagnosis not of sanity, but of "schizophrenia in remission."

Deception is deeply wired into our genes from birth. Most of us have seen the pictures of human babies in wards, wrapped up in swaddling and placed side by side in a row. The surprising feature of these babies is how much they can resemble each other. Or to put it another way, they don't appear to resemble anyone in particular. Babies have characteristically round faces and short pug noses, and most infants of European ancestry have blond hair and blue eyes at birth. So anonymous are human infants at birth that observers cannot match photographs of babies to their parents at a better than chance level. Hospitals frequently put both wrist and leg bands on babies, identifying their parents. This is not to prevent the occasional and terrible cases of baby-snatching from neonatal wards; it is a precaution should one band drop off. We have all heard the sad cases of babies being accidentally switched among parents in hospital rooms and then being brought up by the "wrong" parents, only to discover the distressing truth many years later. Who among us would be able to pick out our own child from a group of newborn infants had we never seen it?

Human babies are probably the most dependent of all species of mammal on their parents for survival. Our offspring require something like twenty-four-hour home care for the first year or so of life. It is extraordinary, then, that they do not come into the world with clear marks or some sort of signal that would allow their parents to recognize them. Lambs and newly born goat kids are far more precocious at birth than are human infants and yet even their mothers can recognize them from their calls. So, why are our babies anonymous at birth? One possibility is that babies have round faces, small chins, and pug noses to fit through the small human birth canal. And yet, there is no reason why babies need blond hair and blue eyes to do this, or why they couldn't have some distinguishing mark or feature. Another possibility is it just doesn't matter that the baby is anonymous: human mothers are not normally separated from their infants, and they know they are the mothers of the infant they gave birth to.

But this is not necessarily true of human fathers (or for that matter most fathers). The baby in his presence might have been fathered by another man, and without his knowing it. This means that a baby might find itself being raised by a domestic father that is not its *biological* father. Were that domestic father to find out, he might not be willing to spend the time and effort caring for that baby. Worse, he might harm or even kill offspring not related to him, as is true of other mammalian fathers. Infanticide by males is common in mammals; it often occurs when a male lion takes over a pride, or a male gorilla ousts the resident silverback. These new males know they are not the fathers of any offspring around them, and have no interest in rearing them for some other male. They don't kill these infants to be spiteful, or King Herod–like, to be sure they get a particular one. Killing a female's offspring will often mean that she comes back into estrus sooner, and the new male will then mate with her and have his own offspring around him instead.

Infanticide by males is rare in humans, but Martin Daly and Margo Wilson have demonstrated that it is around *one hundred* times more likely to occur between male stepparents and their stepchildren than between male biological parents and theirs. This statistic holds even when other factors that might differ between these two kinds of families are taken into account. It must be emphasized that the vast majority of stepparents are not abusive, much less infanticidal, just that both traits are more common in stepparents than biological parents. What scant evidence there is in humans suggests that domestic fathers might not be the biological fathers in 5 to 10 percent of births, without knowing it. Faced with the possibility of reduced investment, or worse, human infants might have fared better if they could hide clues about who their father was and genes for anonymity would have spread. If the domestic father is the biological father, little is lost by this deception. But if the domestic father is not the biological father, potentially much is gained.

Some years ago, I analyzed a mathematical model of this question, which indicated that levels of paternity uncertainty right around the 5–10 percent mark are where anonymity becomes the preferred strategy. We can even speculate that human babies' strategy of deception extends to their behaviors. They should accept their domestic fathers—biological or not—on equal terms, not relying on physical or olfactory cues to influence their responses. Babies might even deceive themselves the better to cover up their deception. It is just possible, then, that human babies have adopted a deliberate strategy of anonymity to conceal their father's identity. Women seem to be aware of men's predicament and the risks this holds for their children, and often collude with their babies in the deception, even if unwittingly. Anthropological accounts reveal that at the birth of a child, the mother, her mother, and other women who might be present are far more likely to say, "It looks like the father," than, "It looks like the mother." Fathers need reassuring; the mothers don't. As time goes on, the risks to the baby decline and they often start to resemble one or the other of their parents, or both. The blond hair and blue eyes so common in Caucasian infants disappear in most of them sometime around the second birthday. As we might expect of a strategy to hide one's identity, the genes that control our adult eye color are switched off at birth (unless the baby does have blue eyes), but are gradually turned on as the baby matures.

divided minds and self-deception

AN IMPORTANT role for self-deception might be to allow the brain to produce a narrative of everyday existence that is somehow consistent, in the face of the gnawing internal conflicts over what to think and how to behave. William Hamilton wrote in the first volume of his *Narrow Roads of Gene Land* that

> Seemingly inescapable conflict within diploid organisms came to me both as a new agonizing challenge and at the same time a release from a personal problem I had had all my life. My own

> conscious and seemingly indivisible self was turning out far from what I had imagined and I need not be so ashamed of my self-pity! I was an ambassador ordered abroad by some fragile coalition [of genes], a bearer of conflicting orders from the uneasy masters of a divided empire. Still baffled about the very nature of the policies I was supposed to support, I was being asked to act, and to act at once—to analyse, report on, and influence the world about me.

What could Hamilton's lament possibly mean? A *diploid* organism is one that like humans and most animals has two copies of each of its genes, one that it received from its mother and the other from its father. These two copies are called *alleles.* Hamilton had come to appreciate that having maternal and paternal alleles meant there is a fundamentally divided parliament of genes locked up inside all of us. Here is why: males and females combine their genes to produce male and female offspring. This means that most of your genes have spent half their evolutionary history living inside a female body and half living inside a male body. If what works best for a female body is not what works best for a male body, then our genes can be tempted to acquire divided loyalties in their different postings. Human males and females have different physiology, morphology, body size, and shape. Human males and females have different life expectancies (females live longer than males, although this might be a reversal of the pattern in our evolutionary past), and most young women mature physically earlier than men. There might also be differences in risk taking and temperament between men and women, but these are less certain and, even if they do exist, could be learned rather than influenced by genes.

So different are the two sexes that it is useful in some respects to think of males and females as slightly different species pursuing divergent lives, only coming together every now and then to mingle their genes. If the metaphor seems strained in humans, consider the case of the group of insects known as the *Strepsiptera*. They are *endoparasites*, meaning parasites that live inside the body of another animal. In a few *Strepsiptera* species, males and females have evolved to live inside the bodies of different hosts—females lives inside an ant, males inside a cricket. Females spend their entire lives dwelling in their host's body, never emerging. They have lost their eyes, wings, antennae, and legs. Males on the other hand must emerge every now and then from their cricket hosts to find these submerged females to mate, and so they retain all these features. How can the same genes be expected to produce such different outcomes?

Natural selection will often have little choice but to mold our genes to do the best job they can by somehow averaging the demands and the payoffs of living in male and female bodies; but that averaging can be tipped toward one sex more than the other. For example, the genes that reside on the chromosomes that determine your sex do not divide their time equally between the sexes, and this can cause them to have sharply divided loyalities. Women have two X chromosomes, one inherited from their father and one from their mother. Men have one X and one Y chromosome, the Y always inherited from their father, the X from their mother. Assuming equal numbers of males and females in a population, this means that Y chromosomes at any given time make up one quarter of the sex chromosomes, the other three quarters being Xs. The Y chromosome only ever resides in men, but X chromosomes divide their time between men and women (technically some parts of the Y can find their way onto the X but this is not important for this example).

Some simple arithmetic tells us that over many generations X chromosomes will spend two thirds of their time in women and one third in men (think of the sex chromosomes in a man and a woman as being numbered from 1 to 4, three of which are X chromosomes and one is a Y chromosome. Then, again assuming equal numbers of men and women, any given X chromosome has a two-thirds chance of being in a woman and a one-third chance of being in a man, because a woman always gets two Xs and a man one). We might expect genes on X chromosomes, then, to be better at making women than at making men even though they can reside in both.

In extreme cases, the loyalties get tipped entirely to one sex, even though the gene can appear in males and females. You inherit a small amount of DNA in the structure known as the mitochondrion that resides in each of your cells. Mitochondria are inherited solely from mothers, even though mothers and fathers both have them. This means that even though the mitochondrion spends exactly half of its time in each sex, this small piece of DNA is at a dead end when it finds itself in a male because it knows it will not be transmitted to his offspring. This puts it in direct conflict with the male's body. There are remarkable instances in the animal kingdom in which genes on the mitochondrion feminize males to such an extent that they become females, thereby ensuring this mitochondrion does get transmitted.

It gets worse. Bacteria known as *Wolbachia* live inside the cells of some ladybird, fruitfly, butterfly, and woodlice species. As with mitochondria, only mothers transmit these *Wolbachia* to their offspring. Now, females of these species lay a large clutch of brother and sister embryos, each carrying the mother's *Wolbachia*. As with the mitochondria, the *Wolbachia* in the male embryos are at a dead end. In some ladybird species, these *Wolbachia* commit suicide by killing the male embryos in which they reside. Why? In ladybirds at least, the surviving sisters (who carry identical copies of the suicidal *Wolbachia*) feast on their dead brothers. This improves the sisters' survival and therefore the survival of the *Wolbachia* in them. Suicide can be a useful policy for spreading copies of your genes.

Our divided loyalties do not begin or end with differences that arise from genes spending differing amounts of time in the sexes. For example, all of our genes have spent a good part of their lives living inside children who try to manipulate their parents in the persistent ways that children do, and another part of their lives as parents trying to resist those charms. Which period of our lives wins? Most people think it must be parents, as they are bigger, stronger, more intelligent, hold the reins of power, and are more devious. Or is it children because their needs are greater? The answer is neither, or more accurately, the answer is that it is the wrong question. Our genes have been selected to influence us in the ways that best suit them at the different times of our lives. That is why the begging and nagging you as a child turns into the scolding and strict parent. As a child it serves your genes to beg, but when you become a parent it often serves those same genes to try to stand firm. Just as was true of the stand-off between deception and the ability to detect it, it is not *you* that wins in the end but your genes. You are just their carrier and the organ that expresses their wishes, and otherwise mostly irrelevant.

A phenomenon known as *genomic imprinting* could mean that the possibilities for a divided mind are far greater than anyone might previously have imagined. Even people who instinctively recoil from the thought that they are just a vehicle for their genes must acknowledge some beauty in the precision of its effects. For the most part, the paired genes that you inherit from your mother and father are identical or nearly so, and normally both the maternal and the paternal copies are "expressed" or used. But the hallmark of genes that are "imprinted" is that some are expressed only when inherited from the mothers, and others are expressed only when inherited from the father, with the other copy in both cases remaining silent. The imprint itself is a chemical tag that gets applied to the gene when the mother makes an egg or the father makes sperm, and its action is to switch the gene off. Now, if for some reason there is a conflict of interest between the same genes, depending upon whether they are inherited from the mother or the father, such that one wants to be on and the other off, genomic imprinting can make that conflict happen.

The best-known cases of genomic imprinting involve genes that influence how large a baby grows in the mother's womb. Normally, at least up to a point, larger babies are healthier. Nevertheless, a mother might wish to limit the growth of her babies *in utero* because a baby that is too large could harm or even kill her during childbirth, or affect her chances of reproducing successfully in the future. All women face this trade-off between what evolutionary biologists call their *current* and *future reproductive success*. Fathers, on the other hand, in an evolutionary sense have been less concerned about how the baby's size might affect the mother. So long as the mother (and thus the baby) survives, men will prefer somewhat larger babies because they tend to fare better. Men won't mind as much—again in an evolutionary sense—if this limits the mother's future reproduction because they can in principle always have children with other women. This results in a conflict of interest between paternal and maternal genes, with the father's genes wanting slightly larger babies than the mother's genes.

Careful genetic studies show that mothers and fathers imprint their genes in just the ways we would predict if they are in conflict over how large they want their babies to be. For instance, we all carry a gene called *insulin-like growth factor 2*, or *Igf2*. *Igf2* produces a protein that, as its name suggests, increases the volume of nutrients the developing embryo receives, causing it to grow larger. If our conflict story is correct, we might expect that fathers will leave their paternal copy of this gene on in the developing fetus, but mothers will switch their maternal copy off by imprinting it. And this is exactly what happens. Another gene, called *Igf2-R*, adds evolutionary intrigue to this story. Standing for insulin-like growth factor 2 *receptor*, this gene captures the product of the *Igf2* gene: it seems to have no other purpose than to nullify *Igf2's* effects. It is as if mothers and fathers have been locked into an evolutionary arms race over the size of the baby and *Igf2-R* exists to help the mother. If so, we would expect the maternal copy of *Igf2-R* to be switched on, but the paternal copy to be imprinted and thereby switched off. And, again, this is exactly what happens. Neither side (mother vs. father) "wins" this evolutionary competition outright; rather, they remain locked into their respective strategies as the best things they can do for their respective genes.

Whether this conflict between the mothers and fathers is also felt within each of their minds is not known, but it could be. Male and female babies grow up to be fathers and mothers, and both sexes will carry these imprinted genes that influenced their growth. Either way, conflicts that might manifest as the sort of divided consciousness Hamilton spoke of can be imagined and might even operate routinely in our minds because of genomic imprinting effects. The founder of psychoanalysis, Sigmund Freud, would have known nothing about genomic imprinting, but it might form the basis of one of his better-known ideas. Freud proposed that a squabbling triumvirate of the *id*, *ego*, and *superego* rule our conscious and subconscious minds. To Freud, your id is your impulsive and self-interested side, your superego is your moral and ethical side, and your ego is left in the middle to arbitrate between these two opposing factions. Freud is rightly criticized for the descriptive, subjective, and elusive nature of these definitions, and yet all of us sense that there is something to them. We have all known the impulsivity and hyperactivity of someone with an overactive id (in Freud's terminology) or the balefully moralizing tone of someone with an overactive superego. One says, "C'mon, just do it!" The other says, "You know, we really shouldn't, it's not right."

To the less Freudian inclined, Freud's id-superego battle is just the familiar tussle between passion and reason. But why is this tussle such a fundamental and unresolved part of us? Extraordinarily, we can construct an expectation for something like id versus superego conflicts from a knowledge of how imprinting works. In most mammal species, upon reaching sexual maturity, one or the other sex—but not both—typically leaves the group it was born into, to find a mate elsewhere. In most mammals it is males that disperse at sexual maturity and these males must then go off on their own, or sometimes in pairs as brothers, leaving their natal territory to seek a new territory in which to live and reproduce. This pattern in the mammals accounts for the familiar stories of male lions taking over prides of lionesses or male gorillas that take over and displace silverback males from the groups they dominate. In birds, it is often the reverse, with females dispersing and having to go off and find a male's territory to join. In humans, the usual mammal pattern is reversed, and so, as with many bird species, it is typically females that leave the group. This is not a prescriptive social statement but rather one based on patterns of female movements in traditional hunter-gatherer societies. Even until recently, we recognize this pattern in the tendency of women to join their husband's household and adopt his surname.

Female dispersal is an unusual trait but one that we also happen to share with chimpanzees, and it seems with Neanderthals. The twelve cannibalized Neanderthal skeletons mentioned earlier were well enough preserved to yield ancient DNA. Analysis of their genes shows that the men were brothers, but the women, as would be expected if they have left their natal territories, were unrelated to each other. Now if, as the chimpanzees and Neanderthals suggest, this difference in which sex leaves the group has been true of our past, then the males in human social groups will tend to be more closely related to each other than they are to the adult women, or the women are to each other. The men are mostly fathers, sons, uncles, nephews, cousins, and brothers; but the women are mostly immigrants from other groups. This difference in average relatedness sets up the potential for a conflict of interest between the genes the mother and father transmit to their offspring. A father might wish his offspring to behave altruistically toward other group members, especially toward children and adult male members of that group. Mothers, on the other hand, being less related to other people in their group, might wish these same offspring to be less altruistic and more self-interested.

And this is where genomic imprinting steps in. If we imagine a gene involved in helping or favoring relatives, including the conscious expression of the emotions and beliefs that support these behaviors, your mother might wish to have her genes switched off to make it less likely they provide aid to genes not related to hers. For the opposite reason, your father might wish to have his switched on. Now since half of your genes come from your mother and half from your father, this sets in place the potential for an enormous tug-of-war between opposing factions of genes—some predisposed to be helpful and cooperative, others less so. If we wish to put this scenario in Freudian terms, there will be a struggle within you between superegolike tendencies to be helpful and idlike tendencies to be more selfish in your outlook. It is then up to your conscience, or ego, to arbitrate.

Next time you feel that tension between the half-crazed side of you that wants to leap recklessly at some opportunity and that annoying other side of you that says, "Whoa!", give a moment's thought to the possibility that it all comes from a genetic tug-of-war inside your brain. I use Freud's ideas here metaphorically and even whimsically, and I don't take these examples to be evidence for his theories. But for purposes of understanding our selves, we can now see the wisdom of Hamilton's statement—there is an inescapable conflict within us, and it might even be played out at the level of our conscious awareness and dispositions toward others. The "I" that we see when we look inside us is perhaps little more than a construction of these interested parties inside our minds, trying to get us to behave in ways that promote them, given the makeup of our social groups.

If this all sounds a bit dubious or incredible, it is worth bearing in mind what Daniel Dennett has called Orgel's "second rule," named after the molecular biologist Leslie Orgel: "evolution is cleverer than you are." And, indeed, a gene has recently been found in mice to have the characteristics predicted if there is conflict between paternal and maternal alleles over how altruistic to be. The gene is called *growth factor receptor-bound protein 10* or *Grb10*, and it is imprinted—paternal copies of this gene are expressed, while the maternal ones receive the imprint and are switched off. It also influences social behavior. When researchers remove the paternal copy of the gene, offspring become more socially dominant—more altruistic in this case—in their interactions with others. This suggests that the paternal copy normally acts to suppress altruism rather than promote it, as in our scenario for humans. Intriguingly, though, in mice it is males who normally disperse from their groups rather than females, so this is the effect we expect. It is too early to know if this finding bears directly on the idea of conflict inside human minds. But it does show that imprinting related to conflict within genomes over how just how altruistic maternal and paternal genes "want" to be can indeed be observed.

consciousness and self-awareness

THE IMPRESSION that we are able to choose freely between different possible courses of action is fundamental to our self-image as beings with free will. But as we have seen, our obsession with free will might be misplaced. Natural selection should have created in us a tendency to do what is good for our survival and reproduction, and not necessarily a tendency to do what we "want" to do. The situation might even be worse for free will than this explanation suggests. Our subjective experience of freedom—or at least more of it than we might imagine—might be little more than an illusion. This is a statement that we all instinctively and intuitively reject, but there is a growing evidence that our actions can be initiated by unconscious mental processes that occur long before we become aware of them.

Sophisticated experiments which measure activity in people's brains while they are asked to perform a task show that our brain can make decisions up to ten seconds before the decision enters our awareness. In one of these experiments, you might be asked to focus your attention on a screen where a stream of letters is presented, such as: *b.....c.....f.....d.....* You are told that when you feel the urge to do so, you should freely decide to push one of two buttons in front of you—one operated by your left-hand index finger, the other by your right—and then press it immediately. After pressing one of the buttons, a screen appears that replays the last few letters in the stream you just viewed. You are then asked to indicate which letter was being displayed when the idea to push the button came into your mind.

People who participate in these tasks use the left and right response buttons about equally often. This is important because it shows they are making choices. But the surprise is that people's brains seem to know what they are going to do before the owners of those brains are made aware of the decision. This unsettling conclusion was apparent from the patterns of the volunteers' brain activity. These patterns changed five to ten seconds *before* the letter the volunteers had indicated was being displayed on the screen when the idea to push the button came into their minds. Imagine yourself in this setting: You might have got the idea to push the button when *d* was on the screen, but your brain activity would tell a different story. It would show changes happening earlier, say, when *b* was on the screen. The researchers know that this was when the decision was made to press the button because the patterns of brain activity can predict which of the buttons—left or right—you would later push. At the time these events were happening, you would not have yet had an inkling of what you were going to do.

These are careful neurological studies by leading scientists using the latest neuro-imaging techniques. They are not the sort of crackpot research studies—such as those on so-called subliminal messages—

that unnerved moviegoers in the 1950s, worried that they had been manipulated by some "hidden persuader," in Vance Packard's memorable phrase. His compelling and worrying book of the same title convinced millions they were powerless to escape instructions being inserted into their minds—to buy Coca-Cola and popcorn—by messages flashed so quickly onto movie screens that their conscious minds were not even aware of them. Campaigners opposed to the use of subliminal messages in advertising produced advertisements showing hordes of slightly manic people, manipulated by the invisible hands of subliminal messages, streaming to buy soft drinks and popcorn. But later research showed Packard's claims to be false. Truly subliminal messages—messages that appear for such a short period that we don't perceive them—don't influence us, or if they do, the effects are weak. It is only when they are shown long enough for us to become aware of them—that is, they are no longer *subliminal*—that we rush out and buy the Coca-Cola or the popcorn. We are being persuaded, but the persuaders are not hidden.

What, then, could the results of the neuro-imaging studies possibly mean? Are we mere observers being diligently, if belatedly, informed of what our brains are up to? The simple answer to that question is almost certainly yes. To a degree that far exceeds what we might have thought, we cannot merely assume that "we" initiate the thought processes that lead to the decisions we make, such as to buy Coca-Cola or popcorn. Instead, our subconscious brains might set to work on problems that it later gives us answers to, and when it does so, this feels like free choice or an action "we" chose. In fact, whenever some thought just seemingly pops into your brain, it might be the result of the sort of subconscious thinking that occurred in the button-pushing experiments.

Our environments routinely give us clues to be thinking about certain things, but the clues and our thoughts about them might sit mostly beneath our awareness. Driving along a motorway, your brain will be subconsciously monitoring your speed and stability, where other cars are and where you are, and all the while readying itself to alert you to the turning or exit you should take. When you choose a seat at a cinema, or select an ice cream flavor, pick out a book at a bookstore or the color to paint the wall of your living room, the decisions might have been made sometime before they were made available to you. How widespread this is is surely one of the most fundamental questions of our inner existence.

The neurological results coincide with a radical proposal from psychology: that we do not have the privileged access or knowledge of our inner selves that we might imagine, and that most of us simply take for granted. René Descartes is best known for his statement, *Cogito, ergo sum*, his philosophical leap of faith for the reality of being. Descartes believed that if there was something that was thinking, then at least we could be sure there was something that existed. But being aware of its existence doesn't ensure that that thinking thing can know itself. Indeed, the social psychologist Daryl Bem proposed in the late 1960s that we come to know our own attitudes, emotions, and character at least in part by inferring them from observing our own behaviors rather than from introspection. Bem thought that when it came to knowing ourselves, we were often in a position no different from an outside observer. Like them, we arrive at inferences about ourselves from watching what we do and from seeing how others react to us.

If this seems far-fetched, who among us can say how we would react in an emergency medical situation, on the battlefield, encountering a burglar, or having to watch abdominal surgery? If you *can* answer one or more of these questions, ask yourself if it is because you have been in one of these situations and so have had a chance to observe how you would react. I once had to have an injection in my forearm to test for tuberculosis—the standard tests that are done to people who have travelled in parts of the world where this disease is still prevalent. I stood and watched as the needle was inserted under my skin. The next thing I knew I was lying on my back on the floor with my feet up in the air, resting on a nurse's shoulders, while another nurse crouched down to attend to the open wound on my forehead. While I was contentedly watching the needle go in, my body had decided to black out. When I fell to the floor, my forehead hit a large piece of equipment. After I came to, the agitated nurses demanded to know why I hadn't warned them. Feebly, all I could say was that I didn't know. Now I do, and I sit down to get any injections. But I only know to do this because I watched what my body did.

Even our inventiveness might be largely a subconscious, or at least a non-verbal, process. Where do our new and imaginative thoughts come from? Why do we use the metaphor of a light bulb switching on to describe them, unless we have a sense that they just seem to come from nowhere out of the dark depths of our minds? In the 1940s, the great mathematician Jacques Hadamard interviewed creative people in an attempt to describe how they came up with their ideas. For Hadamard, creativity did not lie in consciousness, but in the long unconscious work of our minds incubating thoughts, which even once formulated, must pass through what Hadamard thought of as a sort of aesthetic filter before they reach consciousness. Einstein was Hadamard's most celebrated contributor, and for him creative thinking took on a physical property. Hadamard quotes a letter from Einstein in which he described how he could physically feel when thoughts were emerging in his mind, but could not put them into words:

> the words or the language, as they are written or spoken, do not seem to play any role in my mechanism of thought. The physical entities which seem to serve as elements in thought are certain signs and more or less clear images . . . this seems to be the essential feature in productive thought before there is any connection with logical construction in words or other kinds of signs which can be communicated to others.

There is a growing belief that many of our moral decisions might be made before we become aware of them, as if we have an innate moral sense. One suggestion is that we might have an ancient affective or emotional system in our brains that makes split-second decisions about things that have moral content, and then it presents those decisions to a younger, more recently evolved cognitive or deliberative side of our brains that might have been bolted on in the last 100,000 to 200,000 years of our evolution. The evidence for this is that people presented with moral dilemmas can often quickly tell you how they would behave but often struggle to explain why.

Here is an example. A runaway train carriage on a rail line is about to run over and kill five people. You can save them by pushing a button that will divert the carriage onto another track. The trouble is there is a person on that track who will be killed by the carriage. Should you push the button? Most people say yes to this question, even if feeling slightly uneasy about doing so. But now consider a different scenario. Once again, a runaway carriage is bearing down on five people. You can save them, but to do so you must push someone else standing nearby into the carriage's path. Should you do it? Most people say no. When asked to explain the differences, many people are simply dumbfounded—at a loss to provide a rational explanation for what they say they feel instinctively they should do, or why they feel differently about the two situations. Others can eventually provide explanations but much too slowly to have used them to save anyone.

If there is a strand that links these two scenarios, it is that we have an instinctive reluctance to cause harm to others directly, especially when, as in these examples, they pose no threat to us. It is a disposition that might put a brake on our more violent tendencies, and this is something that should be valuable, at least in most circumstances, in our social groups. It is an instinct that also appears, to a far larger extent than most people would have guessed, to be hard-wired into our brains, and they—our brains—act without directly consulting our conscious minds. It could be that our brains get us to make moral decisions and we don't really even know how they arrived at them. Structures in our brains that support this moral decision making will have spread in our evolutionary past, so long as the actions they influence serve our well-being. Much of our so-called moral nature might be just this—dispositions and the behaviors they bring about, some acquired or burnished by learning, others part of our genetic makeup—linked by promoting actions that work for us in the peculiar outlines of our social systems.

Where do these features of our minds leave consciousness, and what is its role? Here is one possibility. Life in a complex social environment such as our own requires us to make decisions about a fluid and constantly shifting situation. The role of a conscious mind might be to weigh up alternative courses of action sent to us from our subconscious minds. This role might be given over to our conscious mind rather than allowing our brains to follow some rule of thumb or algorithm because the social contingencies multiply endlessly—"If I do this, that might occur, and then she might do this or he might do that and that would lead to this. . . ." If not merely a problem of contingencies, the rapid pace of cultural change means that new situations constantly arise and these require conscious deliberation rather than less flexible subconscious rules. We see this today in the field of the social management of technology: should we allow elderly men and women to have children; should we make it possible for someone to clone themselves, or for that matter someone else? Should a woman be allowed to gestate someone else's baby for them? It might be that precisely because of the complexity and novelty of the social relations we regularly engage in, we need something that works in real time and is flexible.

Giving consciousness this role in deliberating and updating us on a real-time basis is a scenario particularly apt for our moral decision making, even if the conscious part appears to happen after the decision has been made by our subconscious minds. After-the-event moral reasoning might help us to understand the connections between external events we can witness, and our own emotions. We can then use these to help us predict others' feelings and emotions, and, importantly, how they might behave. We can frequently observe the same things as others can; it is just that we cannot have direct access to how they feel about those things. Simulating what might be going on in their minds and then comparing the outcomes of our simulations to their actions might be useful. Is someone likely to be bothered, amused, enraged, or euphoric at some set of external events? Analyzing the links between our own emotions and those same external events may therefore give us a better-developed theory of mind that we can put to use in real time as we encounter others. It might be precisely an inability or an impoverished ability to conduct such simulations that plagues people with autism and its milder manifestation in Asperger's syndrome: patients who routinely say they don't know what is going on in the minds of the people around them.

A simpler explanation for consciousness looks to the properties of successful ideas themselves. The ideas that we carry around in our heads are predominantly those that have in our past been good at getting themselves transmitted from one mind to another. Catchy songs like "Fly Me to the Moon," or phrases like "Watch out," or, "Mind your head," or useful pieces of knowledge, such as "Train conductors stop working after 10 p.m.," or "The angle of Polaris from the horizon can be used [at least in the northern hemisphere] to work out your latitude," are more likely to get themselves transmitted than dull or incorrect ones. Countless millions of ideas probably never even see the light of day—or perhaps a better metaphor would be are never given a hearing—because they don't get us to talk about them. Others do, and they are the ones that are, on balance, more likely to be transmitted. Perhaps, then, our ideas in the form of active memes created consciousness as a way to get us to think about and transmit them! Who knows? Maybe it is even our memes constantly agitating and clamoring to be heard that creates the cacophony of "stimulus-independent thought" we saw at the start of Part III.

But there is something unsatisfactory about all of these scenarios for consciousness, and it is this: why is it necessary to conduct deliberations "consciously"? Why is it necessary for a meme to "pop into consciousness" for us to tell someone else about it? These explanations for consciousness beg the question they are meant to answer by assuming the value of consciousness they set out to explain. (As an aside, the meaning of "begging the question" has been changing over recent years so that now many people use it to mean "demanding to be answered." But its original meaning to philosophers was "an answer that assumes in its premise the proposition it sets out to explain.") They assume that consciousness makes us more likely or better able to think about something or act on it. But *why* do we think consciousness improves deliberation, decision making, or for that matter the transmission of memes? Maybe it does, but if we are willing to assume this, there is nothing really to explain.

To see why this assumption is not as obvious as you might think, consider that the game of chess is surely an extreme case of mulling things over before deciding how to act, and of infinite and evanescent possibilities. But this is a game at which computers now routinely beat humans and no one would say that the computers are conscious. When in the 1990s Garry Kasparov played against, and was finally beaten by, the IBM Deep Blue computer, it was his realization that the machine was not conscious that he found most distressing. Kasparov explained that chess is a game of warfare in which terrifying your opponents—striking fear into their hearts—with moves they don't understand or have not seen coming is a vital part of a winning tactic in a grandmaster's game. But computers have no fear; they don't mind losing; and they don't get tired.

It might be objected that computers play chess differently from humans, and this might be true. But this still doesn't tell us why what we think of as our conscious awareness must be conscious to be effective. Here is a suggestion that does not so obviously suffer from begging the question about consciousness, but must be regarded as little more than speculation. Perhaps consciousness arises as a true "sixth sense," albeit a virtual one (our other five senses conventionally being touch, hearing, smell, sight, and taste). Like the Persian "King's Eyes," who were charged with keeping the King informed, perhaps our "consciousness" keeps our hungry-for-knowledge subconscious mind informed of an ever-changing and socially complex outside world that it cannot see. What we *perceive* as consciousness is just a byproduct of the vast amount of brain activity required to produce this sixth sense, and then manage all the continuous cross-talk between it and our subconscious minds, all the while updating the sixth sense with the new perceptions flowing in. Consciousness, or the "I" we see inside us, might just be an artifact of the "post-processing" step that tries to summarize and make sense of the material flowing in, and manage the disagreements between it and what is "downstairs."

For example, our social world changes continually, so that a former ally might have just a moment ago become a competitor. When we search our subconscious mind for how to accommodate these changed circumstances, it might get the wrong item off the memory shelf. We have to send it back, updating it with the new information. Listen to St. Augustine musing in his *Confessions* in the fourth century AD about what he called the "palaces of my memory":

> I come to the fields and spacious palaces of my memory, where are the treasures of innumerable images, brought into it from things of all sorts perceived by the senses. There is stored up, whatsoever besides we think, either by enlarging or diminishing, or any other way varying those things which the sense hath come to; and whatever else hath been committed and laid up, which forgetfulness hath not yet swallowed up and buried. When I enter there, I require what I will to be brought forth, and something instantly comes; others must be longer sought after, which are fetched, as it were, out of some inner receptacle; others rush out in troops, and while one thing is desired and required, they start forth, as who should say, "Is it perchance I?" These I drive away with the hand of my heart, from the face of my remembrance; until what I wish for be unveiled, and appear in sight, out of its secret place. Other things come up readily, in unbroken order, as they are called for; those in front making way for the following; and as they make way, they are hidden from sight, ready to come when I will. All which takes place when I repeat a thing by heart.

Students of the Great Apes might complain this explanation for our consciousness could equally apply to the apes' complex social circumstances, and we should also grant them consciousness. Perhaps we should, but even so, there might be two differences between us and the Great Apes that challenge this objection. One is that our social world is even more complex than that of a Great Ape, including social exchange and the extended forms of cooperation we have seen in earlier chapters. A computational state that keeps the "I" centerstage might be particularly valuable for reminding our subconscious minds to put our social system to best use. But the other is even more fundamental: our minds have discovered language. We alone have a symbolic code for translating our subconscious thoughts from whatever form they might take into the same audible (or tactile) language that we use to communicate with others. It might not be an accident that for most of us consciousness is expressed in our native language. Perhaps it is this aspect of our virtual sixth sense that tips our awareness over into something we can label as "I" or "me."

truth and the difficulty of knowing what to do

THE WORD "truth" is heavily laden with difficult philosophical baggage, but as a shorthand we can take it colloquially to mean knowing the right answer, or knowing what really happened in some situation, or knowing the best course of action, or the best solution to some problem. If we take this as a working definition of truth, then we probably have precious little access to it. The American baseball player and coach Casey Stengel famously advised: "Never make predictions, especially about the future." It is good advice. In the 1950s, the president of IBM, Thomas Watson, Jr., is reported to have said, "I think there's a world market for about five computers." Ken Olson, president of Digital Equipment Corporation in 1977, believed that "There is no reason anyone would want a computer in their home." It is rumored that one publishing executive returned a manuscript to J. K. Rowling, saying that "children just aren't interested in witches and wizards anymore," and that an MGM internal memo about *The Wizard of Oz* said, "That rainbow song's no good. Take it out."

For most of us, much of everyday life is a series of easy decisions that we think we know how to make. But for many of the most important things we do, and most of the important decisions we need to make, we don't have and might not even be able to acquire the information we need to be confident of making the right or best decision. It might also be that our best action depends on what others do. Should I fight those people who live in the next valley and who keep stealing my sheep? What lure should I use on my fishing line in this stream? Is that snake poisonous or is it one of those that just looks like a poisonous one? Is that berry edible? How much should I offer for a house I am thinking of buying? Should I pay more or less than I am into my pension fund? What is the best car for me? Which computer should I buy? How strict should I be with my children? Should I invest in that stock or buy a government bond? Which is the best airline? Should I marry this person?

An amusing but potentially serious manifestation of not knowing what to do or how to behave is called "collective ignorance." You are in a crowded elevator that comes to a halt between floors. Maybe it is just a temporary problem, but maybe not. What should you do? Not wishing to appear foolish or anxious, you look to others for clues. But of course the others are in the same position as you and they are looking to you for the same clues. The result is that everyone inadvertently sends the message to do nothing and you all stand there in silence. It is the position we all occasionally find ourselves in when a fire alarm goes off, a subway train comes grinding to a halt between stations, or, in a big city, we pass someone lying in the street. Should we help them, or are they just some drunk passed out from their own exuberance? But collective ignorance is also why stock markets can rise and fall with such exciting or jarring urgency—few investors know what to do so they just follow what others are doing.

These are questions about whether we should copy others or try to figure out best solutions on our own. As the most intelligent species on the planet, we might think that not only can we work out good solutions, but that doing so rather than relying on others is our best strategy. A simple thought experiment posed by Alan Rogers leads to a different and surprising conclusion. Rogers asks us to imagine a group of people who live in a constantly changing environment such that new problems continually arise that require new solutions. Over time, these people—we can call them *innovators*—work out solutions for surviving and reproducing on their own. This takes time and effort, and they occasionally make mistakes. But they can be expected to maintain a more or less steady level of health and well-being as their innovations just keep up with changes to their environment.

But now imagine that someone is introduced to this group who merely imitates or copies these innovators. This imitator or *social learner* would not have to spend the time and energy trying to work out solutions to problems posed by the environment, and would not suffer the inevitable losses of making the occasional error. This tells us that a social learner who copies what others do, introduced into an environment of innovators, would survive and prosper better than the innovators. Over time, the imitators will therefore increase in number until at some point the population of people is made up mostly of them. But now consider what happens. Once imitators become common, they will frequently copy each other. This is fine so long as it works, but mistakes in copying will creep in, and the imitators will have no way to correct them. The environment will also continue to change. So now the imitators will begin to suffer losses and ill-health because they are employing obsolete solutions.

We learn from this that neither all innovation nor all copying will ever take over in society: in the language of Chapter 3, neither innovation nor copying is an evolutionarily stable strategy. There is also a hint of something we have seen before: that only a handful of innovators is needed. But if this is true, and most of us do copy others, whom should we copy, and how much? Perhaps your neighbor has just purchased a new car; you have been thinking of buying one as well. Should you get the same one? Kevin Laland posed these questions more formally in a computer tournament organized to understand social learning. Elizabeth Pennisi in *Science* describes how people were asked: "Suppose you find yourself in an unfamiliar environment where you don't know how to get food, avoid predators, or travel from A to B. Would you invest time working out what to do on your own, or observe other individuals and copy them? If you copy, who would you copy? The first individual you see? The most common behavior? Do you always copy, or do so selectively? What would you do?"

A young boy I put this question to replied by saying he would copy the most overweight people. There is something to this, especially in the evolutionary setting of being a hunter-gatherer. If body weight is an indication that you have been good at getting food, maybe you are doing something right. It was just this logic that we used to speculate on the meaning of the Venus statues. But Laland wanted to know how we decide whom to copy when we only have access to what others are doing. Entrants to his tournament had to write a computer program that would somehow juggle the alternatives of someone trying to innovate or work out for themselves the best course of action, versus copying or imitating others, and if the latter, whom to imitate. The computer programs operated in a kind of *in silico* social environment in which they could "see" the choices that other programs had made, and thus what behaviors they were displaying. These programs then competed against each other inside a large supercomputer.

Competing strategies ranged from those that relied strongly on innovation to those that always copied others. Startlingly, the winning strategy in Laland's tournament exclusively copied others—it never innovated! By comparison, a strategy that relied almost exclusively on innovation finished ninety-fifth out of one hundred contenders. This is a result that flies in the face of all expectations, but the strategy of always copying works for two very simple but profound reasons. One is that when others around us make decisions and act on them, they have little choice but to demonstrate the best strategy in their repertoire: when you do something, you will typically do what you think is in your best interest. This presents imitators with a set of alternatives from which the truly bad ones have probably already been filtered out. The second is that by virtue of being alive and available to copy, those whom we imitate are survivors, and so what they are doing must be reasonably good.

Remarkably, it matters less exactly whom you copy or precisely what than that you copy rather than try to innovate. Laland's computer tournament, therefore, also lays bare the social implications of learning from others. Our ability to copy and imitate is why our culture can accumulate knowledge and technology. But the winning strategy in the tournament acted like a social parasite, plagiarizing the hard-won knowledge and strategies of others, and thereby avoiding any of the costs of having to try out new ideas on its own. Indeed, its parasitical nature was revealed when Laland ran the winning strategy alone and it performed badly. Just as Alan Rogers's thought experiment would have led us to expect, if no one is innovating, then copiers will end up copying each other, and this will mean that many bad strategies will be copied and maintained.

We have seen this before: social learning—imitation and copying—is visual theft. It is unavoidably steeped in conflict and cooperation because knowledge itself becomes a valuable commodity that might otherwise grant an advantage to the person you visually steal it from. If I can perform some behavior that you wish to learn, I might wish to hide it or even modify it in your presence, or perhaps trade it for some of your knowledge. For your part, you might wish to conceal your interest, act deceptively or furtively, hoping that I will let down my guard. We see these conflicts of interest—and the deceptions they produce—manifesting themselves in patent applications and patent law, industrial and even national espionage, and outright theft. But we also see them in the reluctance, for example, to share old family recipes, reveal where our favorite fishing spot is, where to find the best mushrooms, or what bait we use to catch fish. Deception, competition, and exploitation are built into us because most of us rely on copying others most of the time.

Even when we have access to the so-called facts, we often misuse them, and this too might be because copying has played an important role throughout our history. We know that we are highly susceptible to contagion, false beliefs, neuroses—especially medical and psychological—and conspiracy theories. Why we should be is surprising because our brains have surely evolved to judge risks, to assess likelihoods or probabilities, to defend our minds against undue worry, and to infer what others are thinking. But our minds probably evolved to make these judgments drawing on the experiences of small groups of people—most probably throughout our history the small number of people in our tribe. The trouble is that now we are often confronted with vastly more information about risks, from newspapers and radio or the Internet, and yet we don't always make the best use of it.

We misuse it because our brains assume that the rate at which these things come to our attention from all over the world is the same as the rate in our local area. It is a case of doing *bad mathematics*. In the past, my assessment of the risk of being blown up by a terrorist, or of getting swine flu, or of my child being snatched by a pedophile on the way to school, was calculated from averaging the input of information I received mainly from my small local group, because these were the people I spoke to or heard from, and these were the people whose actions affected me. What the Internet does—and what mass communication does more generally—is to sample those inputs from the 6.8 billion people on Earth. But without my being aware of it, my brain is still considering that the inputs arose from my local community, because that is the case its assessment circuits were built for.

The bad mathematics occurs because my brain assumes a small denominator (the bottom number in a fraction, and here that number is the number of people in my village), but it is using the inputs from the whole world as its numerator (the top number of a fraction). The answer it produces to the question of how likely something is to happen is, then, way too big. So, when I hear every day of children being snatched, my brain gives me the wrong answer to the question of risk: it has divided a big number (the children snatched all over the world) by a small number (the tribe). Call this the "Madeleine McCann effect." We all witnessed months of coverage of this sad case of a kidnapping of a young girl in Portugal that occurred in 2007—as of this writing still unresolved. Although the worry this caused in the rest of us is trivial compared to what the McCanns have suffered, it was probably largely misplaced. But even knowing this, it is hard to shake the feeling that our children are at risk, and this just shows us how deep are the biases in our decision making.

The effects of the bad mathematics don't stop with judging risks. Doing the mathematics wrong means that contagion can leap across the Internet. Contagion arises when people perceive that the numerator (input from Internet) grows far more rapidly than the denominator (village or tribe). Our tendency to copy others just reinforces this perception. Once contagion starts on the Internet, everyone's copying means that the bad mathematics make it explode. The same happens with conspiracy theories: if it seems everyone is talking about something, it must be true!

But this is just the wrong denominator again, because in fact "most" people are not talking about that thing, it is just that the ones who are choose to appear on the Internet (or radio phone-ins, etc.). Neuroses and false beliefs are buttressed: we all worry about our health and in the past would look around us and find that no one else is worrying or ill. But consult the Internet and you might find tens of thousands—maybe more—people are worrying, and they've even developed Web sites to talk about their worry. The 2009 swine flu pandemic turned out to be a damp squib, but you wouldn't have known that from the frenzy at the time.

All of these problems arise because we seldom have access to the truth, and we normally arrive at some guess as to what it is by copying others. The conclusion from tournaments such as Laland's that the number of innovators can be small might be surprising, but when we look around us, this is indeed what we see: successful inventors and entrepreneurs are rare and efforts to find them in television reality shows or to produce them in the classroom only serve to reinforce the point. And this is because for most of what we do, we cannot simply work out in our minds the best course of action, but social learning can sample (or steal) from others' good luck or occasional good judgment. We see an awareness of this even at the highest levels of technical competition in, for example, yachting events such as the America's Cup or racing events such as Formula 1. Boats and cars are often shrouded to conceal, until the very last moment, if ever, the complex shape of a rudder, or the bewildering configuration of airflow across an engine.

A feebleness about knowing what to do has evidently been true throughout our evolution, not just now when we have complicated things like yachts and Formula 1 cars, but also computers and derivative financial products to reach decisions about. What *is* the best way to shape a hand ax, or to make an arrowhead? And how would you know if you stumbled upon the right answer? You wouldn't until you or someone else tried it. And the difficulty even now of answering these comparatively simple questions has meant that we have evolved to be good at what we can be good at: to take advantage of a sort of natural tournament of cultural selection played out in front of us every day, and which presents us with good solutions. Most of us are copiers. Natural selection has seized on the power of copying to make our minds very good at working within what cultural systems have to offer.

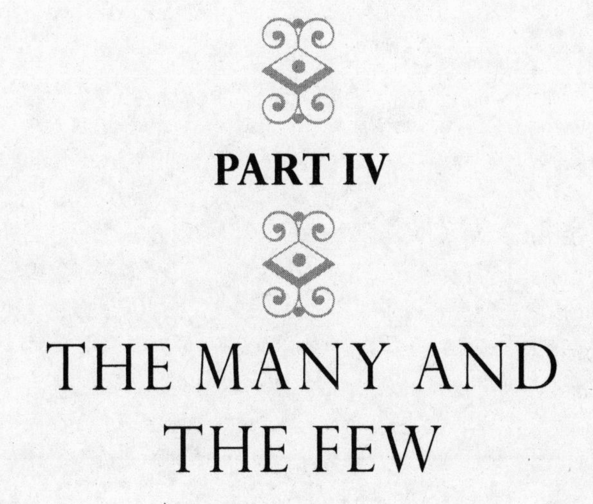

PART IV

THE MANY AND THE FEW

Prologue

TAXIING TO THE terminal at Hong Kong airport, you notice a point across the bay where there is a forest of thin, white structures standing hundreds of feet tall. They have a peculiar monolithic appearance. They do not move or make any sound, and if you arrive at one of those times of the year in Hong Kong when the weather is hot and the air is hazy with pollution and humidity, these white structures look like some giant fungi that has sent up its fruiting bodies from the steaming forest floor, ready to disperse its spores.

But they are not fruiting bodies, at least not of fungi. These stalks have been made by humans. They are high-rise apartment towers, which house tens of thousands of people. And their remarkable feature is that they serve the same purpose as the fungal stalk does for its spores: both are vehicles that carry and promote the survival and reproduction of their inhabitants. But we are not fungi, or even ants, bees, or termites, so how is it that so many of us can live so tightly packed like this, reliant on such a small number of others to govern our lives?

CHAPTER 10

Termite Mounds and the Exploitation of Our Social Instincts

That large groups of humans can be led by a small number of elite for the same reasons as termites, ants, bees, and wasps

a dilemma

TERMITES' MOUNDS and ants' nests can house millions of individuals toiling in dark, cramped, and steamy conditions on behalf of a queen who lives a life devoted almost entirely to reproduction. Most of us instinctively recoil from such a scene as not being part of our nature. Yet it was vividly depicted in the dystopian view of the city of Los Angeles in the film *Blade Runner*, a crowded, teeming, drizzly place full of anonymous strangers. In cities all over the world millions of people live and work side by side ruled by a small elite, and in countries such as China and India over 1 billion people fall under the rule of a few. When we marvel at the purposeful and yet orderly behavior of a colony of ants, scurrying in and out of their nest, some carrying objects, others scouting for prey or invaders, we need not cast our imagination very far to think of construction workers on a large building site or laborers building a pyramid in ancient Egypt. We attend sporting events and musical performances in stadiums at which tens of thousands of us remain for hours only inches from each other, all following the actions of a few on the field or stage. There is something both strange and remarkable about this behavior: hypersocial and hyper-orderly. Apart from the social insects, no other animals can work together in such large numbers. Imagine a construction site or a sports stadium filled with tens of thousands of hyenas, or baboons, or even dogs, a species we have bred in our image.

We are able to live and work among others in our millions. And yet this poses a dilemma for one of the main ideas of this book: nothing in our evolutionary history specifically prepared us for this. If humans evolved a tribal nature that revolves around life in relatively small and exclusive cooperative societies, how do we explain the enormous social groupings of the modern world in which so many can be so willingly led by so few? The growth of human populations happened far too quickly for biological changes to our nature to have kept up. Until perhaps 10,000 years ago, all humans lived in small hunter-gatherer bands. The invention of agriculture changed all that as having the capacity to *produce* rather than simply gather food meant larger numbers of people could reside in the same place. Small bands of maybe ten to three hundred people gradually came to be replaced by tribes that were effectively bands of bands. Tribes gave way to chiefdoms, in which for the first time in our history societies became centralized. There was stratification by class and the chief sat at the top of a formal hierarchy of authority. Chiefdoms eventually gave way in turn to large city-states such as Jericho (in modern-day Palestine) and Çatal Hüyük in Turkey, or the Mesopotamian cities of Ur and Babylon. These were later succeeded by fledgling nation-states.

The forces propelling this growth were many, but mainly of three sorts—protection, economic well-being, and reproductive output. People were, in a word, better off, even if it is by now well established that we were often less healthy in these large groupings. But being better off does not alone tell us why it worked. Were we to provide 10,000 dogs, hyenas, or even apes with unlimited food and protection, we would not get the happy outcome we might have sought. Paul Seabright in *The Company of Strangers* suggests that human societies have been able to grow large because we have acquired an ability to trust strangers. We pay our taxes to unknown bureaucrats, buy things made in foreign lands and from people we do not know, walk past strangers in the street and even allow them into our homes without fear of being robbed or killed. We are able do these things because we have evolved rules and dispositions that allow us to exchange goods and services with people we have never met. And indeed, we have seen in earlier chapters how cooperation and trust can arise. The looming shadow of future encounters with the same people softens our tendencies to cheat them; we acquire reputations and learn those of others; and we count on the knowledge that you and those you would do business with bring to every exchange a sense of fairness that protects you and them from exploitation.

But if these rules remind us of anything, it is that we cannot possibly ever know enough about strangers per se to trust them. In fact, we have seen that there is reason to believe we have a hard-wired wariness of strangers. Rather, what we have acquired throughout our brief evolution is a taste for the benefits of cooperation and some rules that can make it work in the right circumstances. Thus, when we do appear to trust strangers, it is probably because they are not really strangers—we know or think we know something about them, the institutions they work for, or we think there are institutions such as the police, banks, or insurance companies ready to protect us from them. When a man knocks at my door asking to read my electricity meter, if I do let him in, it is with a mild apprehension. And even then I only do it because I know that it is my electrical company's practice to send such people around and that my meter hasn't been read for awhile. If this man knocked late at night, looked threatening, or it wasn't my company's practice to send such people around, it is doubtful I would let him in. Even when I do so, it is only after I have asked for his ID and sized him up, making a quick calculation as to whether I could overpower him should he try to rob me. It will also help if I haven't heard anything in the local news about thieves or muggers who masquerade as meter-checkers as a way of gaining access to people's homes.

When the waitress puts my credit card into a restaurant's electronic scanner, I allow it not because I trust strangers, but because I observe others doing it, or have been told by people I do trust that others have used their credit cards at this establishment, or because I happen to know the restaurant has been there for some time. Still, I often feel a slight anxiety, wondering if some cloning device has been fitted to the scanner and I will receive word the next day that a large loan has been taken out against my card. Reminded that my bank will not charge me for purchases I have not made, I go ahead with the transaction anyway. And when I use the services of taxi drivers, banks, airline pilots, the police, and eBay, it is not that I trust them per se, but that I notice over long periods of time that in general airplanes are flown well, the police are not on the take, taxi drivers don't take advantage of their passengers (on the whole), the reputation comments on eBay seem helpful, and my bank is fair with my money (or is it?). But even this is only true in parts of the world where these various services do work, or where I am familiar with the local culture. Many cities have "no-go" areas. Until recently, it was common in many parts of Africa to avoid putting your money into a bank—the widespread belief, often confirmed in practice, being that you wouldn't get it back.

We learn from this that our capacity to live and work in large societies exploits the tactics we have acquired throughout our evolution for making cooperation work, and even then we begin with the most tentative of exchanges. So, if nothing in our evolutionary history specifically prepared us to live in large societies, almost everything about the way culture works does. Mathematicians call a mechanism *scale-free* if it doesn't change as the size or scale of the group or phenomenon it is applied to changes. This chapter examines evidence that the large social groupings that began to emerge around 10,000 years ago did so by exploiting key evolved features of our cooperative behavior and psychology that happen to be scale-free. Our language, our diffuse and indirect style of cooperation and exchange based on reputation, our ability to specialize, and even our willingness to suspend disbelief—thereby making it more likely we might accept some chief as God's representative on Earth—can all act relatively unfettered by the size of the group in which we reside. Having these scale-free cultural mechanisms meant that our societies could automatically grow to a larger size without having to invent new mechanisms beyond those that were in place by perhaps 160,000 to 200,000 years ago when our species arose.

Even these scale-free mechanisms cannot on their own explain why we chose to live in larger societies; they merely made it possible. Instead, we need to look for properties of our societies that make them not only an easy but also a productive thing to be a part of. Here, it turns out that our larger societies could naturally emerge by taking advantage of three properties they all seem to share: one is that they emerge from *local rules*; the second is that there can be some surprising *efficiencies* of larger groupings; and the third is *social viscosity*, or our tendency to maintain local ties within a larger society. The first of these allows larger societies simply to emerge so long as they pay their way; the second tells us how they pay their way; and the third shows us how our tribal psychology can still operate in a larger society. Revealingly, it is also these three features that oppressive and dictatorial regimes attack or exploit to break down a society and hold it within their grip.

local rules and the emergence of self-organization

WHO DESIGNS the societies we live in? The answer of course is that no one does. No one has a blueprint for the final product, no one has the whole picture in mind, and no one ever has. Instead, our societies naturally emerge from the players within them following what we can think of as local rules, and we should be grateful for this. To see why planning anything as complex as a society is out of the question, just consider what must happen for you to take a plane journey. Some time before leaving home you probably phoned for a taxi to take you to the local bus station. At the bus station you buy a ticket to take you to the airport. At the airport you have to check in and then pass through security and passport checks, perhaps buy some food at a restaurant. And finally you will board a plane and be taken to your destination. Once there, you again pass through various checkpoints, collect your bags, pass through more checkpoints, and then find some form of transport to your hotel.

When we do all this, it feels routine, if annoying and prone to jostling and delays. And yet, consider the apparatus that has to exist just to get you to your destination. Someone got up that morning ready to process your phone call for a taxi; someone else had started the day ready to drive you to the bus station. At the station the bus driver was ready to load your bag and take you to the airport. Once there, a phalanx of people had arisen that morning ready to check you in, handle your bags, perform security checks, and examine your passport. While waiting in the departure lounge, you buy coffee from someone who arose that morning ready to make it for you. The pilot and all the crew and maintenance staff also began their day preparing to carry you as a passenger, as did a small army of people at the other end, including finally the taxi driver who drops you off at your hotel.

We might not appreciate that all of this happens, and normally happens surprisingly well, because no one is actually preparing to do a single thing for you. Instead, everyone is following local rules. Those rules have emerged from a long trial-and-error process of people trying out their small part of the larger picture of a division of labor and exchange of goods and services. Someone drives a taxi, someone else a bus, and so on. The system has grown bit by bit in response to demands, and not a one of the people involved has to know or care very much about you. The systems are not perfect—remember the baby Jesus was born in a stable because there was no room in the inn—but all the things that happen just to get you to the airport are only the tip of the iceberg of things that must happen to make these systems possible. We haven't even ventured down to the layer of the people who designed and then built the planes, buses, taxis, and roads that you used in your journey; the people who dug the raw materials out of the ground to make them; and the people who shipped them around the world to refiners and manufacturers.

Local rules and the complex interdependent systems that emerge from them have been with us throughout the history of biological and cultural evolution. One of the remarkable discoveries of the field of study known as *complex systems* is how order, or what physicists term a *lack of entropy*, can be created out of seeming randomness by individuals or agents following a small number of very simple local rules, and without anyone specifying in advance what the outcome will be. Such systems are said to be self-organizing or self-assembling, and often have so-called emergent properties that were not part of any of the rules. The study of these emergent properties teaches us that it is the local rules themselves, not the finished product, that natural selection or some other selective process has sculpted to make the complex structures. The proof is that these agents never build quite the same structure—such as an ant's nest, or even a large city—twice, as they would were they, say, making a model airplane from a fixed set of instructions.

For example, to build a single mound of wood chips in an environment consisting of randomly scattered chips, a hypothetical group of termites has only to follow one simple rule: wander in an area and if you find a wood chip, pick it up, unless you are already carrying one. In that case, drop the chip you are carrying and walk off. To convince yourself this rule works, imagine an area in which wood chips have been scattered randomly. At first, none of the termites will be carrying a chip, and when they encounter one, they will pick it up. They will then wander until they find another chip, at which point they will drop the one they are carrying, making a "mound" of two chips. Simultaneously, this will be occurring all over the area, so that small heaps of two chips each will dot the landscape. The termites that have dropped a chip set off again wandering until they find another. They might pick one up from a heap of two or encounter a lone chip and pick it up. They continue wandering, and some will bump into a heap of two, where they will drop their chip. The first signs of order are already appearing out of randomness.

If this process is allowed to run for a long time, a smaller and smaller number of increasingly larger mounds will spontaneously emerge. This is because early on, and just by chance, there will be variation in the size of the mounds. Termites wandering around carrying chips will tend to encounter the larger mounds more often, and in that way the mounds become "attractors" for yet more chips. True, termites are more likely also to pick up chips from the larger mounds, but for the reason just stated these chips will tend to be dropped at larger mounds. Eventually, one of these mounds will attract all the chips. At that point, empty-mouthed termites pick up chips from the single mound but then wander aimlessly until they bump back into it and deposit the chips again.

With one simple rule we have got our hypothetical termites to create a highly non-random structure. We might be tempted to praise them for their insight and cooperation, but none of the termites set out to make a mound, and none of them had a blueprint of the mound in their minds. The final "design" is not a design at all but emerges from collections of individuals acting as agents following simple local rules. We know this because were we to set them the same task again, a different mound would appear, and it would be in a different place. But the mound would look much the same as the first one, and the process that led to it would be the same. You can see this for yourself because computer programs that allow you to simulate this termite behavior can easily be found on the Internet.

Now, a heap of chips of wood is not a termite mound, but it is a good start. The tall monolithic mounds of the compass termites we saw in Chapter 2 are made of mud, not wood chips. But these too emerge from agents—possibly hundreds of thousands or more of them—following local rules that vary, depending upon circumstances in the environment. Maybe one simple rule that initiates the building of a mound is to pick up a bit of mud and drop it whenever you encounter another drop. Just as with the wood chips, this will begin to construct larger and larger mounds. Maybe another rule gets workers to hollow them out. Maybe inside one of these mounds a rule instructs workers to "drop some mud on the spot where you feel a blast of cold air." This is a rule that will help to regulate the temperature or patch up holes that could let in rain or intruders. The termites don't need to "know" what they are doing, they just need to follow the local rules, making their behavior contingent on what is going on around them. Natural selection acts on these rules because termites that carry genes "for" these rules build better mounds, and are therefore more likely to survive and reproduce, sending their mound-building strategies into their offspring.

One of the earliest and most pleasing demonstrations of complex behaviors emerging from agents following local rules was Craig Reynolds's simulation of the motions of flocks of birds as they fly around in the evening sky feeding on insects. The fluid and flowing motions of these flocks wheeling around the sky, sometimes separating and then coming back together, avoiding collisions with each other, looks to be a supreme act of purposeful cooperation on the wing. But Reynolds achieved a surprisingly realistic simulation by assigning the individual birds just three simple rules: one is to stay near to and steer in the same direction as your nearest neighbor; the second is to follow the main heading of the group; and the third is to avoid crowding. Add to these rules a small amount of randomness to individuals' behaviors, and flocks of "boids," as Reynolds called them, elegantly and sublimely fly around computer screens. No one bird is directing the flock and the birds are not actively cooperating to produce it. It emerges from the simple rules.

The wonder of what biologists call our *ontogeny*, or the sequence of development of our bodies from a single fertilized egg, is that these immensely complex structures get built without anyone in charge, no little homunculus directing traffic or reading from a blueprint. This happens even though every one of our cells carries the same instructions encoded in our DNA. And yet, some will end up as teeth, others as eyes, or parts of your brain or kidneys, heart or liver. How do they know what to do? We know they don't follow a predetermined plan because identical twins carry exactly the same genetic instructions but their bodies are not identical. Thus, we don't develop by building first a skeleton and then having parts added the way one might assemble a car or other complex piece of machinery. Rather, our bodies emerge gradually and piecemeal as our cells repeatedly divide and assume different forms in different parts of the growing body in response to differing local conditions.

Thus, early on in our embryonic development, we are but a growing and mostly undifferentiated ball of cells called the *blastocyst* attached to the uterine wall via the placenta. It is the cells in this blastocyst that are sometimes called *stem cells* and they are described as *totipotent* because up until this point any of them can eventually become any part of your body. Some might become teeth, or parts of a kidney, eye, or liver, or they might become a fingernail or part of your brain. Gradually, as these totipotent cells continue to divide, they commit themselves in response to local conditions to become one of several broad classes of cells in our bodies. Local conditions might vary among the cells in the blastocyst because some cells just by chance are in the outermost layer of the ball, others are near to the middle, and others are in between.

When the cells begin to commit themselves, they switch from being totipotent to being *pluripotent*. They still don't know their fates but they do have a better idea of what they might be. For instance, a cell might know that it is destined to become part of the central nervous system but whether it will be part of the brain or spinal cord is still unclear. Later yet in development—and it is thought again in response to local conditions—cells commit to particular fates such as becoming a nerve cell, a heart muscle cell, a kidney cell, an eye cell, or perhaps a liver cell. Then, the cells within each of these different tissues (again by following purely local rules) know when to stop making more liver cells, or more cells in the kidneys, or eyes. If too many of a particular kind of cell are made—perhaps too many liver cells—a phenomenon called *apoptosis* removes the excess, and even this occurs in response to local conditions.

This process of development, or ontogeny, normally works remarkably well, and it is wondrous because it just happens on its own. It produces objects of unimaginable complexity, greatly exceeding the most sophisticated objects humans build using their minds. And where our human-made objects are often prone to catastrophic failure—as when spacecraft explode or nuclear power stations melt down—the process of our ontogeny is remarkably reliable, capable of turning out our complicated bodies over and over in every species on Earth. Nevertheless, in mammals, once our cells are committed to a particular fate, they cannot be anything else. This can make car crashes, ski accidents, gunshot wounds, and growing old a nuisance because our cells do not always know how to grow back from these insults. They don't always know how to grow back because some wounds, or simply the ravages of time, require cells to go back to an earlier cell fate to repair the damage, but this is what they cannot do.

If cell biologists could learn how to reverse committed cells back to their totipotent or even pluripotent stage, they could perform something close to miracles, which is why the field of stem cell research is pursued with such vigor. With a knowledge of how to control cell fates, scientists could make the science fiction writers look dull. The limbs and organs, nerves and body parts that they could regrow would be real, making bionic attachments like the one Anakin Skywalker gets fitted with in *Star Wars* after a light-saber accident seem primitive. It is a technology that will make transplants obsolete or just temporary, and conditions like heart disease will be treatable by growing new hearts. Nerve damage and paralysis will be reversible, and some brain diseases will become treatable. Some of these things are already happening as scientists inch by inch figure out how to reprogram cells. The current reliance on embryos to supply stem cells for this kind of research—and the moral and political debates this can produce—is a consequence of our ignorance of how to produce stem cells ourselves. Ironically, some animals—lizards, newts, and crabs, for example—already know how to do this. We know this because they can regrow their limbs, making bones, nerve cells, muscles, and blood vessels from scratch. This might just be another example of Leslie Orgel's second rule: "evolution is cleverer than you are."

As we have seen in nearly every chapter of this book, many of the features and processes that make our bodies work have been rediscovered by cultural evolution. We can see this again in how the individual developmental trajectory of our lives replays or reenacts the nature of cellular ontogeny. We have to come into the world *culturally totipotent*, or individually capable of enough flexibility in our ontogeny to inhabit and make use of the body we have been calling the cultural survival vehicle, and without knowing in advance what it will be like. Like the cells that construct the early part of our embryonic lives, we come into the world not committed to any particular fate (we speculated in Chapter 3 that we might differ in some innate predispositions and talents, and if we do, we could say that our personal development begins at about the point that individual cells become pluripotent). And like them, we acquire a language, customs, beliefs, and specializations as we go, in response to local circumstances. Some of us become lawyers, others artists or boatmakers, construction workers, salespeople, or mechanics. The local circumstances that influence our fates could be as simple as a local shortage of a particular skill, the presence of a teacher, something our parents told us or something about the labels they attached to us, or a current demand, such as the requirements that impending warfare might bring. The societies that emerge from this process, like an old building, are potentially immortal as we come and go, but the structure remains intact.

Like cells committed to a particular fate, once we have committed to a specialization in life, it can be difficult to return us to an earlier time of our lives when we had more options. Programs to teach people new trades or skills are like attempts to restore a cell's totipotency or pluripotency, and just as it is difficult to reprogram cells, sayings like "You can't teach an old dog new tricks" remind us that it becomes far harder to learn a new trade or some new physical skill, or even a second language, later in life. We might be tempted to attribute these difficulties of reprogramming simply to getting old. Surely age does play a role, but this merely raises the question of why we are built this way. The answer might be that in our past, our social environments have been stable enough that once committed to some trajectory, we would not need to change. Our brains might have evolved for earlier times, when they would not be asked to start over and do something new, such as learn a new language, or switch from being a herder to a fisherman.

Our ontogenetic wiring might be designed to crystallize from around the time of adolescence because in our past what you might have been doing at that time of your life was probably a good predictor of what you would do for the rest of your life. But the cheetahlike pace of cultural evolution in our modern world has taken this stability away, exposing the inflexibility in our development. Entire trades vanish as cultural developments render them obsolete or inefficient.

How many chimneysweeps, glass blowers, shoemakers, textile dyers, or hat makers do you know? It is not hard to imagine what we would be like if we had a different kind of ontogeny—something that made us more like the lizards, but at a social level. If we could reverse years of learning and specialization and take people back to an earlier state, the results might be as spectacular as acquiring a new limb. The demands that modern cultures put on us to have this flexibility might already be favoring people whose wiring makes them more flexible and able to adapt and change throughout their lives.

If we think about it, we realize that the way our bodies develop, and our own development within society, have to be based around local rules, and not predetermined fates. There could not be any little homunculus inside us planning our bodies. It is not just that there is too much to know; this little homunculus would also have to be able to predict the future. There could not be a predetermined number of cells for each particular function in a body; or, if there were, this would strikingly limit a body's ability to adapt to new or changing conditions. There are exceptions. The worm known as *C. elegans* does normally assign the same number of cells to each function. But this creature is constructed from a mere 959 of them and is about 1 millimeter long; it doesn't have a wide range of behaviors; and it lives in a fairly constant environment in soils where all of them do the same thing.

Equally, and for the same reasons, no one could plan our societies, and there could not be a predetermined number of people in a society with a predetermined number in each of many different occupations. What if all of a sudden the society needed more of one particular commodity? When societies have been designed—Moore's Utopia, Pol Pot's Year Zero, the vast sprawling social housing estates of Western European social democracies, or the hugely controlling and interfering theocracies of the world—they have usually proved far from utopian. And human-engineered societies such as planned cities might be "optimal" in some respects but often fail to meet our expectations in other ways. Natural and cultural selection have found a set of rules that our minds haven't yet discovered. Orgel's second rule yet again?

If the properties of self-organization tell us how it is we can produce large societies and develop within them merely by following local rules, they cannot tell us why we allow them to grow so large and be led by so few. So we need to know why large societies of people in their millions or even billions have been useful when our pattern and our instincts throughout our history have been to live in small tribal societies. The answers will follow from the other two characteristics of these self-organizing systems mentioned earlier. One is that they can often become more efficient as they grow larger; the other is that they frequently acquire smaller structures within the larger whole.

the economies of scaling

ECONOMIES OF scale arise when some things change faster or slower than some measure of the population, such as the number of people. If they change faster and it is something you want, or they change more slowly and it is something you want to avoid, these are both good outcomes. Prior to the invention of agriculture, all human groups were hunter-gatherers. It is a reasonable guess that there were few economies of scale in hunter-gatherer groups, at least once the group got beyond some minimum size that made it viable to survive. This minimum size might be the number of people required to ensure that there is a variety of skills in the group and sufficient numbers such that if someone is injured or dies, others can step in and take their place. With fewer than this number, the group not only loses some its skills, it is also vulnerable to chance events and bad luck. The minimum size might also be influenced by the numbers needed to compete with or fight other groups.

When a hunter-gatherer group got much larger than this minimal size, it might actually have become less efficient. Imagine the land surrounding the hunter-gatherer group's village or settlement as spreading out in an expanding circle. (We don't suppose that the territory literally forms a circle, it is just that this makes the example easy to imagine.) Everything the group needs to survive must be obtained from that circle: hunters must find sufficient game, foragers must find sufficient plants, and everyone must find water and materials for constructing shelters. Evolutionary biologists call this "central place foraging," and it is a good description of most hunter-gatherers. As the numbers of people in a hunter-gatherer group increases, the old circle will no longer provide enough in the way of resources, and so people will have to range out over a wider area. This means that now everyone, on average, has to walk just that little bit further every day to acquire the food and resources they need. As groups get very large, this daily foraging or hunting route might get so long that it makes more sense to divide the settlement into two, placing a new one some distance away from the first.

This simple limitation of central place foraging might be why there were never any large-scale hunter-gatherer societies, at least as far as anyone knows: hunting-and-gathering positively works against any attempts to centralize and enlarge society. But compare this situation to what happens once agriculture, horticulture, or perhaps fishing technology is invented. Now everyone can produce at least some of what they need from a relatively small area of land. It is even possible that a subset of people can produce sufficient food for the remainder of the group, exchanging the food they make for goods the others provide. The group can then increase in size without requiring that each new person go further to find what they need to live.

With food production industrialized this way, societies can benefit from other economies of scale that seem to arise naturally within larger groups. Modern cities are in many ways more efficient than collections of smaller towns or villages that might together have the same overall number of people. Measures of wealth creation, such as wages and gross domestic product, but also inventions and patents, increase more rapidly than population size: cities seem disproportionately either to attract or produce creative people and industries. On the other hand, measures of infrastructure, such as numbers of gas stations, the length of electrical cables, and road surface area increase more slowly than population size; this is good as it means that in these ways cities are more efficient. So consistent are the economies of scale in cities that some tout them as the emerging "green" centers of the modern world. They may in one place belch out huge quantities of pollution, but it is a smaller quantity than were the city to be broken into a smaller collection of towns and villages.

Economies of scale dilute one of the chief reasons we first saw for a society to split into two: that individuals can promote their long-term reproductive success better by dividing a group than by staying together. Indeed, economies of scale provide an incentive to remain together, yielding more wealth and resources, and reducing important risks. Large numbers of people mean there are more backups should someone be lost or injured, and the economies of scale might mean that the society can afford an army. A centralized society can distribute food efficiently and it can devote resources—resources that it has because of economies of scale—to curb violence, maintain order, or protect people from fraud or theft. In short, these larger societies can establish institutions on the back of their economies of scale because these economies mean there is more wealth to go around. People are quite happy for it to be used collectively because they benefit from it.

We can be sure the incentives from these various economies of scale must have been large or at least steady enough to accumulate to large amounts, because as soon as humans started living in larger societies, they would suffer from two kinds of predation. One was predation from other humans who wanted to sack and plunder towns for their wealth. Archaeological digs show that as soon as large settlements began to form, sometime around 10,000 years ago, so too did defensive walls. Perhaps the most famous story in Western literature—Homer's *Iliad*—could be summarized as the story of a particularly thick defensive wall. Early town builders also sought out positions in the landscape—for example, on top of a hill or cliff—that granted a natural defensive advantage, even though hilltops were inconvenient for most other things, such as collecting water.

The second kind of predation came from within society as higher densities of people meant greater transmission of infectious diseases. It is a simple rule of epidemiology that greater densities of people can support not only more diseases but nastier ones. All a disease needs to do to survive and grow in a population is infect more than one additional person before the body it is in dies or acquires immunity and kills the disease. When people live close together, diseases can be very nasty indeed and still achieve this magic number. Where people are further apart, diseases must become more benign: they have to keep you alive long enough that you will bump into enough people not yet infected to be certain of transmitting the infection to at least one of them. Anyone who has spent time in Africa knows that it is in large cities that you are most likely to get ill. Out in the sparse countryside one's health often spontaneously and delightfully returns.

Our determination to live in cities in spite of these costs shows that natural selection does not maximize well-being, but rather reproduction. Our cities survived because those who lived in them survived better and they produced more children. Had they not, cities would have vanished as quickly as they appeared, as people fled back to the Arcadian countryside. Still, economies of scale are not a free ride to prosperity. Calculations of the growth of cities suggest they rely on a steady stream of innovations to provide new economies as they need continually to reset their carrying capacities. This might be why historical records show that cities can change their sizes relative to other cities abruptly, either by growing on the back of some new innovation or by failing to innovate and collapsing. The collapse of the great Mayan civilizations and perhaps those of the Anasazi in the southwestern United States might have owed as much to a failure to produce technological innovations to increase food production as to any internal decline. But so long as technological innovations maintain economies of scale, it becomes a simple matter to grow the group into a larger society, and eventually into the large metropolitan cities of the modern world.

Even so, why have humans so willingly allowed themselves to be taken over in these large societies by a small number of leaders who often wear absurd costumes, demand taxes and the performance of ridiculous rituals, and make questionable claims to be connected to the gods? Large societies probably do benefit from some kinds of centralization and planning, despite the power of local rules to build them. They benefit from systems for distributing goods, systems for resolving disputes, and, not least, systems for organizing society around shared goals. If these are left to local rules, an *ochlocracy* or rule of the rabble can emerge, and when that happens, the economies of scale begin to leak away. The alternative is to allow for some degree of centralization; but once this happens, power ebbs away from the many and into the hands of the few closest to the apparatus of that centralization. It happens easily because most people are just following local rules anyway, and so long as the few can deliver on their promises to distribute wealth, curb violence, and reduce outside threats, they stand a chance of being retained. A different and provocative answer to this question arises from something we saw in chapters 7 and 9—that there is reason to expect that most of us are followers by nature anyway. Given the power of social learning, we expect that there will be only a small number of innovators among us. If this is true, and we can crudely translate innovators and imitators to leaders and followers, then many of us might be content to let someone else do the leading.

These considerations can explain why we accept the few as our leaders. But what about their sometimes ridiculous or even bizarre rituals and costumes? Here we must remind ourselves that large societies don't exist for the good of the masses, but because we as individuals get more out of them than we would from a smaller grouping or even no grouping at all. So, there is no expectation that the leaders that emerge in the new chiefdoms and nation-states will be fair-minded, equitable, or even moral. We expect that the most charismatic, socially shrewd, sociopathic, and powerful will rise to the top and do everything they can to enshrine their rule, even becoming kleptocrats along the way to whatever extent they can get away with it. Wearing the plumes of a rare bird, or in our modern world building obscenely large palaces—as Romania's Nicolae Ceauşescu did in the middle of Bucharest in the latter part of the twentieth century, or in earlier cases such as the Alhambra, Versailles, or St. Peter's in Rome—is a good advertisement that you have time, power, and resources to waste. This won't be lost on people with our evolved psychology who instinctively recognize that some of that excess power might be used to protect them, but could also be turned on them in vicious acts of suppression. None of this should surprise us. One of the themes of this book is that the cooperative enterprise of society is always finely balanced between the benefits that derive from cooperation on the one hand and the benefits that derive from trying to subvert the system toward your own gain without being caught or overpowered.

The spectacle of elderly, frail, sociopathic, and even senile leaders presiding over military dictatorships probably owes more to individuals and institutions protecting themselves than to any power the so-called leader might have. Dictators shrewdly surround themselves with sycophantic supporters—including their armies and police—and reward them with homes, access to food, and better freedom of movement than is available to others. These perquisites give the sycophants reason to keep their leader in power, and opponents who dare to poke their heads above the parapets are quickly whisked away, despatched, or disappeared. Wily dictators routinely provoke low-level factional disputes by denying any one group exclusive access to power, and so individuals know that once their leader falls, there will be a struggle for power among the various factions, with uncertain outcomes, or worse, death to the losers. Social institutions look after themselves because the people in them, like everyone else, have their self-interest in mind. This is true from the highest levels of government down through organizations and right into your local church or bridge club. If society is ticking over at some minimal level, it might be safer to put up with it than to try to replace it.

For many people, the most compelling question about the disintegration of nation-states that coincided with the collapse of communism in Eastern Europe and the Soviet Union in 1989, or the unrest in many North African and Middle Eastern states in 2011, was why didn't it happen sooner? But revolutions take lots of provocation, and you might just lose your life. If you jump ship too soon and join the opposition before its time, you will probably be tracked down and killed when the revolution fails. On the other hand, if you wait too long to join the opposition, you will be branded a traitor to the revolution and killed. No one will know when to make their move and no one will want to discuss the possibility with others, fearing they might be betrayed. The result is an unnerving and deadly version of the collective ignorance we saw in Chapter 7.

viscosity and the maintenance of small worlds

WE HAVE seen throughout this book that the fundamental feature of human societies is our cooperation and ability to work together as a group. But we have also seen that this cooperation depends upon meeting people repeatedly and on knowing others' reputations. Both of these features fall away rapidly in large groups. And yet at least two aspects of large social groups mean that this doesn't happen nearly as fast as it might. One is that even in large societies, individuals are surprisingly well connected to each other; and the other is that people tend to move around far less than they could.

In 1967, the psychologist Stanley Milgram sent a package to each of 160 randomly chosen people living in Omaha, Nebraska. He asked them to forward the package to a friend or acquaintance whom they thought would bring it closer to a particular final individual, a stockbroker working in Boston, Massachusetts. No one necessarily knew this stockbroker, and he wasn't a famous local or national figure. The study came to be known as the "small world" experiment because on average the packages crossed the hands of only around six people, counting the first and last person, on the way to the stockbroker. This led to the popular notion of "six degrees of separation"—the idea that we are all separated from each other by no more than around six other people (later the subject of John Guare's play by the same name). There were many criticisms of Milgram's study, but later studies of e-mail and instant-messaging networks have confirmed his early results that the world is smaller than it might seem. You can get a sense of this idea yourself by imagining how you might get a message to someone you don't know residing in some foreign land. Do you know anyone there? Do you know someone who might know someone there? Maybe try the embassy? The world begins to shrink as you consider the possibilities.

Other features of our social networks, loosely thought of as our circle of friends, also support Milgram's original findings. In many real social networks, the probability that two people chosen at random will have some link—perhaps friendship or belonging to the same club—is greatly increased if those two people have a mutual acquaintance. This is the same as saying that your friends are much more likely to know one another than two people chosen at random. We expect this sort of clustering and connectivity among people if they don't move around too much. An analysis of over 1 million records of the movement of U.S. dollar bills using the "Where's George?" Web site (the dollar bill features George Washington, the first American president) revealed that over half the time they turned up within six miles of where they had last been recorded. Even this short distance is probably an overestimate as the dollar bills would not always have been recorded by everyone who handled them.

These measures tell us that our large societies are not very well *mixed*. Either we do not naturally move very far in our every day lives or people have tendencies toward what has been called social viscosity—the formation of smaller clubs, cliques, and other subgroupings. Whichever is correct, we tend to know those around us far better than we might expect from the size of the societies we live in, and this effectively reduces the size of those societies. No one actually lives in a society of billions, or millions or thousands, or perhaps even hundreds, for that matter.

Think about how many different people you see each day—people with whom you might have some interaction, not those you just pass in the street. Our tendency to social viscosity means our knowledge of each other is accurate; it makes it easier to trust in reciprocal exchanges; and we can be aware of each other's reputations. This is probably why we have this tendency toward making groups-within-groups in the first place: it is a natural consequence of our ancient psychology for living in small groups. Even large social networks often display a peculiar architecture in which a few popular individuals have links to a large number of other people, while most people have far fewer links. These gregarious people provide short routes that can link many pairs of people, and thereby effectively reduce the size of the population, turning what is a wide world into a rather smaller one.

If our social habits and instincts allow our societies to grow in an almost unlimited way, it shouldn't surprise us that, historically, authoritarian and dictatorial regimes have succeeded in controlling people by removing the elements of the cooperative society on which these instincts act. Dictatorial regimes—the most obvious contemporary example being North Korea—remove trust between individuals, and this destroys cooperation, returning people to a dependent and psychologically infantile state. Once a secret police force or surveillance measures have penetrated the society so deeply that even family members betray one another, the fabric of the cooperative society collapses because reciprocity networks and personal reputations can no longer be relied on. People have little choice but to fall in line, seeking rewards directly from the state because it is the only source.

Dictatorial regimes also remove people's social ontogeny by reserving the right to direct them into any occupation at any given time—like trying to reprogram a cell—thereby losing the efficiencies that might come from specialization. Mao Zedong's reeducation camps are but one infamous example. These same regimes also skillfully play segments of society off against each other by encouraging tribal rivalries, and thereby weakening any opposition that might arise from widespread cooperation among people. Having abolished precisely those features of our social lives that have motivated our cooperative societies throughout our history, it should not surprise us that no authoritarian or planned state has ever flourished.

otherness and the dilution of social ties

IT CAN be difficult to overestimate the importance of the social group or cultural survival vehicle to our history, evolution, and psychology. It is a powerful, sometimes febrile, and often startlingly fragile force, depending for its success on a sense of togetherness. That sense derives from two sources. One is the genetic relatedness among family members that we and every other animal makes use of; but the other is, as we have seen earlier, the uniquely human sense of social or cultural relatedness that makes our cooperation work. When either of these two sources of information signal to us that we have exceeded some minimum amount, we are prompted to behave well toward each other; but even slight perceived differences in social relatedness can end in xenophobia, racism, and extreme violence.

And this leaves us with a dilemma. We live at an extraordinary time in the history of humanity, a time in the twenty-first century when there are still people living a Stone Age existence in the depths of the Amazonian rain forest, who have never been contacted by outsiders, and who know nothing about the modern world. And yet the differences between us and them are smaller than the technological comforts of modern life might have us believe. We could pluck a newborn child from one of these uncontacted tribes and happily raise him or her in one of our societies because our social rules have remained the same even as our technology has changed. And it is these social rules that are being put under strain as globalization and ever more culturally differentiated societies mean that humans are increasingly bombarded by outward signals that their average level of cultural relatedness is lower than has been true for much of their evolutionary history, and even perhaps their own personal history.

We should expect this to bring about changes in the way people behave. Large societies of people will be brought together who have little common cultural identity of the sort that historically has prompted our cultural altruism. If the success of modern society up to this point is anything to go by, new heterogeneous societies will increasingly depend upon clear enforcement of cultural or democratically derived rules to maintain stability, and will creak under the strain of smaller social groupings seeking to disengage from the whole. One early harbinger of a sense of decline in social relatedness might be the increasing tendencies of people to avoid risk, to expect safety, to be vigilant about fairness, to require and to be granted "rights." These might all be symptoms of a greater sense of self-interest, brought about perhaps by decline in the average amount of "togetherness" we feel. In response, we naturally turn inward, effectively reverting to our earlier evolutionary instincts, to a time when we relied on kin selection or cooperation among families for our needs to be met.

How is it that words like "race" and "ethnicity" so easily—even if inadvertently and wrongly—find their way into discussions of migration, multiculturalism, and globalization? The answer is clear, and it has nothing to do with racism or ethnicity and everything to do with statistics and our nature. Humans evolved to live in small isolated groups and are finely tuned to seek people of common values and allegiances. Shared markers of what people perceive as race and ethnicity, then, often come to be taken as statistical markers of common values, and that precious oil called "trust" naturally flows. We do this calculation, and our mouths speak it, without thinking. It must be stressed that there is no necessity whatsoever in these connections, and more often than not they will be wrong, and with hurtful and damaging consequences. But they are connections we are all too prone to make.

We are like this because all that is required for it to have been a successful strategy throughout our history is that markers of common ethnicity were a better-than-chance predictor of common culture, and thus common goals and values. In the long term, individuals playing better-than-chance strategies will outcompete those that don't, however distasteful we might find those strategies in the modern world. This is not to say that the "ethnic-marker-equals-common-value" strategy is the best one, or even desirable; just that it might have worked at the individual level and not been surpassed in a general way throughout our history. The language of multiculturalism slides into ethnic and racial categories so easily as a result of the all-too-human—even if all-too-fallible—search for common values because, in the end, our success as a species has come from cooperation.

Does this make us all racists, bigots, or xenophobes deep down? No, we are far too clever for that. The very feature of our social existence that makes us unique—our ability to cooperate with others—makes us uniquely among the animals capable of moving beyond the divisive politics of race, ethnicity, and multiculturalism. Were we as mindless as apes and ants, this would be impossible: they are racists and xenophobes, and unlike us, this is inflexibly hard-wired into them. Their behavior is based almost exclusively on common genetic ancestry. Ours is not. What our history has demonstrated is that we humans will get along with anyone who wishes to play the cooperative game with us. The returns of cooperation, trade, and exchange that derive from that part of our nature have historically trumped our guesswork based on markers of ethnicity or other features. And they always will.

It would be glib to suggest that the answers to 200,000 years of our history are this easy; but our nature can point the way. That way is not necessarily to seek to use our incomparable intelligence to rebel against the dictates of our genes. The message of this book is that our genes have created in us a machine capable of greater inventiveness and common good than any other on Earth. The key is to provide or somehow create among people stronger clues of trust and common values than might otherwise be suggested by the highly imprecise markers of ethnicity or cultural differences that we have used throughout our history, and then to encourage the conditions that give people a sense of shared purpose and shared outcomes. That is the recipe that carried us around the world beginning around 60,000 years ago, and it still works. Looking around the great cosmopolitan cities of our world, it is hard to avoid the conclusion that this is already happening.

References

Introduction

Ambrose, S. H. 1998. Late Pleistocene human population bottlenecks, volcanic winter, and differentiation of modern humans. *J. Hum. Evol.*, 34, 623–51.

Boyd, R., and P. J. Richerson. 2005. *The Origin and Evolution of Cultures.* Oxford: Oxford University Press.

Dawkins, R. 1976. *The Selfish Gene.* Oxford: Oxford University Press.

———. 1978. *The Extended Phenotype.* Oxford: Oxford University Press.

———. 1995. *River Out of Eden.* London: Weidenfeld & Nicolson.

Hamilton, W. D. 1996. *Narrow Roads of Gene Land: The Collected Papers of W. D. Hamilton.* Vol 1: *Evolution of Social Behavior.* New York: W. H. Freeman, p. 330.

Li, H., and R. Durbin. 2011. Inference of human population history from individual whole-genome sequences. *Nature*, 475, 493–96.

McBrearty, S., and A. Brooks. 2000. The revolution that wasn't: A new interpretation of the origin of modern human behaviour. *J. Hum. Evol.*, 39, 453–563.

Richerson, P. J., and R. Boyd. 2005. *Not By Genes Alone.* Chicago: University of Chicago Press.

Thornton, A., J. Samson, and T. Clutton-Brock. 2010. Multi-generational persistence of traditions in neighbouring meerkat groups. *Proc. Royal Soc. London, B.* doi: 10.1098/rspb.2010.0611.

Winterhalder B., and E. A. Smith. 1992. *Evolutionary Ecology and Human Behavior.* New York: Aldine de Gruyter.

Part I: Mind Control, Protection, and Prosperity

Prologue

Blackmore, S. 1999. *The Meme Machine*. Oxford: Oxford University Press.

Dawkins, R. 2003. *A Devil's Chaplain*. London: Weidenfeld & Nicolson, chap. 3.2, "Viruses of the Mind."

———, and J. R. Krebs. 1979. Arms races between and within species. *Proc. Royal Soc. London, B*, 205, 489–511.

Dennett, D. C. 1991. *Consciousness Explained*. Boston: Little, Brown, p. 207.

———. 1995. *Darwin's Dangerous Idea: Evolution and the Meaning of Life*. New York: Simon & Schuster.

Moore, J. 2002. *Parasites and the Behavior of Animals*. Oxford: Oxford University Press.

Sapolsky, R. 2003. Bugs in the brain. *Scientific American*, 94–97.

Wilson, E. O. 1975. *Sociobiology: The New Synthesis*. Cambridge, MA: Harvard University Press, p. 3.

———. 1978. *On Human Nature*. Cambridge, MA: Harvard University Press, p. 167.

Chapter 1: The Occupation of the World

Barbujani, G., and R. R. Sokal. 1990. Zones of sharp genetic change in Europe are also linguistic boundaries. *Proc. Natl Acad. Sci., USA*, 87, 1816–19.

Brown, P., et al. 2004. A new small-bodied hominin from the Late Pleistocene of Flores, Indonesia. *Nature*, 431, 1055–61.

Cavalli-sforza, L. L., et al. 1988. Reconstruction of human evolution: Bringing together genetic, archaeological, and linguistic data. *Proc. Natl Acad. Sci., USA*, 85, 6002–06.

Dent, S. 2003. *The Language Report*. Oxford: Oxford University Press.

Diamond, J. 1997. *Guns, Germs, and Steel*. New York: W. W. Norton & Company.

Duncan, D., and K. Burns. 1997. *Lewis and Clark: The Journey of the Corps of Discovery*. New York: Alfred A. Knopf, pp. 50–51.

Goebel, T. 2007. The missing years for modern humans. *Science*, 315, 194–96.

Goodall, J., and E. V. Lonsdorf, 2002. "Culture in Chimpanzees," in M. Pagel, ed., *Encyclopedia of Evolution*. New York: Oxford University Press, pp. E29–E38.

Gray, R. D., A. J. Drummond, and S. J. Greenhill. 2009. Language phylogenies reveal expansion pulses and pauses in Pacific settlement. *Science*, 323, 479–83.

Green, R. E., et al. 2010. A draft sequence of the Neandertal genome. *Science*, 328, 710–22.

Harrison, G. A. 1995. *The Human Biology of the English Village*. Oxford: Oxford University Press.

Holden, C., and R. Mace. 1999. Sexual dimorphism in stature and women's work: A phylogenetic cross-cultural analysis. *Amer. J. Phys. Anthropol.*, 110, 27–45.
Kinzler, K. D., E. Dupoux, and E. S. Spelke. 2007. The native language of social cognition. *Proc. Natl Acad. Sci., USA,* 104, 12577.
Kirch, P. V. 2000. *On the Road of the Winds.* Berkeley: University of California Press.
Klein, R. G., and B. Edgar. 2002. *The Dawn of Human Culture.* New York: John Wiley & Sons.
Krause, J., et al. 2010. The complete mitochondrial genome of an unknown hominin from southern Siberia. *Nature*, online, March 24, 2010. doi:10.1038/nature08976.
Kulick, D. 1992. *Language Shift and Cultural Reproduction: Socialization, Self, and Syncretism in a Papua New Guinean Village.* Cambridge: Cambridge University Press, p. 2.
Li, J. Z., et al. 2008. Worldwide human relationships inferred from genome-wide patterns of variation. *Science*, 319, 1100–04.
Mace, R., and M. Pagel. 1995. A latitudinal gradient in the density of human languages in North America. *Proc. Royal Soc. London, B*, 261, 117–21.
———. 2004. The cultural wealth of nations. *Nature*, 428, 275–78.
Marean, C. W., et al. 2007. Early human use of marine resources and pigment in South Africa during the Middle Pleistocene. *Nature*, 449, 905–08.
Mellars, P. 2006. Why did modern human populations disperse from Africa ca. 60,000 years ago? A new model. *Proc. Natl Acad. Sci., USA*, 103, 9381.
Novembre, J., et al. 2008. Genes mirror geography within Europe. *Nature*, 456, 98–101.
Olivieri, A., et al. 2006. The mtDNA legacy of the Levantine early Upper Palaeolithic in Africa. *Science*, 314, 1767–70.
Pagel, M., and R. Mace. 2004. The cultural wealth of nations. *Nature*, 428, 275–78.
Powell, A., et al. 2009. Late Pleistocene demography and the appearance of modern human behavior. *Science*, 324, 1298–1301.
Reich, D., et al. 2010. Genetic history of an archaic hominin group from Denisova Cave in Siberia. *Nature*, 468, 1053–60.
Renfrew, C. 1987. *Archaeology and Language: The Puzzle of Indo-European Origins.* Cambridge, UK: Cambridge University Press, p. 368.
Ridley, M. 2010. *The Rational Optimist: How Prosperity Evolves.* London: The Fourth Estate, Prologue, p. 1.
Rosenberg, N. A., et al. 2002. Genetic structure of human populations. *Science*, 298, 2381–85.
Sokal, R. R., et al. 1990. Genetics and language in European populations. *Amer. Naturalist*, 135, 157–75.
Stringer, C., and P. Andrews. 2005. *The Complete World of Human Evolution.* London: Thames & Hudson. Cf. also *In Our Time*, BBC Radio 4, June 17, 2010.
Tomasello, M. 1999. *The Cultural Origins of Human Cognition.* Cambridge, MA: Harvard University Press.

Wade, N. 2006. *Before the Dawn: Recovering the Lost History of Our Ancestors*. New York: Penguin Press.

Wells, S. 2002. *The Journey of Man: A Genetic Odyssey*. London: Penguin Press.

Chapter 2: Ultra-sociality and the Cultural Survival Vehicle

Dawkins, R. 1978. *The Extended Phenotype*. Oxford: Oxford University Press.

Ember, C. R. 1978. Myths about hunter-gatherers. *Ethnology*, 17, 439–48.

Hawking, S. Reported by the BBC, April 25, 2010; http://news.bbc.co.uk/2/hi/uk_news/8642558.stm.

Hill, K., and A. M. Hurtado. 1995. *Aché Life History: The Ecology and Demography of a Foraging People*. Hawthorne, NY: Aldine de Gruyter.

Junger, S. 2010. *War*. London: HarperCollins/Fourth Estate.

Keeley, L. H. 1996. *War Before Civilization: The myth of the Peaceful Savage*. Oxford: Oxford University Press.

LeBlanc, S. A., and K. E. Register. 2003. *Constant Battles: The Myth of the Peaceful, Noble Savage*. New York: St. Martin's Press.

Maynard Smith, J., and E. Szathmáry. 1995. *The Major Transitions in Evolution*. New York: W. H. Freeman.

Strassmann, J. E., Y. Yong Zhu, and D. C. Queller. 2000. Altruism and social cheating in the social amoeba *Dictyostelium discoideum*. *Nature*, 408, 965–67.

Trivers, R. L. 1971. The evolution of reciprocal altruism. *Quart. Rev. Biol.*, 46, 35–57.

Weinberg, S. 2003. *Facing Up: Science and Its Cultural Adversaries*. Cambridge, MA: Harvard University Press, p. 242.

Wilson, D. S. 2003. *Darwin's Cathedral: Evolution, Religion, and the Nature of Society*. Chicago: University of Chicago Press.

Wynne-Edwards, V. C. 1962. *Animal Dispersal in Relation to Social Behaviour*. Edinburgh: Oliver & Boyd.

Chapter 3: The Domestication of Our Talents

Baron-Cohen, S. 2003. *The Essential Difference: Men, Women and the Extreme Male Brain*. London: Allen Lane Press.

Bouchard, T. J., Jr., et al. 1990. Sources of human psychological differences: The Minnesota study of twins reared apart. *Science*, 250, 223–28.

Bouchard, T. J., Jr., and M. McGue. 2003. Genetic and environmental influences on human psychological differences. *J. Neurobiol.* 54, 4–45.

Dawkins, R. 1980. "Good Strategy or Evolutionarily Stable Strategy?" in G. W. Barlow and J. Silverberg, eds., *Sociobiology: Beyond Nature/Nurture?* Denver, CO: Westview Press.

Duarte, C. M., N. Marbá, and M. Holmer. 2007. Rapid domestication of marine species. *Science*, 316, 382–83.

Gladwell, M. 2009. *Outliers: The Story of Success*. New York & London: Penguin Books.

Hawks, J., et al. 2007. Recent acceleration of human adaptive evolution. *Proc. Natl Acad. Sci., USA*, 104, 20753–58.

Henshilwood, C. S., et al. 2002. Emergence of modern human behavior: Middle Stone Age engravings from South Africa. *Science*, 295, 1278–80.

Mace, R. 1993. Transitions between cultivation and pastoralism in Sub-Saharan Africa. *Current Anthropol.*, 34, 363–82.

McGue, M., and T. J. Bouchard, Jr. 1988. Genetic and environmental influences on human behavioral differences. *Ann. Rev. Neuroscience*, 21, 1–24.

Mithen, S. 1996. *The Prehistory of the Mind: A Search for the Origins of Art, Religion and Science*. London: Thames & Hudson.

Rentfrow, P. J., L. R. Goldberg, and R. Zilca. 2010. Listening, watching, and reading: The structure and correlates of entertainment preferences. *J. Personal.* doi: 10.1111/j.1467-6494.2010.00662.x.

Ricardo, D. 1817. *On the Principles of Political Economy and Taxation*. London: John Murray.

Saunders, P. 2000. *Unequal But Fair? A Study of Class Barriers in Britain*. London: Civitas (Institute for the Study of Civil Society), p. 78.

Sinervo, Barry, and C. M. Lively. 1996. The Rock-Paper-Scissors Game and the evolution of alternative male strategies. *Nature*, 380, 240–43.

Stout, M. 2005. *The Sociopath Next Door*. New York: Random House.

Wolf, M., et al. Life-history trade-offs favor the evolution of animal personalities. *Nature*, 447, 581–86.

Chapter 4: Religion and Other Cultural "Enhancers"

Atran, S. 2002. *In Gods We Trust: The Evolutionary Landscape of Religion*. New York: Oxford University Press.

Barnes, J. 2008. *Nothing to Be Frightened Of*. New York: Alfred A. Knopf, p. 78.

Bloom, P. 2007. Religion is natural. *Development. Sci.*, 10, 147–51.

Boyer, P. 2002. *Religion Explained: The Human Instincts that Fashion Gods, Spirits and Ancestors*. New York: Vintage Books.

Dawkins, R. 2006. *The God Delusion*. London & New York: Bantam Books, p. 31.

Dennett, D. C. 2006. Breaking the Spell: Religion as a Natural Phenomenon. London: Allen Lane Press.

Gombrich, E. H. 1984. *The Story of Art*. 14th ed. London: Phaidon Press, p. 4.

Hoffman, J. 2011. Q&A: Illuminating the dark. *Nature*, 473, 30.

Humphrey, N. Humphrey's Law of the Efficacy of Prayer, at http://www.edge.org/q2004/page2.html.

Hume, D. 1757/1956. *The Natural History of Religion*. Palo Alto, CA: Stanford University Press, pp. 27, 11, 29.

Kahneman, D., and A. Tversky. 1973. On the psychology of prediction. *Psychol. Rev.*, 80, 237–57.
Kierkegaard, S. 1843/2003. *Fear and Trembling*. New York: Penguin Books.
Levinson, O. 2002. "Art: An Adaptive Function?" in Pagel, ed., *Encyclopedia of Evolution*.
Locke, J. 1690/1979. *An Essay Concerning Human Understanding*. Oxford: Oxford University Press, p. 151.
Luria, A. 1968. *The Mind of a Mnemonist: A Little Book About a Vast Memory*. Cambridge, MA: Harvard University Press.
McLuhan, M. 1964/1994. *Understanding Media: The Extensions of Man*. New York: MIT Press, p. 7.
Mithen, S. 1999. *The Prehistory of the Mind: The Cognitive Origins of Art, Religion and Science*. London: Thames & Hudson, p. 20.
———. 2006. *The Singing Neanderthals: The Origins of Music, Language, Mind, and Body*. Cambridge, MA: Harvard University Press, p. 2.
Pinker, S. 2003. *How the Mind Works*. New York: Penguin Books.
Plato. *The Dialogues of Plato*, vol. 3, trans. B. Jowett. 1875/1953. Oxford: Oxford University Press, p. 184.
Pooke, G., and D. Newell. 2007. *Art History: The Basics*. London: Routledge.
Renard, J. 1925/2008. *Journal of Jules Renard*, ed. L. Bogan and E. Roget. Portland, OR: Tin House Books, p. 87.
Ross, D., J. Choi, D. Purves. 2007. Musical intervals in speech. *Proc. Natl Acad. Sci., USA*, 104, 9852–57.
Skinner, B. F. 1948. Superstition in the pigeon. *J. Experiment. Psych.*, 38, 168–72.
Stark, R. 1990. Micro foundations of religion: A revised theory. *Sociolog. Theory*, 17, 264–89.
Trivers, R. 1985. *Social Evolution*. Menlo Park, CA: Benjamin Cummings.
Veblen, T. 1899/1994. *The Theory of the Leisure Class*. New York: Penguin Books.
Wade, N. 2009. *The Faith Instinct: How Religion Evolved and Why It Endures*. New York: Penguin Press.
Wilson, D. S. 1990. Species of Thought: A comment on evolutionary epistemology. *Biol. and Philos.*, 5, 39.
———. 2003. *Darwin's Cathedral: Evolution, Religion and the Nature of Society*. Chicago: University of Chicago Press.
Wright, R. 2009. *The Evolution of God: The Origins of Our Beliefs*. New York: Little, Brown.
Zahavi, A., and A. Zahavi. 1999. *The Handicap Principle: A Missing Piece of Darwin's Puzzle*. Oxford: Oxford University Press.

Part II: Cooperation and Our Cultural Nature

Prologue

Sagan, C. See www.youtube.com/watch?v=PdYMLq7NY_M.

Chapter 5: Reciprocity and the Shadow of the Future

Axelrod, R. 1984. *The Evolution of Cooperation*. New York: Basic Books.

———, and W. Hamilton. 1981. The evolution of cooperation. *Science*, 211, 1396.

Basu, K. B. 1984. *The Less-Developed Economy: A Critique of Contemporary Theory*. Oxford: Blackwell Publishers.

Bowles, S., et al. 2006. Group competition, reproductive levelling, and the evolution of human altruism. *Science*, 314, 1569–72.

Boyd, R. R., and P. Richerson. 2005. *The Origin and Evolution of Cultures*. Oxford: Oxford University Press, pp. 52–65.

Dawkins, R. 1976. *The Selfish Gene*. Oxford: Oxford University Press.

Duncan, D., and K. Burns. 1997. *Lewis and Clark: The Journey of the Corps of Discovery*. New York: Alfred A. Knopf, p. 54.

Hamilton, W. D. 1964. The genetical evolution of social behaviour, I and II. *J. Theoret. Biol.*, 7, 1–16 and 17–31.

Hornstein, H., E. Fisch, and M. Holmes. 1968. Influence of a model's feelings about his behavior and his relevance as a comparison other on observers' helping behavior. *J. Personal. and Soc. Psychol.*, 10, 220–26.

Keeley, L. H. 1996. *War Before Civilization: The Myth of the Peaceful Savage*. Oxford: Oxford University Press, pp. 114–15.

Levitt, S. D., and J. A. List. 2008. *Homo economicus* evolves. *Science*, 319, 909–10.

List, J. 2002. Testing neo-classical competitive market theory in the field. *Proc. Natl Acad. Sci., USA*, 99, 15827–30.

Maynard Smith, J. 1979. Hypercycles and the origin of life. *Nature*, 280, 445–46.

Nowak, M., and K. Sigmund. 1993. A strategy of win-stay, lose outperforms generous tit for tat in the prisoner's dilemma game. *Nature*, 364, 56–58.

Ridley, M. 1996. *The Origins of Virtue*. Harmondsworth, UK: Viking Penguin.

Trivers, R. L. 1971. The evolution of reciprocal altruism. *Quart. Rev. Biol.*, 46, 35–57.

———. 2006. "Reciprocal Altruism Thirty Years Later," in P. M. Kappeler and C. P. van Schaik, eds., *Cooperation in Primates and Humans: Mechanisms and Evolution*. Berlin: Springer-Verlag, pp. 77, 79.

Wilkinson, G. S. 1984. Reciprocal food-sharing in the vampire bat. *Nature*, 308, 181–84.

Wilson, E. O. 1978. *On Human Nature*. Cambridge, MA: Harvard University Press, p. 157.

Chapter 6: Green Beards and the Reputation Marketplace

Choi, J. K., and S. Bowles. 2007. The coevolution of parochial altruism and war. *Science*, 318, 636–40.

Cosmides, L., and J. Tooby. 1992. "Cognitive Adaptations for Social Exchange," in J. H. Barkow, L. Cosmides, and J. Tooby, eds., *The Adapted Mind*. New York: Oxford University Press.

Dawkins, R. 1976. *The Selfish Gene*. Oxford: Oxford University Press, p. 89.

Dunbar, R. I. M. 1996. *Grooming, Gossip and the Evolution of Language*. Cambridge, MA: Harvard University Press.

Flack, J. C., et al. 2006. Policing stabilizes construction of social niches in primates. *Nature*, 439, 426–29.

Hawkes, K. 1993. Why hunter-gatherers work: An ancient version of the problem of public goods. *Current Anthropol.*, 34, 341.

Keller, L., and K. G. Ross. 1998. Selfish genes: A greenbeard in the fire ant. *Nature*, 394, 573–75.

Lehman, L., and M. Feldman. 2008. War and the evolution of belligerence and bravery. *Proc. Royal Soc. London, B*. doi:10.1098/rspb.2008.0842.

Marlowe, F. W. 2010. *The Hadza: Hunter-Gatherers of Tanzania*. Berkeley: University of California Press, p. 234.

Nowak, M. A., and K. Sigmund. 2005. Evolution of indirect reciprocity. *Nature*, 437, 1291–98.

Queller, D. C., et al. 2003. Single-gene greenbeard effects in the social amoeba *Dictyostelium discoideum*. *Nature*, 299, 105–06.

Sassoon, S. 1918. "Suicide in the Trenches," *Counter Attack and Other Poems*. New York: E.P. Dutton, p. 31.

Trivers, R. 2006. "Reciprocal Altruism Thirty Years Later," in Kappeler and van Schaik, eds., *Cooperation in Primates and Humans: Mechanisms and Evolution*.

Chapter 7: Hostile Forces

Alexander, R. D. 1979. *Darwinism and Human Affairs*. Seattle: University of Washington Press.

Atkins, P. 1981. *The Creation*. San Francisco: W. H. Freeman, p. 3.

Basalla, G. 1988. *The Evolution of Technology*. Cambridge, UK: Cambridge University Press.

Boyd, R., and P. J. Richerson. 2005. *The Origin and Evolution of Cultures*. Oxford: Oxford University Press.

Bramble, D. M., & D. E. Lieberman. 2004. Endurance running and the evolution of *Homo*. *Nature*, 432, 345–52.

Darwin, C. 1888. *The Descent of Man and Selection in Relation to Sex*, 2nd ed. London: John Murray, p. 57.

Dawkins, R. 1976. *The Selfish Gene*. Oxford: Oxford University Press.

———. 1986. *The Blind Watchmaker: Why the Evidence of Evolution Reveals a Universe Without Design*. New York: W. W. Norton & Company.

Diamond, J. 1995. "The Evolution of Human Inventiveness," in M. P. Murphy and L. A. J. O'Neill, eds. *What Is Life? The Next Fifty Years. Speculations on the Future of Biology*. Cambridge, UK: Cambridge University Press.

Edgerton, D. 2011. In praise of Luddism. *Nature*, 471, p. 27.

Enard, W., et al. 2002. Molecular evolution of FOXP2, a gene involved in speech and language. *Nature*, 418, 869–72.

Gombrich, E. H. 1936. *A Little History of the World*. New Haven, CT: Yale University Press, p. 5.
Humphrey, N. K. 1976. "The Social Function of Intellect." in P. P. G. Bateson and R. A. Hinde, eds., *Growing Points in Ethology*. Cambridge, UK: Cambridge University Press.
Ingram, C. J. E., et al. 2009. Lactose digestion and the evolutionary genetics of lactase persistence. *Hum. Genetics*, 124, 579–91.
Jolly, A. 1966. Lemur social behaviour and primate intelligence. *Science*, 153: 501–6.
Kittler, R., M. Kayser, and M. Stoneking. 2003. Molecular evolution of *Pediculus humanus* and the origin of clothing. *Current Biol.*, 13, 1414–17. Erratum 2004, vol. 14, doi:10.1016/j.cub.2004.12.024.
Lee, R. 1969. Quotation taken from R. L. Trivers. 1971. The evolution of reciprocal altruism. *Quart. Rev. Biol.*, 46, p. 45.
Pagel, M., and W. Bodmer. 2003. A naked ape would have fewer parasites. *Biology Letters. Proc. Royal Soc. London, B, Suppl.*, 270, S117–S119. doi: 10.1098/rsbl.2003.0041.
Paley, W. 1802/2006. *Natural Theology*, intro. and notes by M. D. Eddy and D. M. Knight. Oxford: Oxford University Press, p. 1.
Polybius. *The Histories* (Book 1). Loeb Classical Library Edition, trans. W. R. Paton. 1975. Cambridge, MA: Harvard University Press, p. 57.
Richerson, P. J., and R. Boyd. 2005. *Not By Genes Alone*. Chicago: University of Chicago Press.
Rogers, A. R. and S. Wooding. 2004. Genetic variation at the MCIR Locus and the time since loss of human body hair. *Current Anthropol.*, 45, 105–24.
Ruff, C. B., E. Trinkhaus, and T. W. Holliday. 1997. Body mass and enchephalization in Pleistocene *Homo*. *Nature*, 387, 173–76.
Stone, L. 1983. Interpersonal violence in English society, 1300–1980. *Past and Present*, 101, 22–33.
Tishkoff, S. A., et al. 2007. Convergent adaptation of human lactase persistence in Africa and Europe. *Nature Genetics*, 39(1), 32–40.
Wrangham, R. 2009. *Catching Fire: How Cooking Made Us Human*. New York: Basic Books.

Part III: The Theatre of the Mind

Prologue

Dawkins, R. 1995. *River Out of Eden*. London: Weidenfeld & Nicolson, pp. 1–2.
Hume, D. 1740/1967. *A Treatise of Human Nature*. Oxford: Oxford University Press, p. 163.
Mason, M. F. 2007. Wandering minds: The default network and stimulus-independent thought. *Science*, 315, 393–95.

Chapter 8: Human Language—The Voice of Our Genes

Atkinson, Q. D., et al. 2008. Languages evolve in punctuational bursts. *Science*, 319, 588.

Christiansen, M. H., and N. Chater. 2008. Language as shaped by the brain. *Behav. Brain Sci.*, 31, 489–558.

Darwin, C. 1888. *The Descent of Man . . .*, 2nd ed. London: John Murray, p. 59.

Diamond, J. 1995. "The Evolution of Human Inventiveness," in Murphy and O'Neill, eds., *What Is Life? The Next Fifty Years. Speculations on the Future of Biology*, p. 49.

Encode Project Consortium. 2007. Identification and analysis of functional elements in 1% of the human genome by the ENCODE pilot project. *Nature*, 447, 799–815. doi:10.1038.

Fitch, W. T. 2010. *The Evolution of Language.* Cambridge and New York: Cambridge University Press.

Gray, R. D., and Q. D. Atkinson. 2003. Language-tree divergence times support the Anatolian theory of Indo-European origin. *Nature*, 426, 435–39.

Kinglake, A. W. 1845. *Eothen, or, Traces of Travel Brought Home from the East.* 4th ed. London: John Oliver, chap. 17, p. 267.

Lieberman, E., et al. 2007. Quantifying the evolutionary dynamics of language. *Nature*, 449, 713–16.

Mattick, J. S. 2003. Challenging the dogma: The hidden layer of non-protein-coding RNAs in complex organisms. *BioEssays*, 25, 930–39.

Orgel, L., and F. H. C. Crick. 1980. Selfish DNA: The ultimate parasite. *Nature*, 284, 604–7.

Pagel, M. 2000. "The History, Rate, and Pattern of World Linguistic Evolution," in C. Knight, M. Studdert-Kennedy, and J. Hurford, eds., *The Evolutionary Emergence of Language.* New York: Cambridge University Press, pp. 391–416.

———. 2008. Rise of the digital machine. *Nature*, 452, 699.

———. 2009. Human language as a culturally transmitted replicator. *Nature Reviews Genetics*, 10, 405–15.

———. 2009. Words That Last. *National Geographic*, Science feature, December issue.

———, Q. D. Atkinson, and A. Meade. 2007. Frequency of word-use predicts rates of lexical evolution throughout Indo-European history. *Nature*, 449, 717–20.

Piantodosi, S.T., H. Tily, and E. Gibson. 2011. Word lengths are optimized for efficient communication. *Proc. Natl Acad. Sci., USA.* doi: 10.1073/pnas.1012551108.

Pinker, S. 1994. *The Language Instinct.* New York: William Morrow.

Renfrew, C. 1987. *Archaeology and Language: The Puzzle of Indo-European Origins.* Cambridge: Cambridge University Press.

Rostand, E. 1898. *Cyrano de Bergerac: A Play in Five Acts*, trans. G. Thomas and M. F. Guillemard. Cambridge, MA: Harvard University Press, p. 36.

Ruhlen, M. 1994. *On the Origin of Languages: Studies in Linguistic Taxonomy*. Stanford, CA: Stanford University Press, p. 117.
Swadesh, M. 1952. Lexico-statistic dating of prehistoric ethnic contacts. *Proc. Amer. Phil. Soc.*, 96, 453–63.
Zipf, G. K. 1949. *Human Behavior and the Principle of Least Effort*. Reading, MA: Addison-Wesley.

Chapter 9: Deception, Consciousness, and Truth

Augustine, St., Bishop of Hippo. 1853. *Confessions of S. Augustine*, rev. from a former translation by E. B. Pusey. Vol. 1 of library of the fathers of the Holy Catholic Church. Oxford: John Henry Parker, 1853.
Barnes, J. 2008. *Nothing to Be Frightened Of.* New York: Alfred A. Knopf, p. 228.
Bem, D.J. 1967. Self-perception: An alternative interpretation of cognitive dissonance phenomena. *Psychol. Rev.*, 74, 183–200.
———, and A. Allen. 1974. On predicting some of the people some of the time: The search for cross-situational consistencies in behavior. *Psychol. Rev.*, 81, 506–20.
Cashdan, E., and R. Trivers. 2002. "Self-deception," in Pagel, ed., *Encyclopedia of Evolution*.
Daly, M., and M. I. Wilson. 1984. "A Sociobiological Analysis of Human Infanticide," in G. Hausfater and S. B. Hrdy, eds., *Infanticide: Comparative and Evolutionary Perspectives*. New York: Aldine Press.
Dawkins, R. 1976. *The Selfish Gene*. Oxford: Oxford University Press.
Garfield, A. S., et al. 2011. Distinct physiological and behavioral functions for parental alleles of imprinted *Grb10*. *Nature*, 469, 534–538.
Graham, C., and D. Haig. 1991. Genomic imprinting and the strange case of the insulin-like growth factor II receptor. *Cell*, 64, 1045–46.
Hadamard, J. 1945/1973. *The Mathematician's Mind: The Psychology of Invention in the Mathematical Field*. Princeton: Princeton University Press, p. 143 (1945).
Hamilton, W. D. 1996. *Narrow Roads of Gene Land*. Vol. 1: *Evolution of Social Behavior*. New York: W. H. Freeman, pp. 134–35.
Hauser, M. D. 2006. *Moral Minds*. New York:Ecco/HarperCollins.
Hume, D. 1740/1967. *A Treatise of Human Nature*. Oxford: Oxford University Press.
Laland, K. 2010—see Rendell, et al.
LaLueza-Fox, C. 2011. Genetic evidence for patrilocal mating behavior among Neandertal groups. *Proc. Natl Acad. Sci., USA*, 108, 250–253.
Mead, S., et al. 2009. A novel protective prion protein variant that colocalizes with kuru exposure. *New England J. Med.*, 361, 2056–65.
Montaigne, Michel de. 1580/1958. *The Essays*, trans. D. M. Frame. Palo Alto, CA: Stanford University Press, chap. 31, "Of Cannibals," p. 158.
Packard, V. 1957. *The Hidden Persuaders*. Brooklyn, NY: Ig Publishing.

Pagel, M. 1997. Desperately concealing father: A theory of parent-infant resemblance. *Animal Behav.*, 53, 973–81.
———. 1999. Mother and father in surprise genetic agreement. *Nature*, 397, 19–20.
Pennisi, E. 2010. Conquering by copying. *Science*, 328, 167.
Rendell, L., et al. 2010. Why copy others? Insights from the Social Learning Strategies Tournament. *Science*, 328, 208–13.
Rogers, A. R. 1989. Does biology constrain culture? *Amer. Anthropol.*, 90, 819–31.
Rose, D. 2009. Systemic NHS failures allowed cannibal Peter Bryan to kill twice. *The Times* (UK) *online*, September 3, 2009, www.timesonline.co.uk/tol/news/uk/crime/article6820206.ece.
Rosenhan, D. L. 1973. On being sane in insane places. *Science*, 179, 250–58.
Soon, C. S., et al. 2008. Unconscious determinants of free decisions in the human brain. *Nature Neuroscience*, 11, 543–45. doi:10.1038/nn.2112.
Trivers, R. 1985. *Social Evolution*. Menlo Park, CA: Benjamin Cummings, chap. 16.
———. 2000. The elements of a scientific theory of self-deception. *Annals N.Y. Acad. Sci.*, 907, 114–31.
Ubeda, F., and A. Gardner. 2010. A model for genomic imprinting in the social brain: Juveniles. *Evolution*, 64, 2587–2600.

Part IV: The Many and the Few

Chapter 10: Termite Mounds and the Exploitation of Our Social Instincts

Batty, M. 2006. Rank clocks. *Nature*, 444, 592–96.
Bettencourt, L. M. A., et al. 2007. Growth, innovation, scaling, and the pace of life in cities. *Proc. Natl Acad. Sci., USA*, 104, 7301–6.
Brockman, D., L. Hufnagel, and T. Geisel. 2006. The scaling laws of human travel. *Nature*, 439, 462–65.
Currie, T. E., et al. 2010. Rise and fall of political complexity in island South-East Asia and the Pacific. *Nature*, 467, 801–4.
Diamond, J. 2005. *Collapse: How Societies Choose to Fail or Succeed*. New York: Viking Press.
Ohtsuki, H., et al. 2006. A simple rule for the evolution of cooperation on graphs and social networks. *Nature*, 441, 502–5.
Reynolds, C. W. 1987. Flocks, herds, and schools: A distributed behavioral model. *Computer Graphics*, 21(4) (SIGGRAPH '87 Conference Proceedings), 25–34.
Seabright, P. 2004. *The Company of Strangers: A Natural History of Economic Life*. Princeton: Princeton University Press.
Travers, J., and S. Milgram. 1969. An experimental study of the small world problem. *Sociometry*, 32, 425–43.

Bibliography

Alexander, R. D. 1979. *Darwinism and Human Affairs.* Seattle: University of Washington Press.

Ambrose, S. H. 1998. Late Pleistocene human population bottlenecks, volcanic winter, and differentiation of modern humans. *J. Hum. Evol.*, 34, 623–51.

Atkins, P. 1981. *The Creation*. San Francisco: W. H. Freeman.

Atkinson, Q. D., et al. 2008. Languages evolve in punctuational bursts. *Science*, 319, 588.

Atran, S. 2002. *In Gods We Trust: The Evolutionary Landscape of Religion*. New York: Oxford University Press.

Axelrod, R. 1984. The Evolution of Cooperation. New York: Basic Books.

———, and W. Hamilton. 1981. The evolution of cooperation. *Science*, 211, 1396.

Barbujani, G., and R. R. Sokal. 1990. Zones of sharp genetic change in Europe are also linguistic boundaries. *Proc. Natl Acad. Sci., USA*, 87, 1816–19.

Barnes, J. 2008. *Nothing to Be Frightened Of.* New York: Alfred A. Knopf.

Baron-Cohen, S. 2003. *The Essential Difference: Men, Women and the Extreme Male Brain*. London: Allen Lane Press.

Basalla, G. 1988. *The Evolution of Technology*. Cambridge, UK: Cambridge University Press.

Basu, Kaushik, B. 1984. *The Less-Developed Economy: A Critique of Contemporary Theory*. Oxford: Blackwell Publishers.

Batty, M. 2006. Rank clocks. *Nature*, 444, 592–96.

Bem, D. J. 1967. Self-perception: An alternative interpretation of cognitive dissonance phenomena. *Psychol. Rev.*, 74, 183–200.

———, and A. Allen. 1974. On predicting some of the people some of the time: The search for cross-situational consistencies in behavior. *Psychol. Rev.*, 81, 506–20.

Bettencourt, L. M. A., et al. 2007. Growth, innovation, scaling, and the pace of life in cities. *Proc. Natl Acad. Sci., USA*, 104, 7301–6.

Blackmore, S. 1999. *The Meme Machine*. Oxford: Oxford University Press.

Bloom, P. 2007. Religion is natural. *Development. Sci.*, 10, 147–51.

Bouchard, T. J. Jr., et al. 1990. Sources of human psychological differences: The Minnesota study of twins reared apart. *Science*, 250, 223–28.

Bouchard, T. J., Jr., and M. McGue. 2003. Genetic and environmental influences on human psychological differences. *J. Neurobiol.*, 54, 4–45.

Bowles S., et al. 2006. Group competition, reproductive leveling, and the evolution of human altruism. *Science*, 314, 1569–72.

Boyd, R., and P. Richerson. 2005. *The Origin and Evolution of Cultures*. Oxford: Oxford University Press.

Boyer, P. 2002. *Religion Explained: The Human Instincts That Fashion Gods, Spirits and Ancestors*. New York: Vintage Books.

Bramble, D. M., & D. E. Lieberman. 2004. Endurance running and the evolution of *Homo*. *Nature*, 432, 345–52.

Brockman, D., L. Hufnagel, and T. Geisel. 2006. The scaling laws of human travel. *Nature*, 439, 462–65.

Brown, P., et al. 2004. A new small-bodied hominin from the Late Pleistocene of Flores, Indonesia. *Nature*, 431, 1055–61.

Cashdan, E., and R. Trivers. 2002. "Self-Deception," in M. Pagel, ed., *Encyclopedia of Evolution*. New York: Oxford University Press.

Cavalli-sforza, L. L., et al. 1988. Reconstruction of human evolution: Bringing together genetic, archaeological, and linguistic data. *Proc. Natl Acad. Sci., USA*, 85, 6002–6.

Choi, J. K., and S. Bowles. 2007. The coevolution of parochial altruism and war. *Science*, 318, 636–40.

Christiansen, M. H., and N. Chater. 2008. Language as shaped by the brain. *Behav. Brain Sci.*, 31, 489–558.

Cosmides, L., and J. Tooby. 1992. "Cognitive Adaptations for Social Exchange," in J. H. Barkow, L. Cosmides, and J. Tooby, eds., *The Adapted Mind*. New York: Oxford University Press.

Currie, T. E., et al. 2010. Rise and fall of political complexity in island South-East Asia and the Pacific. *Nature*, 467, 801–4.

Daly, M., and M. I. Wilson. 1984. "A Sociobiological Analysis of Human Infanticide," in G. Hausfater and S. B. Hrdy, eds., *Infanticide: Comparative and Evolutionary Perspectives*. New York: Aldine Press.

Darwin, C. 1859. *On the Origin of Species by Means of Natural Selection*. London: John Murray.

———. 1888 *The Descent of Man and Selection in Relation to Sex*, 2nd ed. London: John Murray.

Dawkins, R. 1976. *The Selfish Gene*. Oxford: Oxford University Press.

———. 1978. *The Extended Phenotype*. Oxford: Oxford University Press.

———. 1980. "Good Strategy or Evolutionarily Stable Strategy," in G. W. Barlow and J. Silverberg, eds., *Sociobiology: Beyond Nature/Nurture?* Boulder, CO: Westview Press.

———. 1986. *The Blind Watchmaker: Why the Evidence of Evolution Reveals a Universe Without Design*. New York: W. W. Norton & Company.
———. 1995. *River Out of Eden*. London: Weidenfeld & Nicolson.
———. 2003. *A Devil's Chaplain*. London: Weidenfeld & Nicolson.
———. 2006. *The God Delusion*. London & New York: Bantam Books.
——— and J. R. Krebs. 1979. Arms races between and within species. *Proc. Royal Soc. London, B, Biol. Sci.*, 205, 489–511.
Dennett, D. C. 1991. *Consciousness Explained*. Boston: Little, Brown.
———. 1995. *Darwin's Dangerous Idea: Evolution and the Meaning of Life*. New York: Simon & Schuster.
———. 2006. *Breaking the Spell: Religion as a Natural Phenomenon*. London: Allen Lane.
Dent, S. 2003. *The Language Report*. Oxford: Oxford University Press.
Diamond, J. 1995. "The Evolution of Human Inventiveness," in M. P. Murphy and L. A. J. O'Neill, eds., *What Is Life? The Next Fifty Years. Speculations on the Future of Biology*. Cambridge, UK: Cambridge University Press.
———. 1997. *Guns, Germs, and Steel*. New York: W. W. Norton & Company.
———. 2005. *Collapse: How Societies Choose to Fail or Succeed*. New York: Viking Press.
Duarte, C. M., N. Marbá, and M. Holmer. 2007. Rapid domestication of marine species. *Science*, 316, 382–83.
Dunbar, R. I. M. 1996. *Grooming, Gossip, and the Evolution of Language*. Cambridge, MA: Harvard University Press.
Duncan, D., and K. Burns 1997. *Lewis and Clark: The Journey of the Corps of Discovery*. New York: Alfred A. Knopf.
Edgerton, D. 2011. In praise of Luddism. *Nature*, 471, 27–29.
Ember, Carol R. 1978. Myths about hunter-gatherers. *Ethnology*, 17, 439–48.
Enard, W., et al. 2002. Molecular evolution of FOXP2, a gene involved in speech and language. *Nature*, 418, 869–72.
Encode Project Consortium. 2007. Identification and analysis of functional elements in 1% of the human genome by the ENCODE pilot project. *Nature*, 447, 799–815. doi:10.1038.
Fitch, W. T. 2010. *The Evolution of Language*. Cambridge and New York: Cambridge University Press.
Flack, J. C., et al. 2006. Policing stabilizes construction of social niches in primates. *Nature*, 439, 426–29.
Garfield, A. S., et al. 2011. Distinct physiological and behavioral functions for parental alleles of imprinted *Grb10*. *Nature*, 469, 534–38.
Gladwell, M. 2009. *Outliers: The Story of Success*. New York: Penguin Books.
Goebel, T. 2007. The missing years for modern humans. *Science*, 315, 194–96.
Gombrich. E. H. 1936. *A Little History of the World*. New Haven, CT: Yale University Press.
———. 1984. *The Story of Art*. 14th ed., London: Phaidon Press.
Goodall, J., and E. V. Lonsdorf. 2002. "Culture in Chimpanzees," in M. Pagel, ed., *Encyclopedia of Evolution*. New York: Oxford University Press.

Graham, C., and D. Haig. 1991. Genomic imprinting and the strange case of the insulin-like growth factor II receptor. *Cell*, 64, 1045–46.

Gray, R. D., and Q. D. Atkinson. 2003. Language-tree divergence times support the Anatolian theory of Indo-European origin. *Nature*, 426, 435–39.

Gray, R. D., A. J. Drummond, and S. J. Greenhill. 2009. Language phylogenies reveal expansion pulses and pauses in Pacific settlement. *Science*, 323, 479–83.

Green, R. E., et al. 2010. A draft sequence of the Neandertal genome. *Science*, 328, 710–22.

Hadamard, J. 1945/1973. *The Mathematician's Mind: The Psychology of Invention in the Mathematical Field*. Princeton: Princeton University Press.

Hamilton, W. D. 1964. The genetical evolution of social behaviour, I and II. *J. Theor. Biol.*, 7, 1–16 and 17–31.

———. 1996. *Narrow Roads of Gene Land: The Collected Papers of W. D. Hamilton*, Vol. 1: *Evolution of Social Behavior*. New York: W. H. Freeman.

Harrison, G. A. 1995. *The Human Biology of the English Village*. Oxford: Oxford University Press.

Hauser, M. D. 2006. *Moral Minds*. New York: Ecco/HarperCollins.

Hawkes, K. 1993. Why hunter-gatherers work: An ancient version of the problem of public goods. Current Anthropol., 34, 341–61.

Hawking, S. Reported by the BBC, April 25, 2010; http://news.bbc.co.uk/2/hi/uk_news/8642558.stm.

Hawks, J., et al. 2007. Recent acceleration of human adaptive evolution. *Proc. Natl. Acad. Sci., USA*, 104, 20753–58.

Henshilwood, C. S., et al. 2002. Emergence of modern human behavior: Middle Stone Age engravings from South Africa. *Science*, 295, 1278–80.

Hill, K., and A. M. Hurtado. 1995. *Aché Life History: The Ecology and Demography of a Foraging People*. Hawthorne, NY: Aldine de Gruyter.

Hoffman, J. 2011. Q&A: Illuminating the dark. *Nature*, 473, 30.

Holden, C., and R. Mace. 1997. Phylogenetic analysis of the evolution of lactose digestion in adults. *Hum. Biol.*, 69, 605–28.

———. 1999. Sexual dimorphism in stature and women's work: A phylogenetic cross-cultural analysis. *Amer. J. Phys. Anthropol.* 110, 27–45.

Hornstein, H., E. Fisch, and M. Holmes. 1968. Influence of a model's feelings about his Behavior and his relevance as a comparison other on observers' helping behavior. *J. Personal. and Soc. Psych.*, 10, 220–26.

Hume, D. 1740/1967. *A Treatise of Human Nature*. Oxford: Oxford University Press.

———. 1757/1956. *The Natural History of Religion*. Palo Alto, CA: Stanford University Press.

Humphrey, N. K. 1976. "The Social Function of Intellect," in P. P. G. Bateson and R. A. Hinde, eds., *Growing Points in Ethology*. Cambridge, UK: Cambridge University Press.

———. Humphrey's Law of the Efficacy of Prayer at www.edge.org/q2004/page2.html.

Ingram, C. J. E., et al. 2009. Lactose digestion and the evolutionary genetics of lactase persistence. *Hum. Genetics*, 124, 579–91.

Jolly, A. 1966. Lemur social behaviour and primate intelligence. *Science*, 153, 501–6.

Junger, S. 2010. *War.* London: HarperCollins/Fourth Estate.

Kahneman, D., and A. Tversky. 1973. On the psychology of prediction. *Psycho. Rev.*, 80, 237–57.

Keeley, L. H. 1996. *War Before Civilization: The Myth of the Peaceful Savage.* Oxford: Oxford University Press.

Keller, L., and K. G. Ross. 1998. Selfish genes: A greenbeard in the fire ant. *Nature*, 394, 573–75.

Kierkegaard, S. 1843/2003. *Fear and Trembling.* New York: Penguin Books.

Kinglake, A. W. 1845. *Eothen, or, Traces of Travel Brought Home from the East.* 4th ed. London: John Oliver.

Kinzler, K. D., E. Dupoux, and E. S. Spelke. 2007. The native language of social cognition. *Proc. Natl Acad. Sci., USA*, 104, 12577–80.

Kirch, P. V. 2000. *On the Road of the Winds.* Berkeley: University of California Press.

Kittler, R., M. Kayser, and M. Stoneking. 2003. Molecular evolution of *Pediculus humanus* and the origin of clothing. *Current Biol.*, 13, 1414–17. Erratum, 2004, vol. 14 doi:10.1016/j.cub.2004.12.024.

Klein, R. G., and B. Edgar. 2002. *The Dawn of Human Culture.* New York: John Wiley & Sons.

Krause, J., et al. 2010. The complete mitochondrial genome of an unknown hominin from southern Siberia. *Nature*, online, March 24, 2010. doi:10.1038/nature08976.

Kulick, D. 1992. *Language Shift and Cultural Reproduction: Socialization, Self, and Syncretism in a Papua New Guinean Village.* Cambridge, UK: Cambridge University Press.

LaLueza-Fox, C. 2011. Genetic evidence for patrilocal mating behavior among Neandertal groups. *Proc. Natl Acad. Sci., USA*, 108, 250–53.

LeBlanc, S. A., and K. E. Register. 2003. *Constant Battles: The Myth of the Peaceful, Noble Savage.* New York: St. Martin's Press.

Lehman, L., and M. Feldman. 2008. War and the evolution of belligerence and bravery. *Proc. Royal Soc. London, B.* doi:10.1098/rspb.2008.0842.

Levinson, Orde. 2002. "Art: An Adaptive Function?" in M. Pagel, ed., *Encyclopedia of Evolution.* Oxford: Oxford University Press.

Levitt, S. D., and J. A. List. 2008. *Homo economicus* evolves. *Science*, 319, 909–10.

Li, J. Z., et al. 2008. Worldwide human relationships inferred from genome-wide patterns of variation. *Science*, 319, 1100–1104.

Lieberman, E., et al. 2007. Quantifying the evolutionary dynamics of language. *Nature*, 449, 713–16.

List, J. 2002. Testing neo-classical competitive market theory in the field. *Proc. Natl Acad. Sci., USA*, 99, 15827–30.

Locke, J. 1690/1979. *An Essay Concerning Human Understanding.* Oxford: Oxford University Press.

Luria, A. 1968. *The Mind of a Mnemonist: A Little Book About a Vast Memory*. Cambridge, MA: Harvard University Press.

Mace, R. 1993. Transitions between cultivation and pastoralism in sub-Saharan Africa. *Current Anthropol.*, 34, 363–82.

———, and M. Pagel. 1995. A latitudinal gradient in the density of human languages in North America. *Proc. Royal Soc. London, B*, 261:117–21.

Marean, C. W, et al. 2007. Early human use of marine resources and pigment in South Africa during the Middle Pleistocene. *Nature*, 449:905–8.

Marlowe, F. W. 2010. *The Hadza: Hunter-Gatherers of Tanzania*. Berkeley: University of California Press.

Mason, M. F. 2007. Wandering minds: The default network and stimulus-independent thought. *Science*, 315, 393–95.

Mattick, J. S. 2003. Challenging the dogma: The hidden layer of non-protein-coding RNAs in complex organisms. *BioEssays*, 25, 930–39.

Maynard Smith, J. 1979. Hypercycles and the origin of life. *Nature*, 280, 445–46.

———, and E. Szathmáry. 1995. *The Major Transitions in Evolution*. New York: W. H. Freeman.

McBrearty, S., and A. Brooks. 2000. The revolution that wasn't: A new interpretation of the origin of modern human behaviour. *J. Hum. Evol.*, 39, 453–563.

McGue, M., and T. J. Bouchard, Jr. 1988. Genetic and environmental influences on human behavioral differences. *Annual Rev. Neuroscience*, 21, 1–24.

McLuhan, M. 1964/1994. *Understanding Media: The Extensions of Man*. New York: MIT Press.

Mead, S., et al. 2009. A novel protective prion protein variant that colocalizes with kuru exposure. *New England J. of Med.*, 361, 2056–65.

Mellars, P. 2006. Why did modern human populations disperse from Africa ca. 60,000 years ago? A new model. *Proc. Natl Acad. Sci., USA*, 103, 9381.

Mithen, S. 1996. *The Prehistory of the Mind: A Search for the Origins of Art, Religion and Science*. London: Thames & Hudson.

———. 2006. *The Singing Neanderthals: The Origins of Music, Language, Mind, and Body*. Cambridge, MA: Harvard University Press.

Montaigne, Michel de. 1580. *The Essays*, trans. D. M. Frame. 1958. Palo Alto, CA: Stanford University Press.

Moore, J. 2002. *Parasites and the Behavior of Animals*. Oxford: Oxford University Press.

Novembre, J., et al. 2008. Genes mirror geography within Europe. *Nature*, 456, 98–101.

Nowak, M., and K. Sigmund. 1993. A strategy of win-stay, lose outperforms generous tit for tat in the prisoner's dilemma game. *Nature*, 364, 56–58.

———. 2005. Evolution of indirect reciprocity. *Nature*, 437, 1291–98.

Ohtsuki, H., et al. 2006. A simple rule for the evolution of cooperation on graphs and social networks. *Nature*, 441, 502–5.

Olivieri, A., et al. 2006. The mtDNA legacy of the Levantine early Upper Palaeolithic in Africa. *Science*, 314, 1767–70.

Orgel, L., and F. H. C. Crick. 1980. Selfish DNA: The ultimate parasite. *Nature*, 284, 604–7.

Packard, V. 1957. *The Hidden Persuaders*. Brooklyn, NY: Ig Publishing.

Pagel, M. 1997. Desperately concealing father: A theory of parent-infant resemblance. *Anim. Behav.*, 53, 973–81.

———. 1999. Mother and father in surprise genetic agreement. *Nature*, 397, 19–20.

———. 2000. "The History, Rate, and Pattern of World linguistic Evolution," in: C. Knight, M. Studdert-Kennedy, and J. Hurford, eds. *The Evolutionary Emergence of Language*. New York: Cambridge University Press.

———. 2008. Rise of the digital machine. *Nature*, 452, 699.

———. 2009. Human language as a culturally transmitted replicator. *Nature Reviews Genetics*, 10, 405–15.

———, Q. D. Atkinson, and A. Meade. 2007. Frequency of word-use predicts rates of lexical evolution throughout Indo-European history. *Nature*, 449, 717–20.

———. and W. Bodmer. 2003. A naked ape would have fewer parasites. *Biology Letters* (*Proc. Royal Soc. London B., Suppl*, 270, S117–S119). doi 10.1098/rsbl.2003.0041.

———, and R. Mace. 2004. The cultural wealth of nations. *Nature*, 428, 275–78.

Paley, W. 1802/2006. *Natural Theology*, intro. and notes by M. D. Eddy and D. M. Knight. Oxford: Oxford University Press.

Pennisi, E. 2010. Conquering by copying. *Science*, 328, 165–67.

Piantodosi, S. T., H. Tily, and E. Gibson. 2011. Word lengths are optimized for efficient communication. *Proc. Natl Acad. Sci., USA*. doi 10.1073/pnas.1012551108.

Pinker, S. 1994. *The Language Instinct*. New York: William Morrow.

———. 2003. *How the Mind Works*. New York: Penguin Books.

Plato. *The Dialogues of Plato*, vol. 3, trans. B. Jowett. 1875/1953. 2nd ed. Oxford: Oxford University Press.

Polybius. *The Histories* (Book 1). Loeb Classical Library Edition, trans. W. R. Paton. 1975. Cambridge, MA: Harvard University Press.

Pooke, G., and D. Newell. 2007. *Art History: The Basics*. London: Routledge.

Powell, A., et al. 2009. Late Pleistocene demography and the appearance of modern human behavior. *Science*, 324, 1298–1301.

Queller, D. C., et al. 2003. Single-gene greenbeard effects in the social amoeba *Dictyostelium discoideum*. *Nature*, 299, 105–6.

Reich, D., et al. 2010. Genetic history of an archaic hominin group from Denisova Cave in Siberia. *Nature*, 468, 1053–60.

Renard, J. 1925/2008. *Journal of Jules Renard*, ed. L. Bogan and E. Roget. Portland, OR: Tin House Books.

Rendell, L., et al. 2010. Why copy others? Insights from the Social Learning Strategies Tournament. *Science*, 328, 208–13.

Renfrew, C. 1987. *Archaeology and Language: The Puzzle of Indo-European Origins*. Cambridge, UK: Cambridge University Press.

Rentfrow, P. J., L. R. Goldberg, and R. Zilca. 2010. Listening, watching, and reading: The structure and correlates of entertainment preferences. *J. Personal.* doi: 10.1111/j.1467-6494.2010.00662.x.

Reynolds, C. W. 1987. Flocks, herds, and schools: A distributed behavioral model. *Computer Graphics*, 21(4) (SIGGRAPH '87 Conference Proceedings), 25–34.

Ricardo, D. 1817. *On the Principles of Political Economy and Taxation*. London: John Murray.

Richerson, P. J., and R. Boyd. 2005. *Not By Genes Alone*. Chicago: University of Chicago Press.

Ridley, M. 1996. *The Origins of Virtue*. Harmondsworth, UK: Viking Penguin.

———. 2010. *The Rational Optimist: How Prosperity Evolves*. London: Fourth Estate.

Rogers, A. R. 1989. Does biology constrain culture? *Amer. Anthropol.*, 90, 819–31.

———, and S. Wooding. 2004. Genetic variation at the MCIR Locus and the time since loss of human body hair. *Current Anthropol.*, 45, 105–24.

Rose, D. 2009. Systemic NHS failures allowed cannibal Peter Bryan to kill twice. *The Times* (UK) online, September 3, 2009; www.timesonline.co.uk/tol/news/uk/crime/article6820206.ece.

Rosenberg, N. A., et al. 2002. Genetic structure of human populations. *Science*, 298, 2381–85.

Rosenhan, D. L. 1973. On being sane in insane places. *Science*, 179, 250–58.

Ross, D., J. Choi, and D. Purves. 2007. Musical intervals in speech. *Proc. Natl Acad. Sci., USA*, 104, 23.

Rostand, E. 1898. *Cyrano de Bergerac: A Play in Five Acts*, trans. G. Thomas and M. F. Guillemard. Cambridge, MA: Harvard University Press.

Ruff, C. B., E. Trinkhaus, and T. W. Holliday. 1997. Body mass and encephalization in Pleistocene *Homo*. *Nature*, 387, 173–76.

Ruhlen, M. 1994. On the Origin of Languages: Studies in Linguistic Taxonomy. Stanford, CA: Stanford University Press.

Sagan, C. See www.youtube.com/watch?v=PdYMLq7NY_M.

Sapolsky, R. 2003. Bugs in the brain. *Scientific Amer.*, 94–97.

Sassoon, S. 1918. "Suicide in the Trenches," *Counter Attack and Other Poems*. New York: E. P. Dutton.

Saunders, P. 2000. *Unequal But Fair? A Study of Class Barriers in Britain*. London: Civitas (Institute for the Study of Civil Society).

Seabright, P. 2004. *The Company of Strangers: A Natural History of Economic Life*. Princeton: Princeton University Press.

Sinervo, Barry, and C. M. Lively. 1996. The Rock-Paper-Scissors Game and the evolution of alternative male strategies. *Nature*, 380, 240–43.

Skinner, B. F. 1948. Superstition in the pigeon. *J. Experiment. Psych.*, 38, 168–72.

Sokal, R. R., et al. 1990. Genetics and language in European populations. *Amer. Naturalist*, 135, 157–75.

Soon, C. S., et al. 2008. Unconscious determinants of free decisions in the human brain. *Nature Neuroscience*, 11, 543–45. doi:10.1038/nn.2112.

Stark, R. 1990. Micro foundations of religion: A revised theory. *Sociolog. Theory*, 17, 264–89.

Stone, L. 1983. Interpersonal violence in English society, 1300–1980. *Past and Present*, 101, 22–33.
Stout, M. 2005. *The Sociopath Next Door.* New York: Random House.
Strassmann, J. E., Y. Yong Zhu, and D. C. Queller. 2000. Altruism and social cheating in the social amoeba *Dictyostelium discoideum. Nature*, 408, 965–67.
Stringer, C., and P. Andrews. 2005. *The Complete World of Human Evolution.* London: Thames & Hudson. Cf. also In Our Time, BBC Radio 4, June 17, 2010.
Swadesh, M. 1994. *On the Origin of Languages: Studies in Linguistic Taxonomy.* Stanford, CA: Stanford University Press.
Thornton, A., J. Samson, and T. Clutton-Brock. 2010. Multi-generational persistence of traditions in neighbouring meerkat groups. *Proc. Royal Soc. London, B.* doi: 10.1098/rspb.2010.0611.
Tishkoff, S. A., et al. 2007. Convergent adaptation of human lactase persistence in Africa and Europe. *Nature Genetics*, 39(1), 32–40.
Tomasello, M. 1999. *The Cultural Origins of Human Cognition.* Cambridge, MA: Harvard University Press.
Travers, J., and S. Milgram. 1969. An experimental study of the small world problem. *Sociometry*, 32, 425–43.
Trivers, R. L. 1971. The evolution of reciprocal altruism. *Quart. Rev. Biol.*, 46, 35–57.
———. 1985. *Social Evolution.* Menlo Park, CA: Benjamin Cummings.
———. 2000. The elements of a scientific theory of self-deception. *Annals N.Y. Acad. Sci.*, 907, 114–31.
———. 2006. "Reciprocal Altruism Thirty Years Later," in P. M. Kappeler and C. P. van Schaik, eds., *Cooperation in Primates and Humans: Mechanisms and Evolution.* Berlin: Springer-Verlag.
Ubeda, F., and A. Gardner. 2010. A model for genomic imprinting in the social brain: Juveniles. *Evolution*, 64, 2587–2600.
Veblen, T. 1899/1994. *The Theory of the Leisure Class.* New York: Penguin Books.
Wade, N. 2006. *Before the Dawn: Recovering the Lost History of Our Ancestors.* New York: Penguin Press.
———. 2009. *The Faith Instinct: How Religion Evolved and Why It Endures.* New York: Penguin Press.
Weinberg, S. 2003. *Facing Up: Science and Its Cultural Adversaries.* Cambridge, MA: Harvard University Press.
Wells, S. 2002. *The Journey of Man: A Genetic Odyssey.* London: Penguin Press.
Wilkinson, G. S. 1984. Reciprocal food-sharing in the vampire bat. *Nature*, 308, 181–84.
Wilson, D. S. 1990. Species of thought: A comment on evolutionary epistemology. *Biol. and Philos.*, 5, 37–62.
———. 2003. *Darwin's Cathedral: Evolution, Religion and the Nature of Society.* Chicago: University of Chicago Press.
Wilson, E. O. 1975. *Sociobiology: The New Synthesis.* Cambridge, MA: Harvard University Press.

———. 1978. *On Human Nature*. Cambridge, MA: Harvard University Press.

Winterhalder, B., and E. A. Smith. 1992. *Evolutionary Ecology and Human Behavior*. New York: Aldine de Gruyter.

Wolf, M., et al. Life-history trade-offs favor the evolution of animal personalities. *Nature*, 447, 581–86.

Words That Last. 2009. *National Geographic*, Science feature, December issue.

Wrangham, R. 2009. *Catching Fire: How Cooking Made Us Human*. New York: Basic Books.

Wright, R. 2009. *The Evolution of God: The Origins of Our Beliefs*. New York: Little, Brown.

Wynne-Edwards, V. C. 1962. *Animal Dispersal in Relation to Social Behaviour*. Edinburgh: Oliver & Boyd.

Zahavi, A., and A. Zahavi. 1999. *The Handicap Principle: A Missing Piece of Darwin's Puzzle*. Oxford: Oxford University Press.

Zipf, G. K. 1949. *Human Behavior and the Principle of Least Effort*. Reading, MA: Addison-Wesley.

Index